Lecture Notes in Computer Science 613

Edited by G. Goos and J. Hartmanis

Advisory Board: W. Brauer   D. Gries   J. Stoer

J. P. Myers, Jr.   M. J. O'Donnell (Eds.)

Constructivity in Computer Science

Summer Symposium
San Antonio, TX, June 19-22, 1991
Proceedings

Springer-Verlag

Berlin Heidelberg New York
London Paris Tokyo
Hong Kong Barcelona
Budapest

Series Editors

Gerhard Goos
Universität Karlsruhe
Postfach 69 80
Vincenz-Priessnitz-Straße 1
W-7500 Karlsruhe, FRG

Juris Hartmanis
Department of Computer Science
Cornell University
5149 Upson Hall
Ithaca, NY 14853, USA

Volume Editors

J. Paul Myers, Jr.
Department of Computer Science, Trinity University
715 Stadium Drive, San Antonio, TX 78212, USA

Michael J. O'Donnell
Department of Computer Science, The University of Chicago
Ryerson Hall, 1100 E. 58th Street, Chicago, IL 60637-2745, USA

CR Subject Classification (1991): F.4.1, I.2.3, F.3.2-3, D.3.1

ISBN 3-540-55631-1 Springer-Verlag Berlin Heidelberg New York
ISBN 0-387-55631-1 Springer-Verlag New York Berlin Heidelberg

Typesetting: Camera ready by author/editor
Printing and binding: Druckhaus Beltz, Hemsbach/Bergstr.
45/3140-543210 - Printed on acid-free paper

Preface

Mathematicians have long recognized the distinction between an argument showing that an interesting object exists and a procedure for actually constructing the object. Some reject nonconstructive proof of existence as invalid, but even those who accept nonconstructive proof usually value the additional insight given by a mental construction. Computer science adds a new dimension of interest in constructivity, since a computer program is a formal description of a constructive procedure that can be executed automatically. So, computer science motivates an interest in constructions as objects with useful behaviors, in addition to the mathematical interest in constructions as direct sources of insight. That constructivity has assumed much importance in computer science is reflected in the title of this symposium, mirroring the name of the first colloquium: "Constructivity in Mathematics" (Heyting, 1957).

The Symposium on Constructivity in Computer Science was sponsored by Trinity University, the University of Chicago, and the Association for Symbolic Logic. The symposium drew participation from Canada, France, Germany, the People's Republic of China, Sweden, the United Kingdom, and the United States of America. Topics discussed were quite diverse, including semantics and type theory, theorem proving, logic, analysis, topology, combinatorics, nonconstructive methods in graph theory, and a special track on curriculum and pedagogy.

This volume contains papers presented at the symposium. Preliminary written versions of papers were distributed, in addition to the lectures. The presentations stimulated very lively discussion, and the papers have been revised based on the feedback from those discussions.

Serge Yoccoz presented a paper, *Some Properties and Applications of the Lawson Topology*, which was not available for these proceedings. The paper of Thierry Coquand, who was unable to attend, was delivered by Chetan Murthy.

Acknowledgments

The Midstates Science and Mathematics Consortium of the Pew Charitable Trusts supported activities that led to the planning for the symposium through a 1989 faculty development award to Professor Myers.

Trinity University provided some funding and visa support for participants. And the University's William Knox Holt Continuing Education and Conference Center provided meeting and residential facilities as well as registration support.

The Association for Symbolic Logic, the American Mathematical Society, and the Special Interest Group on Automata and Computability Theory of the Association for Computing Machinery supported the symposium through free announcements in their periodicals.

And we appreciate encouragement and logistical support (e.g., scheduling assistance and mailing lists) from Robert Constable, Herbert Enderton, Ward Henson, Daniel Leivant, Albert Meyer, Yiannis Moschovakis, and Dana Scott.

May 1992

J. Paul Myers, Jr.
Michael J. O'Donnell

Participants

Jawahar Chirimar	University of Pennsylvania
Ching-Tsun Chou	UCLA
Thierry Coquand	Chalmers Tekniska Högskola
John Doner	University of California at Santa Barbara
Michael R. Fellows	University of Victoria
Newcomb Greenleaf	Columbia University
William Halchin	Software Engineering Consultant
Jim Hoover	University of Alberta
Douglas J. Howe	Cornell University
Jiafu Xu	Nanjing University
Jianguo Lu	Nanjing University
Stuart A. Kurtz	The University of Chicago
Michael A. Langston	University of Tennessee
James Lipton	University of Pennsylvania
Andrew Malton	Queen's University
John C. Mitchell	Stanford University
Chetan R. Murthy	Cornell University
J. Paul Myers, Jr.	Trinity University
Maria Napierala	Oregon Graduate Institute
Michael J. O'Donnell	The University of Chicago
Walter Potter	Southwestern University
Ronald E. Prather	Trinity University
Uday S. Reddy	University of Illinois at Urbana-Champaign
Scott F. Smith	The Johns Hopkins University
Ned Staples	Southeastern Massachusetts University
Yong Sun	University of York
Vipin Swarup	The MITRE Corporation
Simon Thompson	University of Kent at Canterbury
J.V. Tucker	University College of Swansea
Klaus Weihrauch	Fern Universität
Judith Wilson	Temple University
Serge Yoccoz	LABRI, Université de Bordeaux
J.I. Zucker	McMaster University

Contents

Connecting Formal Semantics to Constructive Intuitions

Stuart A. Kurtz   John C. Mitchell
The University of Chicago   Stanford University

Michael J. O'Donnell
The University of Chicago

1   Abstract

We use formal semantic analysis to generate intuitive confidence that the Heyting Calculus is an appropriate system of deduction for constructive reasoning. Well-known modal semantic formalisms have been defined by Kripke and Beth, but these have no formal concepts corresponding to constructions, and shed little intuitive light on the meanings of formulae. In particular, the well-known completeness proofs for these semantics do not generate confidence in the sufficiency of the Heyting Calculus, since we have no reason to believe that every intuitively constructive truth is valid in the formal semantics.

Läuchli has proved completeness for a realizability semantics with formal concepts analogous to constructions, but the analogy is inherently inexact. We argue that, in spite of this inexactness, every intuitively constructive truth is valid in Läuchli semantics, and therefore the Heyting Calculus is powerful enough to prove all constructive truths. Our argument is based on the postulate that a uniformly constructible object must be communicable in spite of imprecision in our language, and we show how the permutations in Läuchli's semantics represent conceivable imprecision in a language, independently of the particular structure of the language.

We look at some of the details of a generalization of Läuchli's proof of completeness for the propositional part of the Heyting Calculus, in order to expose the required model constructions and the constructive content of the result. We discuss the reasons why Läuchli's completeness results on the predicate calculus are not constructive.

2   General Introduction

This paper presents a detailed outline of a three-part lecture given by Michael J. O'Donnell at the symposium on Constructivity in Computer Science. The lecture describes collaborative work in progress by the three authors above, attempting to use formal proofs of completeness for the Heyting Calculus to provide *intuitive* confidence that all constructively true propositional formulae are provable. Feedback from the symposium participants, and particularly a very detailed and cogent critique from James Lipton of the University of Pennsylvania, helped substantially in improving the presentation and the scholarship of the work.

The speaker tried to stimulate thinking on a number of side topics, and to connect to many of the issues raised in other papers at the conference, but did not attempt a thorough survey of the area. Some technical improvements to formal systems and proofs of completeness are introduced, but they are all variations of previously known results. The goal of the lecture is thorough understanding of known technical results and their connection to intuitive concepts, rather than new technicalities, or a thorough survey of known results.

Section 3 presents the lecture as an outline, rather than a narrative. Part I introduces the basic intuitions of constructivism, and describes the types of insights that we hope to get from a formal semantic treatment. It describes Kripke's and Beth's formal semantics for constructive logic, and explains why these do *not* give the desired insights. Part II introduces the realizability and formulae-as-types approaches to semantics. It defines Läuchli's version of realizability, based on permutation-invariant functions, and explains permutation invariance as a plausible *necessary condition* for reliable communicability, and therefore for constructibility, of a function. Finally, Part III gives the formal proof that the Heyting Propositional Calculus is complete for Läuchli's realizability semantics.

3 The Lecture

I. First part: Introduction to constructivism, the intuitive use of semantics.

I.A. What is "constructivism"?

I.A.1. We seek a useful formalism for a reasonable constructive philosophy, not a treatment of a particular historical school, such as intuitionism.

——**2.** The basic intuition of constructivism is that, by asserting the proposition α we claim to have a mental construction verifying α. The precise meaning of "construction" is problematic. A number of philosophers and mathematicians have discussed the problem, including Heyting [19], Dummett [7], Beeson [1], Kleene [25].

——**3.** Here are some examples of classically true formulae that are rejected by constructive logic. They are not all equivalent. See [22, 7, 50, 47] for discussion of their various strengths.

I.A.3.a. $\alpha \vee \neg\alpha$ *(excluded middle)*

——**b.** $\alpha \vee (\alpha \Rightarrow \beta)$

——**c.** $\neg\neg\alpha \Rightarrow \alpha$ *(double negation elimination)*

——**d.** $((\alpha \Rightarrow \beta) \Rightarrow \alpha) \Rightarrow \alpha$ *(Peirce's law [36])*

——**e.** $\neg\alpha \vee \neg\neg\alpha$

——**f.** $(\alpha \Rightarrow \beta) \vee ((\alpha \Rightarrow \beta) \Rightarrow \alpha)$

——**g.** $(\alpha \Rightarrow \beta) \vee (\beta \Rightarrow \alpha)$

——**h.** $((\alpha \Rightarrow \beta) \Rightarrow \gamma) \Rightarrow ((\beta \Rightarrow \alpha) \Rightarrow \gamma) \Rightarrow \gamma$

——**i.** $\alpha \vee (\alpha \Rightarrow \beta) \vee \neg\beta$

——**j.** $\neg\alpha \vee \neg\neg\alpha \vee (\alpha \Rightarrow (\neg\beta \vee \neg\neg\beta))$

I.A.4. Formal system: Heyting Propositional Calculus [18]—essentially Classical Propositional Calculus without the law of excluded middle.

I.B. Semantics: the study of meaning.

I.B.1. The use of semantics: justify and explain a formal system of inference by clarifying its connection to *intuitive* meaning. The intensional structure of formal semantics, not the mere extension of the class of true formulae, connects formalism to intuitive meaning[1]. We must inspect carefully and rigorously, but informally, the connection between formal semantics and intuitive meaning, then examine formally the connection between formal semantics and a formal system of inference.

I.B.2. Notation: Given a fixed system of proof, let F be a formal semantic system. F provides a set of possible interpretations for atomic propositional symbols and, for each interpretation, criteria for marking certain formulae "true". Let α be a formula, and let Γ be a set of formulae.

- $\Gamma \vdash \alpha$ means that α is provable when Γ are assumed.

- $\Gamma \models \alpha$ means that α holds *intuitively* whenever the assumptions in Γ hold.

- $\Gamma \models_F \alpha$ means that α is marked true in every F-interpretation for which all assumptions in Γ are marked true. $\models_F$ is the *logical consequence* relation induced by F.

I.B.3. Intuitive vs. formal measures of strength of a formal system.

I.B.3.a. Faithfulness[2]: $\Gamma \vdash \alpha$ implies $\Gamma \models \alpha$.

——**b.** Soundness: $\Gamma \vdash \alpha$ implies $\Gamma \models_F \alpha$.

——**c.** Fullness: $\Gamma \models \alpha$ implies $\Gamma \vdash \alpha$.

——**d.** Completeness: $\Gamma \models_F \alpha$ implies $\Gamma \vdash \alpha$.

I.B.4. To go from soundness for $\models_F$ to faithfulness for $\models$, we need to show intuitively but rigorously that $\Gamma \models_F \alpha$ implies $\Gamma \models \alpha$. That is, $\models_F$ is a *lower bound* for $\models$. This argument is usually very simple.

——**5.** To go from completeness for $\models_F$ to fullness for $\models$, we need to show that $\Gamma \models \alpha$ implies $\Gamma \models_F \alpha$. That is, $\models_F$ is an *upper bound* for $\models$. This argument is usually difficult, and may

---

[1]Tarski's classical semantics [44] are useful largely because they reveal meaning from syntactic structure, although Tarski claimed that he only wanted to mark the true formulae.

[2]Feferman has related concepts of faithfulness and adequacy [9, 1].

require a *different* formal semantic system from the one used for faithfulness.

I.C. Formal classical semantics [44, 51] (truth tables) is thoroughly successful at supporting intuitive fullness and faithfulness.

I.C.1. Every truth table corresponds to a conceivable situation.

————**2.** A truth table verifying Γ but falsifying α yields a *concrete* interpretation of atomics in Γ and α making Γ clearly true and α clearly false. So, completeness clearly supports fullness[3]. We may write out the truth table on the blackboard, and interpret the propositional symbol α as denoting the concrete proposition expressed by, "In the table on the blackboard, the symbol 'α' appears next to the mark 'T'."

————**3.** Classical semantics supports faithfulness once we agree that every proposition is true or false, so every conceivable situation has a truth table.

I.D. Kripke and Beth semantics for constructive logic: We look only at Kripke semantics [29]. Beth semantics [2] has better constructive technical properties, but the same intuitive drawbacks.

I.D.1. Formal definitions[4]:

Definition 1 *A Kripke model for constructive propositional logic is a triple* $\mathfrak{M} = \langle \mathbf{W}, \preceq, \nu \rangle$, *where* $\mathbf{W}$ *is a set of worlds,* $\preceq$ *is a reflexive, transitive binary relation on* $\mathbf{W}$ *called reachability, and* ν *is a function, called satisfaction, from* $\mathbf{W}$ *to valuations on atomic propositional symbols which is closed under reachability, i.e., if* $\upsilon \preceq \mathfrak{w}$, *and if* $\nu_{\upsilon}(\alpha)$, *then* $\nu_{\mathfrak{w}}(\alpha)$.

Definition 2 *Let* $\mathfrak{M} = \langle \mathbf{W}, \preceq, \nu \rangle$ *be a Kripke model. The relation* $\mathfrak{M}, \mathfrak{w} \models_K \alpha$ *is defined inductively for* $\mathfrak{w} \in \mathbf{W}$ *and propositional formulae* α *as follows:*

- *If* α *is an atomic propositional symbol, then* $\mathfrak{M}, \mathfrak{w} \models_K \alpha$ *if and only if* $\nu_{\mathfrak{w}}(\alpha)$;

- $\mathfrak{M}, \mathfrak{w} \models_K (\alpha \wedge \beta)$ *if and only if* $\mathfrak{M}, \mathfrak{w} \models_K \alpha$ *and* $\mathfrak{M}, \mathfrak{w} \models_K \beta$;

- $\mathfrak{M}, \mathfrak{w} \models_K (\alpha \vee \beta)$ *if and only if* $\mathfrak{M}, \mathfrak{w} \models_K \alpha$ *or* $\mathfrak{M}, \mathfrak{w} \models_K \beta$;

- $\mathfrak{M}, \mathfrak{w} \models_K (\alpha \Rightarrow \beta)$ *if and only if for every world* υ *such that* $\mathfrak{w} \preceq \upsilon$, *and such that* $\mathfrak{M}, \upsilon \models_K \alpha$, *we also have* $\mathfrak{M}, \upsilon \models_K \beta$.

- $\mathfrak{M}, \mathfrak{w} \models_K \neg\alpha$ *if and only if there is no world* υ *such that* $\mathfrak{w} \preceq \upsilon$ *and* $\mathfrak{M}, \upsilon \models_K \alpha$.

In addition, we have the following abbreviated forms:

- $\mathfrak{M} \models_K \alpha$ *if and only if* $\mathfrak{M}, \mathfrak{w} \models_K \alpha$ *for every world* $\mathfrak{w}$ *in* $\mathbf{W}$;

- $\models_K \alpha$ *if and only if* $\mathfrak{M} \models_K \alpha$ *for every Kripke model* $\mathfrak{M}$;

- $M \models_K \alpha$ *if and only if* $\mathfrak{M} \models_K \alpha$ *for all* $\mathfrak{M} \in M$; *and*

- $\Gamma \models_K \alpha$ *if and only if* $\{\mathfrak{M} \mid (\forall \beta \in \Gamma)[\mathfrak{M} \models_K \beta]\} \models_K \alpha.$

I.D.2. Examples: countermodels to the formulae of I.A.3 above. Points represent worlds, arrows show reachability. Next to each point is the set of atomic formulae holding in that world. See Figures 1–4.

I.D.3. Restrictions, variations, on Kripke models.

I.D.3.a. Restrict to trees: same formulae α and consequences $\Gamma \models_K \alpha$ are valid, but theories of these models are all disjunctively closed (if $\alpha \vee \beta$ holds in a model, then either α holds or β holds).

————**b.** Restrict to forests: no change.

————**c.** Restrict to discrete, well-founded, partial orders: no change.

————**d.** Allow inconsistent worlds where all formulae are true: invalidate $\neg(\alpha \wedge \neg\alpha)$. To fix, reinterpret $\neg\alpha$ as α holds only where everything holds.

————**e.** Restrict to linear orderings: no countermodels for

- $\neg\alpha \vee \neg\neg\alpha$

- $(\alpha \Rightarrow \beta) \vee ((\alpha \Rightarrow \beta) \Rightarrow \alpha)$

- $(\alpha \Rightarrow \beta) \vee (\beta \Rightarrow \alpha)$

- $((\alpha \Rightarrow \beta) \Rightarrow \gamma) \Rightarrow ((\beta \Rightarrow \alpha) \Rightarrow \gamma) \Rightarrow \gamma$

---

[3]Truth tables show that classical propositional logic is the strongest possible logic without fallacious computation of truth values.

[4]These are the S_4 models of modal logic [28].

Figure 1: Kripke countermodel for
$$\alpha \lor \neg\alpha$$
$$\alpha \lor (\alpha \Rightarrow \beta)$$
$$\neg\neg\alpha \Rightarrow \alpha$$
$$((\alpha \Rightarrow \beta) \Rightarrow \alpha) \Rightarrow \alpha$$

Figure 2: Kripke countermodel for
$$\neg\alpha \lor \neg\neg\alpha$$
$$(\alpha \Rightarrow \beta) \lor ((\alpha \Rightarrow \beta) \Rightarrow \alpha)$$
$$(\alpha \Rightarrow \beta) \lor (\beta \Rightarrow \alpha)$$
$$((\alpha \Rightarrow \beta) \Rightarrow \gamma) \Rightarrow ((\beta \Rightarrow \alpha) \Rightarrow \gamma) \Rightarrow \gamma$$

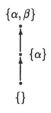

Figure 3: Kripke countermodel for
$$\alpha \lor (\alpha \Rightarrow (\beta \lor \neg\beta))$$

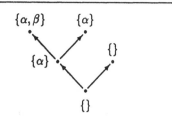

Figure 4: Kripke countermodel for
$$\neg\alpha \lor \neg\neg\alpha \lor (\alpha \Rightarrow (\neg\beta \lor \neg\neg\beta))$$

I.D.3.f. Restrict to (upper-semi-) lattices: no countermodel for $\neg\alpha \lor \neg\neg\alpha$. To fix, allow inconsistent worlds and reinterpret $\neg\alpha$ as α *holds only where everything holds.*

———g. Limit chains to length 1: no countermodel for $\alpha \lor (\alpha \Rightarrow (\beta \lor \neg\beta))$.

———h. Beth's variation [2] changes the criterion for a disjunction holding at a world. Roughly, he lets $(\alpha \lor \beta)$ hold at a world if every path through future worlds is forced to eventually make one of α or β true. For every Beth model we can prove classically, but not constructively, that there is a Kripke model with the same theory. The metamathematics for Beth's semantics is more constructive than for Kripke's. Lipton discusses other variations, combining Kripke semantics with realizability concepts [33].

I.D.4. Intuitive foundations for Kripke semantics.

I.D.4.a. Intuitive meaning of letting $\nu_{\mathfrak{v}}(\alpha)$ in Kripke model: α is constructively *knowable* or *provable* at $\mathfrak{v}$, rather than α is *true* at $\mathfrak{v}$.

———b. Intuitive meaning of $\mathfrak{v} \preceq \mathfrak{w}$ in Kripke model: in $\mathfrak{v}$, one possible future situation is $\mathfrak{w}$.

———c. So, Kripke semantics is intuitively *temporal* and *epistemic*, but there is no direct representative of a *construction,* and no reason to suppose that the temporal and epistemic relations of a Kripke model must be constructively meaningful ones.

———d. Notice that the branching of possible time developments is *crucial* to Kripke semantics. Without branching time, by I.D.3.e above, new formulae, such as $(\alpha \Rightarrow \beta) \lor (\beta \Rightarrow \alpha)$, become valid.

I.D.5. Kripke semantics is insufficient for faithfulness and fullness.

I.D.5.a. From I.D.4 above, we have no intuitive reason to think that *every* Kripke model represents a conceivable temporal development of *constructive* knowledge. The requirement of branching time is particularly suspect: we know that there is only one real future, so why can we not use that knowledge in constructive reasoning?

———b. If there are Kripke models that do not represent conceivable constructive realities, then Kripke semantics may invalidate intuitively true formulae. So, a proof of completeness for Kripke semantics does not support an intuitive claim of fullness[5].

I.D.5.c. It is not even clear that soundness for Kripke semantics supports faithfulness, since the absence of an explicit representation of constructions raises doubt whether the models contain sufficient information to determine truth and falsehood. E.g., suppose that in every world for which α holds, β holds as well, but the knowledge of that fact is not constructive. Is it reasonable to say that $\alpha \Rightarrow \beta$ holds constructively?

I.E. Strategy for developing formal semantics to support faithfulness and fullness of constructive formal logic:

I.E.1. A formal semantics may *approximate* intuitive truth, rather than characterize it precisely.

I.E.1.a. Too many models leads to too few truths, and vice versa[6].

———**b.** Individual models may also allow too many or too few truths to hold.

I.E.2. Define two different formal semantic treatments of constructive logic, approximating intuitive truth below and above. Rather than allowing more or fewer models, make the formal criteria for marking formulae "true" in a given model alternately too strict and too liberal. Let $\Gamma \models \alpha$ indicate that α holds in all conceivable situations where Γ hold, according to *intuitive* constructive semantics.

I.E.2.a. Curry-Howard-realizability ($\models_\lambda$) is a formal *lower bound* on intuitive constructive truth. Show *rigorously and intuitively* that $\Gamma \models_\lambda \alpha$ implies $\Gamma \models \alpha$.

———**b.** Läuchli-realizability ($\models_L$) is a formal *upper bound* on intuitive constructive truth. Show *rigorously and intuitively* that $\Gamma \models \alpha$ implies $\Gamma \models_L \alpha$.

I.E.2.c. The Heyting Calculus (formal system for constructive reasoning) is sound and complete for Curry-Howard-realizability and for Läuchli-realizability. Prove *formally* that $\Gamma \vdash \alpha$ implies $\Gamma \models_\lambda \alpha$ (soundness), and that $\Gamma \models_L \alpha$ implies $\Gamma \vdash \alpha$ (completeness).

———**d.** Then, all of the relations $\vdash$, $\models$, $\models_\lambda$, and $\models_L$ must be extensionally equivalent.

II. Second part: Realizability and formulae-as-types interpretations of constructive logic.

II.A. Basic idea: Each formula has a class of *evidence* that might be given in support of it. Certain evidence constitutes *constructive proof*. A formal semantics selects certain pieces of evidence as *realizers*, which mark the formulae that they support as *true*. The accuracy of a formal realizability semantics is determined by the correlation between its realizers and the intuitively constructive proofs. This correlation has never been made exact, and probably cannot be.

II.A.1. Kleene's realizability [23, 24, 26]: realizers are indexes of computable functions. Completeness fails for Constructive Propositional Logic [39].

———**2.** Curry's and Howard's formulae-as-types [20, 6]: realizers are lambda-definable functions of types appropriate to the formulae that they realize. Formal completeness is easy, but does not support fullness, since the definition of realizers appears to be too strict. Formal soundness is also easy to prove, and supports faithfulness.

———**3.** Läuchli's realizability [30]: realizers are permutation-invariant functions of appropriate types. Läuchli has a formal completeness proof. We argue that it supports fullness.[7]

II.B. Positive Constructive Propositional Logic (no negation). Negation introduces technically superficial, but annoying complications. Think of $\neg\alpha$ as $\alpha \Rightarrow \perp$, where $\perp$ represents contradiction or absurdity, and $\perp \Rightarrow \beta$ holds for all β.

II.B.1. Formal definitions.

II.B.1.a. Formulae are built from atomic propositions, using disjunction ($\vee$), conjunction

---

[5]Dummett [7] has a more thorough critique of Kripke-Beth semantics. Troelstra [48, 47] shows that validity in Beth semantics is the same as intuitive constructive validity, if constructive relations are admitted to depend on an implicit parameter representing a choice sequence, implicitly quantified in a particular way. Such an interpretation of relations begs essentially the same questions as the temporal-epistemic interpretation of logic.

[6]The inverse relation between number of models and number of truths is exploited classically in nonstandard models of arithmetic [40] and Henkin models of higher-order logic [17].

[7]Friedman has an unpublished formal completeness proof for a technically simpler realizability semantics. We see no argument for fullness based on Friedman's result.

$(\wedge)$, and implication $(\Rightarrow)$. Parentheses will be dropped when convenient.

Definition 3 *Assume that there is an unlimited supply of atomic propositional symbols. The collection of propositional formulae is defined inductively as follows:*

- *Every atomic propositional symbol is a propositional formula;*

- *If α and β are propositional formulae, then so are $(\alpha \vee \beta)$, $(\alpha \wedge \beta)$, and $(\alpha \Rightarrow \beta)$.*

II.B.1.b. Proofs are built from labels (x, y, etc.) using pairing ($\langle \bullet, \bullet \rangle$), selection ($\sigma_0$ and σ_1), marking ($\langle 0, \bullet \rangle$ and $\langle 1, \bullet \rangle$), conditional ($\chi$), discharging a labelled assumption ($\lambda \bullet : \bullet . \bullet$), and application (shown by juxtaposition). $x_1 : \alpha_1, \ldots, x_n : \alpha_n \vdash b : \beta$ means that b is a proof of β using only the assumptions $\alpha_1, \ldots, \alpha_n$, and referring to them by the labels $x_1, \ldots, x_n$.

Definition 4 *Assume that there is an unlimited supply of labels for assumptions. The (deterministic) proof formulae are defined inductively as follows:*

(B) *If x is a label, then $x : \alpha \vdash x : \alpha$ for all formulae α;*

($\wedge$I) *If $\Gamma \vdash a : \alpha$ and $\Delta \vdash b : \beta$, then $\Gamma, \Delta \vdash \langle a, b \rangle : (\alpha \wedge \beta)$;*

($\wedge$E) *If $\Gamma \vdash c : (\alpha \wedge \beta)$, then $\Gamma \vdash (\sigma_0 c) : \alpha$, and $\Gamma \vdash (\sigma_1 c) : \beta$;*

($\vee$I) *If $\Gamma \vdash a : \alpha$, then $\Gamma \vdash \langle 0, a \rangle : (\alpha \vee \beta)$, and $\Gamma \vdash \langle 1, a \rangle : (\beta \vee \alpha)$;*

($\vee$E) *If $\Gamma \vdash c : (\alpha \vee \beta)$, $\Delta \vdash d : (\alpha \Rightarrow \gamma)$, and $\Theta \vdash e : (\beta \Rightarrow \gamma)$, then $\Gamma, \Delta, \Theta \vdash (\chi cde) : \gamma$;*

($\Rightarrow$I) *If $x : \alpha, \Gamma \vdash b : \beta$ and Γ does not contain an assumption with label x, then $\Gamma \vdash (\lambda x : \alpha . b) : (\alpha \Rightarrow \beta)$;*

($\Rightarrow$E) *If $\Gamma \vdash a : \alpha$, and $\Delta \vdash b : (\alpha \Rightarrow \beta)$, then $\Gamma, \Delta \vdash (ba) : \beta$;*

(K) *If $\Gamma \vdash a : \alpha$, then $\Gamma, \Delta \vdash a : \alpha$.*

II.B.2. Proofs in lambda notation correspond directly to natural deduction proofs[8].

II.B.2.a. If $\Gamma \vdash b : (\alpha \Rightarrow \beta)$ and $\Delta \vdash a : \alpha$, then the proof (ba) of β from assumptions in Γ, Δ may be rewritten as in Figure 5.

Assume Γ, Δ
    $\ldots b \ldots$ (a proof of $\alpha \Rightarrow \beta$ from Γ)
    $\alpha \Rightarrow \beta$
    $\ldots a \ldots$ (a proof of α from Δ)
    α
    β by *modus ponens*

Figure 5: Natural deduction proof corresponding to (ba)

II.B.2.b. Similarly, if $x : \alpha, \Gamma \vdash b : \beta$, then the proof $(\lambda x : \alpha . b)$ of $\alpha \Rightarrow \beta$ from assumptions in Γ may be rewritten as in Figure 6.

Assume Γ
    x: Assume α
        $\ldots b \ldots$ (a proof of β from α, Γ)
        β
    Discharge assumption x of α
    $\alpha \Rightarrow \beta$ by the *deduction rule*

Figure 6: Natural deduction proof corresponding to $(\lambda x : \alpha . b)$

II.B.3. Proofs in lambda notation may also be read as constructions of functions. Reduction to normal form is an evaluation method [43].

II.C. Formal realizability models vs. intuitive realizability structures (the problem of formalizing evidence).

II.C.1. Formal definition:

Definition 5 *A realizability model is a pair $\langle U, P \rangle$, where U is a set, and P maps atomic propositional formulae to subsets of U. We extend P to all positive propositional formulae inductively as follows:*

---

[8]Normalization of proofs in lambda notation is closely connected to cut elimination. See Gentzen [14], Prawitz [38], Stenlund [43], Girard [16].

- $P(\alpha \wedge \beta) = P(\alpha) \times P(\beta)$ (cross product).

- $P(\alpha \vee \beta) = (\{0\} \times P(\alpha)) \cup (\{1\} \times P(\beta))$ (marked union).

- $P(\alpha \Rightarrow \beta) = P(\beta)^{P(\alpha)}$ (function space).

An element of $P(\alpha)$ is called evidence *for α.*

Notice that **U** contains only the evidence for *atomic* propositional formulae, and we then construct a hierarchy of sets above **U** by cross product, union, and function space constructions to provide evidence for nonatomics.

II.C.2. Relation to intuitive constructive situations.

II.C.2.a. Class of pieces of evidence is formalized as a *set*.

————**b.** Pairs of evidence to support conjunctions are formalized as *ordered pairs*.

————**c.** Marked objects presenting evidence for disjunctions are formalized as *ordered pairs*, using the particular marks of 0 and 1 in the first component.

————**d.** Rules, presenting evidence for implications, are formalized as *functions in extension*.

II.C.3. Variants on realizability semantics:

II.C.3.a. Restrict sets $P(\alpha)$ of evidence to be mutually disjoint: no change.

————**b.** Restrict sets $P(\alpha)$ of evidence for atomic formulae to be finite, alternately infinite, alternately nonempty: no change.

————**c.** Restrict sets $P(\alpha)$ of evidence for atomic formulae to be of finite cardinality less than n: similar to restricting length of chains in Kripke models.

————**d.** Extension to predicate calculus has technical complexities, but no deep problems: Läuchli has one such treatment [30]. Extension to higher-order logic involves models of higher-order typed lambda calculus and is quite subtle, but feasible [3, 15, 46].

II.C.4. Marking certain pieces of evidence as realizers:

II.C.4.a. Intuitive realizability ($\models$): realizers are *uniformly constructible* evidence. This seems to be impossible to formalize satisfactorily. It is analogous to enumerating the total computable functions [34].

————**b.** Classical realizability: every piece of evidence is a realizer. We get classical logic. The empty set represents falsehood, and *every* nonempty set represents truth.

————**c.** Curry-Howard realizability ($\models_\lambda$): realizers are the *lambda-definable* evidence[9]. Easy to prove soundness and completeness, since proofs are lambda terms. Soundness for lambda-realizability supports faithfulness, since it is clear that all lambda-definables are uniformly constructible. Completeness does not support fullness, since it begs the question whether all uniformly constructible evidence is lambda-definable.

————**d.** Läuchli realizability ($\models_\mathcal{L}$): realizers are the *permutation-invariant*[10] pieces of evidence.

II.C.4.d.i. Formal definitions:

Definition 6 *A permutation π on a set **U** setwise stabilizes a subset $S \subseteq U$ if and only if, for all $a \in S$, $\pi(a) \in S$.*

Definition 7 *A Läuchli-model is a triple $\mathfrak{M} = \langle U, P, \Pi \rangle$, where $\langle U, P \rangle$ is a realizability model, and Π is a group of permutations of **U**, setwise stabilizing $P(\alpha)$ for each atomic formula α. P extends to all propositional formulae as in Definition 5. For each formula α, each permutation $\pi \in \Pi$ induces a permutation π_α on $P(\alpha)$ as follows:*

- *If α is atomic and $a \in P(\alpha)$, then $\pi_\alpha(a) = \pi(a)$*

- *If $\langle a, b \rangle \in P(\alpha \wedge \beta)$, then $\pi_{\alpha \wedge \beta}(\langle a, b \rangle) = \langle \pi_\alpha(a), \pi_\beta(b) \rangle$*

- *If $\langle 0, a \rangle \in P(\alpha \vee \beta)$, then $\pi_{\alpha \vee \beta}(\langle 0, a \rangle) = \langle 0, \pi_\alpha(a) \rangle$*

---

[9]Notice that realizability semantics can give meaning to proofs, as well as to formulae. Most formal studies of logic treat proofs as purely syntactic mechanisms for enumerating theorems.

[10]Permutations, with the particular inductive definition of permutations on function spaces in Definition 7 below, are a special case of the *logical relations* used to characterize lambda definability [42, 35].

- If $\langle 1, b \rangle \in P(\alpha \vee \beta)$, then
$\pi_{\alpha \vee \beta}(\langle 1, b \rangle) = \langle 1, \pi_\beta(b) \rangle$

- If $a \in P(\alpha \Rightarrow \beta)$, then $\pi_{\alpha \Rightarrow \beta}(a) = \pi_\beta \circ a \circ \pi_\alpha^{-1}$
($\circ$ *is function composition*)

If Π is the set of all permutations setwise stabilizing $P(\alpha)$ for each atomic α, then $\langle \mathbf{U}, P, \Pi \rangle$ is the full *Läuchli model for $\langle \mathbf{U}, P \rangle$.*

Definition 8 *Let $\mathfrak{M}$ be a Läuchli model.*

- $\mathfrak{M} \models_{\mathcal{L}} a : \alpha$ *if and only if $a \in P(\alpha)$, and $\pi_\alpha(a) = a$ for all $\pi \in \Pi$.*

- $\mathfrak{M} \models_{\mathcal{L}} \alpha$ *if and only if there exists a such that $\mathfrak{M} \models_{\mathcal{L}} a : \alpha$.*

- $\mathcal{M} \models_{\mathcal{L}} \alpha$ *if and only if $\mathfrak{M} \models_{\mathcal{L}} \alpha$ for all $\mathfrak{M} \in \mathcal{M}$.*

- $\Gamma \models_{\mathcal{L}} \alpha$ *if and only if $\{\mathfrak{M} \mid (\forall \alpha \in \Gamma)[\mathfrak{M} \models_{\mathcal{L}} \alpha]\} \models_{\mathcal{L}} \alpha$.*

II.C.4.d.ii. Permutation invariance is a plausible necessary condition for uniform constructibility[11].

II.C.4.d.ii.a. In order to prove a formula, I must communicate a construction to you, using syntactic names for objects rather than the objects themselves.

————*b.* We may not agree perfectly on the assignment of names to objects, and may never detect a disagreement, since we communicate only through names.

————*c.* If assignment of names to objects is one-to-one, then there is a permutation mapping each object that I name to the object that you associate with its name.

————*d.* So, permutation-invariant evidence is simply evidence that may be communicated reliably, in spite of confusion about names. Under reasonable assumptions about the nature of constructions and communication, every uniform construction must be permutation-invariant (but not the converse).

II.C.4.d.ii.e. We must inspect the inductive definition of permutations on $P(\alpha)$ intuitively to insure that it represents accurately the real impact of name confusion.

————*f.* Notice that there might be other sorts of inaccuracy in communication, not modelled by permutations (e.g., the assignment of names to objects might not be one-to-one). This does not matter, since we are only claiming that permutation invariance is *necessary* for uniform constructibility, not *sufficient*.

II.C.4.d.iii. Weaknesses and subtleties in the argument for permuation invariance.

II.C.4.d.iii.a. Our argument is vulnerable to the possibility that not all permutations can arise as the results of conceivable inaccuracies in communication. It is at least plausible that all of the permutations required for the formal proof of completeness do arise naturally.

————*b.* The argument that permutation invariance is necessary for uniform constructibility only supports directly the claim that $\models \alpha$ implies $\models_{\mathcal{L}} \alpha$. To get the stronger, and desirable, result that $\Gamma \models \alpha$ implies $\Gamma \models_{\mathcal{L}} \alpha$ for finite sets Γ, we need the *deduction* property for $\models$ (i.e., $\alpha \models \beta$ implies $\models \alpha \Rightarrow \beta$). To extend the result to infinite Γ, we need the *compactness* property for $\models$ (i.e., $\Gamma \models \alpha$ implies that there is some finite subset $\Gamma' \subseteq \Gamma$ with $\Gamma' \models \alpha$). Both of these properties are highly plausible, and are articles of faith for many thinkers, but we cannot find compelling technical arguments for them.

————*c.* Suppose that, inside the inductive clause defining $P(\alpha \Rightarrow \beta)$, we restrict this set to contain only uniformly constructible functions from $P(\alpha)$ to $P(\beta)$, instead of *all* functions of that type. Then, we get classical logic. So, the realizability interpretation of constructive logic must require constructions to operate on objects that are merely *postulated* to be constructions, and are treated as black boxes. It must not allow constructions to assume that their inputs are also constructions in a predefined formal system.

II.C.4.d.iv. Examples of invariant evidence.

II.C.4.d.iv.a. The identity function is invariant, and realizes $\alpha \Rightarrow \alpha$.

————*b.* The application functional is invariant, and realizes $\alpha \Rightarrow (\alpha \Rightarrow \beta) \Rightarrow \beta$.

---

[11]It is intriguing to think that similar necessary conditions for computability might lead to a new kind of evidence for the Church-Turing thesis [4, 49].

II.C.4.d.v. Invariant functions are not necessarily computable. In particular, an invariant function from $P(\alpha)$ to $P(\beta)$ may map the invariants of $P(\alpha)$ arbitrarily to the invariants of $P(\beta)$.

————**vi.** $\neg\alpha$ may be interpreted as $\alpha \Rightarrow \bot$, where $\bot$ represents absurdity or contradiction. In order to insure that $\bot \Rightarrow \beta$ for all β, we must restrict models so that there is an invariant mapping $P(\bot)$ to $P(\beta)$, for each formula β (if we insure this for atomic β, it will hold for nonatomics by an easy induction). Note that, if we require $P(\bot) = \emptyset$, we have no countermodel for $\neg\alpha \vee \neg\neg\alpha$, since $P(\alpha)$ nonempty implies that $P(\neg\alpha)$ is empty. This treatment of negation is not very satisfying to the intuition. Traditional constructive negation may not be philosophically compatible with the intuitions of realizability semantics.

II.C.5. Converting Läuchli models to Kripke models. Given a Läuchli model for PCPL, it is easy to construct a Kripke model with the same theory.

II.C.5.a. The set of worlds in the Kripke model is the set of subgroups of the permutation group in the Läuchli model.

————**b.** From a given group/world Π, the reachable worlds are exactly the subgroups of Π.

————**c.** The atomic formulae holding at a given world are those whose evidence sets contain an object invariant under all permutations in the group. It is obvious that once set true, an atomic formula remains true in all reachable worlds.

————**d.** By induction on the structure of a formula α, we can prove that the Kripke and Läuchli criteria for α holding at a given world are equivalent. The proof is simple, but highly nonconstructive, using the axiom of choice. For finite, or just well-ordered, Läuchli models, the proof is constructive. Lemma 1 below is the key to the logical equivalence of the Läuchli and Kripke models.

Lemma 1 *Let* $\mathfrak{L} = \langle U, P, \Pi \rangle$ *be a Läuchli model.* (1) $\Rightarrow$ (2) *below. Assuming the* Axiom of Choice, (1) $\Leftrightarrow$ (2). *If $\mathfrak{L}$ is finite, or just well-ordered,* (1) $\Leftrightarrow$ (2) *constructively, without AC.*

1. $\mathfrak{L} \models_{\mathcal{L}} \alpha \Rightarrow \beta$

2. For all subgroups $\Pi' \subseteq \Pi$ such that $\langle U, P, \Pi' \rangle \models_{\mathcal{L}} \alpha$, $\langle U, P, \Pi' \rangle \models_{\mathcal{L}} \beta$

For the propositional calculus, all constructed countermodels can be finite, so the lemma above is constructive.

II.C.5.e. The Kripke model so constructed is always a lattice. Every formula holds at the top. If negation is added, we must use the variant of Kripke models that allows inconsistent worlds (I.D.3.d and I.D.3.f)[12].

————**f.** Since Läuchli models translate to a highly restricted set of Kripke models, a completeness proof for Läuchli semantics is *prima facie* stronger than a completeness proof for Kripke semantics.

III. Third part: Proof of completeness.

III.A. Constructive fallacies can be understood in terms of a *nondeterministic* lambda notation for proofs.

III.A.1. In deterministic proofs, proofs by cases from disjunctions are handled with the conditional operator χ. When $c: (\alpha \vee \beta)$, $a: (\alpha \Rightarrow \gamma)$, and $b: (\beta \Rightarrow \gamma)$, we get $(\chi cab): \gamma$. When χ is interpreted as a function, $(\chi\langle 0, d \rangle ab)$ evaluates to (ad), and $(\chi\langle 1, e \rangle ab)$ evaluates to (be). In nondeterministic proofs, (χcab) is replaced by $[(a(\rho_0 c)), (b(\rho_1 c))]_\gamma$. $(\rho_i c)$ selects the second component of c (the meaningful content from the marked union) only in the case where the first component of c (the mark) is i. When the mark in c is not i, the value is undefined. $[\bullet, \bullet]_\gamma$ selects nondeterministically one of the two values in the brackets, selecting a well-defined proof of γ whenever possible, and being undefined otherwise[13].

————**2.** Formal definitions: Proofs are built from labels (x, y, etc.) using pairing ($\langle \bullet, \bullet \rangle$), selection ($\sigma_0$ and σ_1), marking ($\langle 0, \bullet \rangle$ and $\langle 1, \bullet \rangle$), nondeterministic union over α ($[\bullet, \bullet]_\alpha$), projection of marked unions (ρ_0 and ρ_1), discharging a labelled assumption ($\lambda\bullet: \bullet . \bullet$), and application

---

[12]Models with inconsistent worlds are called *fallible*. They have been used to improve the constructive content of metamathematics [47].

[13]The simulation of conditionals by nondeterministic union and selection is also used in Pratt's Dynamic Logic [37] to make the theory more elegant. Felleisen's Control and Abort primitives [10] appear to accomplish something very similar with a different syntactic structure.

(shown by juxtaposition).

$$x_1\colon \alpha_1, \ldots, x_m\colon \alpha_m \vdash b_1\colon \beta_1, \ldots, b_n\colon \beta_n$$

means that, assuming that each x_i is a proof of α_i, at least one of b_j must be a proof of the corresponding β_j. The choice of j may depend on the unknown values of $x_1, \ldots, x_m$.

Definition 9 *As in Definition 4, assume an unlimited supply of labels for assumptions. The nondeterministic proof formulae are defined inductively as follows:*

(nId) *If x is a label, then $x\colon \alpha \vdash x\colon \alpha$ for all formulae α;*

(n$\wedge$I) *If $\Gamma \vdash a\colon \alpha, \Phi$ and $\Delta \vdash b\colon \beta, \Psi$, then $\Gamma, \Delta \vdash \langle a, b\rangle\colon (\alpha \wedge \beta), \Phi, \Psi$;*

(n$\wedge$E) *If $\Gamma \vdash c\colon (\alpha \wedge \beta), \Phi$, then $\Gamma \vdash (\sigma_0 c)\colon \alpha, \Phi$, and $\Gamma \vdash (\sigma_1 c)\colon \beta, \Phi$;*

(n$\vee$I) *If $\Gamma \vdash a\colon \alpha, \Phi$, then $\Gamma \vdash \langle 0, a\rangle\colon (\alpha \vee \beta), \Phi$, and $\Gamma \vdash \langle 1, a\rangle\colon (\beta \vee \alpha), \Phi$;*

(n$\vee$E) *If $\Gamma \vdash c\colon (\alpha \vee \beta), \Phi$, then $\Gamma \vdash (\rho_0 c)\colon \alpha, (\rho_1 c)\colon \beta, \Phi$;*

(nN) *If $\Gamma \vdash a_0\colon \alpha, a_1\colon \alpha, \Phi$, then $\Gamma \vdash [a_0, a_1]_\alpha\colon \alpha, \Phi$;*

(n$\Rightarrow$I) *If $x\colon \alpha, \Gamma \vdash b\colon \beta$, and Γ does not contain an assumption with label x, then $\Gamma \vdash (\lambda x\colon \alpha \, . \, b)\colon (\alpha \Rightarrow \beta)$;*

(n$\Rightarrow$E) *If $\Gamma \vdash b\colon (\alpha \Rightarrow \beta), \Phi$, and $\Delta \vdash a\colon \alpha, \Psi$, then $\Gamma, \Delta \vdash (ba)\colon \beta, \Phi, \Psi$;*

(nK) *If $\Gamma \vdash \Phi$, then $\Gamma, \Delta \vdash \Phi, \Psi$.*

III.A.3. Equivalence of deterministic and nondeterministic proof. The proof is fairly straightforward induction on the structure of proof formulae.

Theorem 2 *There exists a collection of nondeterministic proof formulae, $a_0^{nd}, \ldots, a_n^{nd}$ such that $\Gamma \vdash a_0^{nd}\colon \alpha_0, \ldots, a_n^{nd}\colon \alpha_n$ if and only if there exists a deterministic proof formula a^d such that $\Gamma \vdash a^d\colon (\alpha_0 \vee \cdots \vee \alpha_n)$.*

III.A.4. Nondeterministic proof notation reveals the constructive fallacies in certain classical arguments. Of course, we do not rule out the possibility of other sound arguments for the same formulae.

III.A.4.a. The classical argument for $\alpha \vee (\alpha \Rightarrow \beta)$ uses the following nonconstructive rule in the "proof" of Figure 7.

(n$\Rightarrow$IC) *If $x\colon \alpha, \Gamma \vdash b\colon \beta, \Phi$, and Γ does not contain an assumption with label x, then $\Gamma \vdash (\lambda x\colon \alpha \, . \, b)\colon (\alpha \Rightarrow \beta), \Phi$.*

Notice how the label/variable x is used as an undischarged assumption in $\langle 0, x\rangle$, and inconsistently as if it had two different types (i.e., as if it labelled the two different formulae α and β) within $\langle 1, (\lambda x\colon \alpha \, . \, x)\rangle$. The basic idea of nondeterministic proof as embodied in (nId) allows the type-inconsistent usage in the second contingent proof in line (2), because the first contingent proof is type-consistent. Then, the generalized implication introduction rule (n$\Rightarrow$IC) allows the type-consistent use of the assumption x in the first contingent proof in line (2) to remain undischarged, because the type-inconsistent use in the second contingent proof *is* discharged. Thus, each contingent proof does one thing right, and one thing wrong, and the rules (n$\vee$I) and (nN) allow them to be combined and treated as correct.

III.A.4.b. The classical argument for $(\alpha \Rightarrow \beta) \vee (\beta \Rightarrow \alpha)$ uses the following constructively fallacious rule in the abbreviated "proof" of Figure 8.

(n$\Rightarrow$ICW) *If $x\colon \alpha, \Gamma \vdash b_0\colon \beta_0, \ldots, b_n\colon \beta_n$, and Γ does not contain an assumption with label x, then $\Gamma \vdash (\lambda x\colon \alpha.b_0)\colon (\alpha \Rightarrow \beta_0), \ldots, (\lambda x\colon \alpha.b_n)\colon (\alpha \Rightarrow \beta_n)$*

The fallacious step is line (3). Notice that, in line (2), a *single* object u assumed to prove $(\alpha \vee \beta)$ was guaranteed to provide either a proof $(\rho_0 u)$ of α or a proof $(\rho_1 u)$ of β. In line (3), u was abstracted *separately* from each of $(\rho_0 u)$ and $(\rho_1 u)$, so in effect there are two *different* u's in line (3), each assumed to prove $\alpha \vee \beta$. There is no way to guarantee that either the first proves α or the second proves β, since it is quite possible that the first proves β and the second proves α. So, line (3) does not intuitively represent a pair of constructions at least one of which must be correct, and therefore does not fulfill the intent of the nondeterministic proof formulae. The remaining steps merely work the fallacy of line (3) into the form $(\alpha \Rightarrow \beta) \vee (\beta \Rightarrow \alpha)$.

1. $x : \alpha \vdash x : \alpha$ (nId).

2. $x : \alpha \vdash x : \alpha, x : \beta$ 1, (nK).

3. $\vdash x : \alpha, (\lambda x : \alpha \,.\, x) : (\alpha \Rightarrow \beta)$ 2, (n$\Rightarrow$IC).

4. $\vdash \langle 0, x \rangle : (\alpha \vee (\alpha \Rightarrow \beta)),$
   $(\lambda x : \alpha \,.\, x) : (\alpha \Rightarrow \beta)$ 3, (nVI).

5. $\vdash \langle 0, x \rangle : (\alpha \vee (\alpha \Rightarrow \beta)),$
   $\langle 1, (\lambda x : \alpha \,.\, x) \rangle : (\alpha \vee (\alpha \Rightarrow \beta))$ 4, (nVI).

6. $\vdash [\langle 0, x \rangle, \langle 1, (\lambda x : \alpha \,.\, x) \rangle] : (\alpha \vee (\alpha \Rightarrow \beta))$ 5, (nN).

Figure 7: Constructively fallacious argument for $\alpha \vee (\alpha \Rightarrow \beta)$.

---

1. $u : (\alpha \vee \beta) \vdash u : (\alpha \vee \beta)$ (nId).

2. $u : (\alpha \vee \beta) \vdash (\rho_0 u) : \alpha, (\rho_1 u) : \beta$ 1, (nVE).

3. $\vdash (\lambda u : (\alpha \vee \beta) \,.\, (\rho_0 u)) : ((\alpha \vee \beta) \Rightarrow \alpha),$
   $(\lambda u : (\alpha \vee \beta) \,.\, (\rho_1 u)) : ((\alpha \vee \beta) \Rightarrow \beta)$ 2, (n$\Rightarrow$ICW).

For brevity, let a denote the first proof formula in line (3), and let b denote the second. Let c and d be proof formulae for the easily derived constructive tautologies of lines (4) and (5) below:

4. $\vdash c : ((\alpha \vee \beta) \Rightarrow \alpha) \Rightarrow (\beta \Rightarrow \alpha)$

5. $\vdash d : ((\alpha \vee \beta) \Rightarrow \beta) \Rightarrow (\alpha \Rightarrow \beta)$

Using these abbreviations, we continue:

6. $\vdash (ca) : (\beta \Rightarrow \alpha),$
   $(db) : (\alpha \Rightarrow \beta)$ 3, 4 (n$\Rightarrow$E); 3, 5 (n$\Rightarrow$E)

7. $\vdash \langle 1, (ca) \rangle : (\alpha \Rightarrow \beta) \vee (\beta \Rightarrow \alpha),$
   $\langle 0, (db) \rangle : (\alpha \Rightarrow \beta) \vee (\beta \Rightarrow \alpha)$ 6, (nVI) twice

8. $\vdash [\langle 1, (ca) \rangle, \langle 0, (db) \rangle] : (\alpha \Rightarrow \beta) \vee (\beta \Rightarrow \alpha)$ 7, (nN)

Figure 8: Constructively fallacious argument for $(\alpha \Rightarrow \beta) \vee (\beta \Rightarrow \alpha)$.

III.B. Proofs of completeness are much simpler with a *sequent calculus* instead of lambda notation for proofs.

III.B.1. Sequents are formal assertions of provability in the form

$$\alpha_1, \ldots, \alpha_m \vdash \beta_1, \ldots, \beta_n$$

This is equivalent to a Beth tableau node [2, 12] with the αs marked T and the βs marked F. The sequent above may also be thought of roughly as asserting $(\alpha_1 \wedge \cdots \wedge \alpha_m) \Rightarrow (\beta_1 \vee \cdots \vee \beta_m)$.

————**2.** The rules in Table 1 generate precisely the valid sequents[14]. The proof is a reasonably straightforward induction on the length of sequent derivation for ($\Rightarrow$), and the syntactic structure of a nondeterministic proof formula for ($\Leftarrow$).

Theorem 3
A sequent relation $\alpha_0, \ldots, \alpha_m \vdash \beta_0, \ldots, \beta_n$ is derivable by the rules in Table 1 if and only if, for all labels $x_0, \ldots, x_m$, there exist nondeterministic proof formulae $b_0, \ldots, b_n$ such that $x_0: \alpha_0, \ldots, x_m: \alpha_m \vdash b_0: \beta_0, \ldots, b_n: \beta_n$ (equivalently, by Theorem 2, there exists a deterministic proof formula a^d such that $x_0: \alpha_0, \ldots, x_m: \alpha_m \vdash a^d: (\alpha_0 \vee \cdots \vee \alpha_n))$.

III.B.3. Sequent rules are best understood *backwards* (bottom to top). E.g., ($\vee$R) says that, in order to prove $\alpha \vee \beta$ or Ψ from assumptions Γ, it suffices to prove $\alpha \vee \beta$ or α or β or Ψ from assumptions Γ.

————**4.** The rules are designed so that, as much as possible, when applied backwards they add subformulae of formulae that already appear, but do not delete any formulae. ($\Rightarrow$RS) is the only exception—it deletes Ψ from the right-hand side. ($\Rightarrow$RW) is the only rule that fails to introduce *both* principal subformulae of some formula that already appears—it introduces only the right subformula β of $\alpha \Rightarrow \beta$.

————**5.** Variations on the sequent rules.

III.B.5.a. The most familiar sequent rules for constructive logic allow only a single formula on the right-hand sides of sequents[15]. Singleton right-hand sides are often thought of as distinguishing constructive logic from classical logic. If the rules of Table 1 are pruned by restricting the Ψ on the right-hand sides of ($\odot$L) rules to be singletons, and by eliminating all but one key formula on the right-hand sides of ($\odot$R) rules (the ($\vee$R) rule requires two versions—one with α above the line, and one with β), we prove the same sequents in a fashion that is more obviously constructive. But, the completeness proof becomes much more difficult. All that is *required* to preserve constructive validity is that the ($\Rightarrow$RS) rule has the singleton β on the right-hand side above the line. The following two examples show how multiple formulae on the right-hand side in ($\Rightarrow$RS) lead to constructive fallacies.

III.B.5.b. Constructively fallacious rules.

III.B.5.b.i. The following constructively fallacious sequent rule is analogous to the fallacious nondeterministic proof rule (n$\Rightarrow$IC) of III.A.4.a. The analogous "proof" of $\vdash \alpha \vee (\alpha \Rightarrow \beta)$ is in Figure 9. This is the standard sequent rule for classical logic.

$$(\Rightarrow\text{RC}) \; \frac{\Gamma, \alpha \vdash \alpha \Rightarrow \beta, \beta, \Psi}{\Gamma \vdash \alpha \Rightarrow \beta, \Psi}$$

---

1. $\alpha \vdash \beta, (\alpha \Rightarrow \beta), \alpha \vee (\alpha \Rightarrow \beta), \alpha$	(Id)
2. $\vdash (\alpha \Rightarrow \beta), \alpha \vee (\alpha \Rightarrow \beta), \alpha$	1, ($\Rightarrow$RC)
3. $\vdash \alpha \vee (\alpha \Rightarrow \beta)$	2, ($\vee$R).

Figure 9: Constructively fallacious sequent derivation for $\vdash \alpha \vee (\alpha \Rightarrow \beta)$.

III.B.5.b.ii. The following constructively fallacious sequent rule is analogous to the fallacious nondeterministic proof rule (n$\Rightarrow$ICW) of III.A.4.b. This rule may be used to prove $(\alpha \Rightarrow \beta) \vee (\beta \Rightarrow \alpha)$, but it is not powerful enough for full classical logic.

---

[14]All of the rules except ($\Rightarrow$RW) are direct translations of Beth's tableau rules [2, 12]. Sequent rules may be viewed as a method for generating closed lambda terms without producing open terms in the induction.

[15]But Beth's tableau rules [2, 12] have multiple formulae marked F, which is really the same thing as multiple formulae on the right of a sequent.

$$(\text{Id})\ \Gamma, \alpha \vdash \alpha, \Psi$$

$$(\wedge\text{L})\ \frac{\Gamma, \alpha, \beta, \alpha \wedge \beta \vdash \Psi}{\Gamma, \alpha \wedge \beta \vdash \Psi} \qquad (\wedge\text{R})\ \frac{\begin{array}{c}\Gamma \vdash \alpha \wedge \beta, \alpha, \Psi \\ \Gamma \vdash \alpha \wedge \beta, \beta, \Psi\end{array}}{\Gamma \vdash \alpha \wedge \beta, \Psi}$$

$$(\vee\text{L})\ \frac{\begin{array}{c}\Gamma, \alpha, \alpha \vee \beta \vdash \Psi \\ \Gamma, \beta, \alpha \vee \beta \vdash \Psi\end{array}}{\Gamma, \alpha \vee \beta \vdash \Psi} \qquad (\vee\text{R})\ \frac{\Gamma \vdash \alpha \vee \beta, \alpha, \beta, \Psi}{\Gamma \vdash \alpha \vee \beta, \Psi}$$

$$(\Rightarrow\text{L})\ \frac{\begin{array}{c}\Gamma, \alpha \Rightarrow \beta \vdash \alpha, \Psi \\ \Gamma, \beta, \alpha \Rightarrow \beta \vdash \Psi\end{array}}{\Gamma, \alpha \Rightarrow \beta \vdash \Psi} \qquad (\Rightarrow\text{RS})\ \frac{\Gamma, \alpha \vdash \beta}{\Gamma \vdash \alpha \Rightarrow \beta, \Psi}$$

$$(\Rightarrow\text{RW})\ \frac{\Gamma \vdash \alpha \Rightarrow \beta, \beta, \Psi}{\Gamma \vdash \alpha \Rightarrow \beta, \Psi}$$

Table 1: Constructive Sequent Rules

$$(\Rightarrow\text{RCW})\ \frac{\Gamma, \alpha \vdash \alpha \Rightarrow \beta_0, \beta_0, \ldots, \alpha \Rightarrow \beta_n, \beta_n}{\Gamma \vdash \alpha \Rightarrow \beta_0, \ldots, \alpha \Rightarrow \beta_n}$$

III.B.5.c. ($\Rightarrow$RW) is redundant, and may be eliminated without loss of power. But, this rule makes the proof of completeness simpler. The combination of ($\Rightarrow$RS) and ($\Rightarrow$RW) yield the most powerful approximation that we can find to ($\Rightarrow$RC), while maintaining constructive validity.

III.C. Formal proof of completeness for the constructive sequent rules.

III.C.1. Basic idea: a procedure that searches for a derivation or a countermodel. Given a sequent to prove, apply the rules of Table 1 *backwards* in an attempt to reduce the given sequent to instances of the axiom (Id). If this attempt fails, the structure of the search is a tree describing a countermodel[16].

Definition 10 *A model* $\mathfrak{M}$ *is a* Läuchli *countermodel for a sequent* $\Gamma \vdash \Psi$ *if and only if* $\mathfrak{M} \models_{\mathcal{L}} \alpha$ *for every* $\alpha \in \Gamma$ *and* $\mathfrak{M} \not\models_{\mathcal{L}} \beta$ *for every* $\beta \in \Psi$. Kripke countermodels *are analogous.*

---

[16]The proof is very similar to the one in Fitting's book [12]. He does not use ($\Rightarrow$RW), so the termination criteria for his search procedure are more subtle.

III.C.2. Application of sequent rules backwards extends the search tree by two types of branching, as well as nonbranching extension. Two types of leaves may be produced.

III.C.2.a. ($\wedge$R), ($\vee$L), and ($\Rightarrow$L) have two sequents above the line, *both* of which must be proved in order to prove the one below the line. This produces *conjunctive* branching in the search tree. Branching is required because only one of the two major subformulae of a certain formula is chosen in each of the sequents above the line.

————b. ($\Rightarrow$RS) requires one of the arbitrarily many implications on a right-hand side to be chosen for reduction. *Any one* of the possible sequents above the line may be proved in order to prove the one below the line. This produces *disjunctive* branching in the search tree. Branching is required because the unchosen formulae on the right-hand side are deleted from the sequent.

————c. ($\wedge$L), ($\vee$R), and ($\Rightarrow$RW) extend paths in the search tree without branching. No branching is necessary, because both major subformulae of a certain formula are added to the sequent, and no formula is deleted.

————d. An instance of (Id) is a *success* leaf.

————e. A sequent for which the application of a rule produces nothing new is a *failure* leaf.

III.C.3. The completeness proof hinges on showing that *disjunctive* branching in the search for a derivation corresponds precisely to *conjunctive* branching in the construction of a countermodel, and vice versa, that success and failure interchange similarly at leaf nodes, and that nonbranching extensions to the search tree preserve countermodels as well as provability.

III.C.4. Formal definitions used in the search procedure.

III.C.4.a. Leaves of the search tree are either *closed*, indicating a successful branch of the search, or *open*, indicating failure.

Definition 11 *A sequent $\Gamma \vdash \Psi$ is* closed *if and only if $\Gamma \cap \Psi \neq \emptyset$; it is* open *otherwise.*

III.C.4.b. By applying rules backward as much as possible through nonbranching extensions and disjunctive branching, we produce *semisaturated* sequents. Only conjunctive branching can produce anything new from a semisaturated sequent.

Definition 12 *A sequent $\Gamma \vdash \Psi$ is* semisaturated *if and only if, for all formulae α and β*

1. $\alpha \wedge \beta \in \Gamma$ *implies $\alpha \in \Gamma$ and $\beta \in \Gamma$;*

2. $\alpha \wedge \beta \in \Psi$ *implies $\alpha \in \Psi$ or $\beta \in \Psi$;*

3. $\alpha \vee \beta \in \Gamma$ *implies $\alpha \in \Gamma$ or $\beta \in \Gamma$;*

4. $\alpha \vee \beta \in \Psi$ *implies $\alpha \in \Psi$ and $\beta \in \Psi$;*

5. $\alpha \Rightarrow \beta \in \Gamma$ *implies $\alpha \in \Psi$ or $\beta \in \Gamma$;*

6. $\alpha \Rightarrow \beta \in \Psi$ *implies $\beta \in \Psi$.*

Definition 13 *A sequent $\Gamma' \vdash \Psi'$ is a* semisaturation *of $\Gamma \vdash \Psi$ if and only if $\Gamma \subseteq \Gamma'$, $\Psi \subseteq \Psi'$, $\Gamma' \vdash \Psi'$ is semisaturated, and for all Γ'' and Ψ'' with $\Gamma \subseteq \Gamma'' \subseteq \Gamma'$ and $\Psi \subseteq \Psi'' \subseteq \Psi'$ such that $\Gamma'' \vdash \Psi''$ is semisaturated, $\Gamma'' = \Gamma'$ and $\Psi'' = \Psi'$. That is, a semisaturation is a (not usually unique) minimal extension of a sequent to semisaturated form.*

III.C.4.c. Conjunctive branching with $(\Rightarrow RS)$ yields the *associates* of a semisaturated sequent.

Definition 14 *Let $\Gamma \vdash \alpha \Rightarrow \beta, \Psi$ be semisaturated, with $\alpha \notin \Gamma$. Then $\Gamma, \alpha \vdash \beta$ is an* associate *of $\Gamma \vdash \alpha \Rightarrow \beta, \Psi$.*

III.C.4.d. Alternation of conjunctive and disjunctive branching eventually yields *saturated* sequents, which must be leaves of the search tree, because backward application of rules yields nothing new.

Definition 15 *A sequent $\Gamma \vdash \Psi$ is* saturated *if and only if it is semisaturated and, in addition,*

7. $\alpha \Rightarrow \beta \in \Psi$ *implies $\alpha \in \Gamma$.*

III.C.5. Crucial properties of semisaturations, associates, saturated sequents.

III.C.5.a. A saturated sequent is either closed and derivable, or it is open and has a countermodel.

Lemma 4 *If $\Gamma \vdash \Psi$ is closed, then it is derivable.*

Proof: Trivial from the basis rule (Id) of Theorem 3.

Lemma 4   □

Lemma 5 *If $\Gamma \vdash \Psi$ is saturated and open, then there is a countermodel[17] $\mathfrak{M}$ for $\Gamma \vdash \Psi$.*

Proof: Let $U = \{0, 1, 2\}$, $P(\alpha) = \{0\}$ for all atomic formulae $\alpha \in \Gamma$, $P(\alpha) = \{1, 2\}$ for all atomic formulae $\alpha \notin \Gamma$, and let $\pi(0) = 0$, $\pi(1) = 2$, $\pi(2) = 1$. Let $\mathfrak{M} = \langle U, p, \{\pi, e\} \rangle$. By an elementary induction on the structure of γ, using saturation at each step, $\gamma \in \Gamma$ implies $\mathfrak{M} \models_c \gamma$, and $\gamma \in \Psi$ implies $\mathfrak{M} \not\models_c \gamma$.

Lemma 5   □

III.C.5.b. Derivations for all semisaturations of a sequent yield a derivation of the given sequent; a countermodel for any semisaturation yields a countermodel for the given sequent.

Lemma 6 *If all semisaturations of $\Gamma \vdash \Psi$ are derivable, then $\Gamma \vdash \Psi$ is derivable.*

---

[17]In fact, the countermodel is essentially a classical one, recoded into a nonclassical semantic system.

Proof: The semisaturations of $\Gamma \vdash \Psi$ are precisely the results of backwards derivation from $\Gamma \vdash \Psi$ using all of the rules of Theorem 3 except ($\Rightarrow$RS).

Lemma 6 □

Lemma 7 *If $\Gamma' \vdash \Psi'$ is a semisaturation of $\Gamma \vdash \Psi$, and $\mathfrak{M}$ is a countermodel for $\Gamma' \vdash \Psi'$, then $\mathfrak{M}$ is a countermodel for $\Gamma \vdash \Psi$.*

Proof: Direct, since $\Gamma \subseteq \Gamma'$ and $\Psi \subseteq \Psi'$.

Lemma 7 □

III.C.5.c. A derivation for any associate of a sequent yields a derivation for the given sequent; countermodels for all associates may be combined into a countermodel for the given sequent.

Lemma 8 *If some associate of $\Gamma \vdash \Psi$ is derivable, then $\Gamma \vdash \Psi$ is derivable.*

Proof: Direct, by the ($\Rightarrow$RS) rule.

Lemma 8 □

The difficult part of the completeness proof is the following lemma.

Lemma 9 *If $\Gamma \vdash \Psi$ is open and semisaturated, and if every associate $\Gamma' \vdash \Psi'$ has a countermodel, then $\Gamma \vdash \Psi$ has a countermodel.*

Proof: For Kripke models, the construction is simple. Let $\mathfrak{R}_1, \ldots, \mathfrak{R}_n$ be countermodels for the n associates of $\Gamma \vdash \Psi$. Combine them together with a new root world to produce $\mathfrak{R}$, as shown in Figure 10.

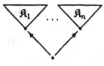

$$\{\gamma \in \Gamma \mid \gamma \text{ is atomic}\}$$

Figure 10: A Kripke countermodel for $\Gamma \vdash \Psi$, constructed from countermodels $\mathfrak{R}_1, \ldots, \mathfrak{R}_n$ for its associates.

At the new root world, let precisely those atomic formulae in Γ hold. Because the sequent rules never delete a formula on the left-hand side, all of these atomic formulae already hold in $\mathfrak{R}_1, \ldots, \mathfrak{R}_n$, so the constructed structure is a well-defined Kripke model.

By induction on the structure of a formula γ, $\gamma \in \Gamma$ implies $\mathfrak{R} \models_K \gamma$, and $\gamma \in \Psi$ implies $\mathfrak{R} \not\models_K \gamma$.

Basis: If γ is an atomic element of Γ, then $\mathfrak{R} \models_K \gamma$ by definition of $\mathfrak{R}$. If γ is an atomic element of Ψ, then $\gamma \notin \Gamma$ by the openness of $\Gamma \vdash \Psi$, and so $\mathfrak{R} \not\models_K \gamma$ by definition.

Induction: The cases ($\vee$L), ($\vee$R), ($\wedge$L), and ($\wedge$R) follow as in Lemma 5. The two interesting cases are ($\Rightarrow$L) and ($\Rightarrow$R).

($\Rightarrow$L) Let $\gamma \equiv \alpha \Rightarrow \beta \in \Gamma$. Now, either $\alpha \in \Psi$ or $\beta \in \Gamma$ by semisaturation.
If $\alpha \in \Psi$, then $\mathfrak{R} \not\models_K \alpha$ by our inductive hypothesis. Furthermore, each $\mathfrak{R}_i \models_K \alpha \Rightarrow \beta$, and so $\mathfrak{R} \models_K \alpha \Rightarrow \beta$ by the definition of $\models_K$. If $\beta \in \Gamma$, then $\mathfrak{R} \models_K \beta$ by our inductive hypothesis, and so $\mathfrak{R} \models_K \alpha \Rightarrow \beta$ trivially.

($\Rightarrow$R) Let $\gamma \equiv \alpha \Rightarrow \beta \in \Psi$. We have $\beta \in \Psi$ by semisaturation. If $\alpha \in \Gamma$, then $\mathfrak{R} \models_K \alpha$ and $\mathfrak{R} \not\models_K \beta$, so $\mathfrak{R} \not\models_K \alpha \Rightarrow \beta$. If $\alpha \notin \Gamma$, then α is α_i and β is β_i for some $\alpha_i \Rightarrow \beta_i \in \Psi_0$. By definition of $\mathfrak{R}_i$, $\mathfrak{R}_i \not\models_K \alpha \Rightarrow \beta$, so $\mathfrak{R} \not\models_K \alpha \Rightarrow \beta$.

If $\mathfrak{R}_1, \ldots, \mathfrak{R}_n$ are lattices, then it is easy to make the constructed Kripke model into a lattice, simply by adding a top element where every atomic formula is set true.

The construction for Läuchli models is more subtle. We can always produce a lattice-shaped Kripke model from each of the component Läuchli models, then combine them into another lattice-shaped Kripke model. But, there is no obvious way to construct the combined lattice as the subgroup structure of some permutation group constructed from the permutation groups in the component Läuchli models. So, we use a simple permutation-group construction that produces a *larger* lattice than the obvious Kripke one, then add objects to the universe that insulate the theory of the model from the effects of the extra subgroups.

For simplicity, we consider only the case $n = 2$. Given Läuchli countermodels $\mathfrak{L}_1 = \langle \mathbf{U}_1, P_1, \Pi_1 \rangle$ and $\mathfrak{L}_2 = \langle \mathbf{U}_2, P_2, \Pi_2 \rangle$ for the associates of $\Gamma \vdash \Psi$, we first take a disjoint union of two copies of each of the universes $\mathbf{U}_1$ and $\mathbf{U}_2$:

$$(\{0,1\} \times \mathbf{U}_1) \cup (\{3,4\} \times \mathbf{U}_2)$$

and a permutation group that permutes each subuniverse $\langle j, \mathbf{U}_i \rangle$ independently according to the appropriate Π_i, and also independently interchanges corresponding elements in copies of the same subuniverse where desired. The subgroups that setwise stabilize each subuniverse are intended to simulate the Kripke submodels $\mathfrak{K}_1$ and $\mathfrak{K}_2$. But, in Läuchli semantics, the models with these subgroups do not act independently, because they share mutual subgroups in which objects in more than one subuniverse are stabilized, and they allow strange subgroups in which nothing is stabilized, but in which permutations in the different subuniverses are linked together. Figure 11 shows the actual subgroup structure of the constructed Läuchli model. The regions marked $\mathfrak{L}_1$ and $\mathfrak{L}_2$ on the sides have exactly the subgroup structure of the given models $\mathfrak{L}_1$ and $\mathfrak{L}_2$. Unfortunately, we also get the three central regions of subgroups as well, containing the problematic subgroups mentioned above. Somehow, we must insulate the theory of the model from the effects of these extra subgroups[18].

The basic idea for insulating the two key submodels from interactions through their mutual subgroups is to add a permutation gizmo that gives *every* formula an invariant piece of evidence in each subgroup that stabilizes something in more than one subuniverse. For the case $n = 2$, merely add two gizmo objects, g_1 and g_2, to the universe, and make them members of the evidence set $P(\alpha)$ for *every* atomic formula α. Let each of the permutations that interchanges copies of universes $\mathbf{U}_1$ and $\mathbf{U}_2$, also interchange g_1 and g_2. All other permutations leave g_1 and g_2 fixed. Then, every subgroup that stabilizes objects in *both* $\mathbf{U}_1$ and $\mathbf{U}_2$ must stabilize g_1 and g_2 as well. So, in the troublesome region at the top of Figure 11, every atomic formula (and therefore every formula) has an invariant, and the whole region acts like a single maximal Kripke world with no effect on the theory of the model. The

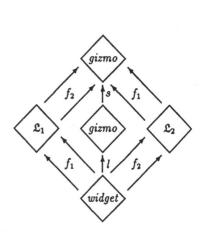

- The leftward transitions marked f_1 eliminate the permutation that interchanges $\langle 0, u_1 \rangle$ and $\langle 1, u_1 \rangle$, for $u_1 \in \mathbf{U}_1$.

- The rightward transitions marked f_2 eliminate the permutation that interchanges $\langle 3, u_2 \rangle$ and $\langle 4, u_2 \rangle$, for $u_2 \in \mathbf{U}_2$.

- The upward transition marked l links the two flipping permutations, allowing both or neither, but not one without the other.

- The upward transition marked s eliminates both flipping permutations.

- The subgroup structures within the left and right regions marked $\mathfrak{L}_1$ and $\mathfrak{L}_2$ are exactly the same as the subgroup structures of those given models.

- In the two central regions marked *gizmo*, all formulae have invariant evidence, because the gizmo points g_1 and g_2 are stabilized.

- In the lower region marked *widget*, a formula α has invariant evidence precisely if the nearest subgroups in both the $\mathfrak{L}_1$ and $\mathfrak{L}_2$ regions have invariant evidence for α.

Figure 11: Subgroup structure of a Läuchli countermodel for $\Gamma \vdash \Psi$, constructed from countermodels for its associates.

[18]Läuchli's proof uses cyclic groups only, and reduces the permutation group theory involved to number theory.

peculiar subgroups in the center of Figure 11, in which neither of U_1 nor U_2 is stabilized, but the permutations that flip the pairs of subuniverses are joined (i.e., both may be applied, or neither, but not one without the other), also provide invariants for all formulae, since they force the gizmo points g_1 and g_2 to be interchanged an even number of times, which leaves them invariant.

Finally, to insulate the key submodels from interactions that link their permutations, we add a large widget, containing the cross product $U_1 \times U_2$, and let each permutation of copies of U_1 and U_2 operate also on the appropriate components of pairs in the cross product. As a result, linking subgroups yield an invariant in $P(\alpha)$ if and only if their two subgroups stabilizing U_1 and U_2 *each* have invariants in $P(\alpha)$ (proof by induction on the structure of α). This widget also provides an invariant object in $P(\alpha)$ for each atomic formula $\alpha \in \Gamma$, making those atomic formulae hold in the constructed Läuchli model. The widget described above prevents the region at the bottom of Figure 11 from affecting the theory of the constructed model.

The Läuchli model resulting from all the considerations above is $\mathfrak{L} = \langle U, P, \Pi \rangle$, defined in Table 2. In this combined model, $\{0,1\} \times U_1$ and $\{3,4\} \times U_2$ produce the copies of the subgroup structure of $\mathfrak{L}_1$ and $\mathfrak{L}_2$, $\{2\} \times U_1 \times U_2$ contains the widget elements, and g_1 and g_2 are the gizmo elements. The proof that the combined Läuchli model $\mathfrak{L}$ is a countermodel to $\Gamma \vdash \Psi$ is left to the reader. Use induction on the structure of a formula, and apply Lemma 1.

Lemma 9 ☐

III.C.6. The derivation/countermodel search procedure[19] for a sequent $\Gamma \vdash \Psi$.

III.C.6.a. If $\Gamma \vdash \Psi$ is *saturated,* then

III.C.6.a.i. if $\Gamma \vdash \Psi$ is closed, then derive $\Gamma \vdash \Psi$ trivially by (Id)

————ii. otherwise $\Gamma \vdash \Psi$ is open, so construct the trivial countermodel.

III.C.6.b. Else if $\Gamma \vdash \Psi$ is *semisaturated,* then run the procedure recursively on each of the associates, and

III.C.6.b.i. if some associate is *derivable,* extend its derivation with one application of ($\Rightarrow$RS) to a derivation of $\Gamma \vdash \Psi$,

————ii. otherwise every associate has a *countermodel,* so construct a countermodel for $\Gamma \vdash \Psi$ as sketched in Lemma 9.

III.C.6.c. Otherwise, $\Gamma \vdash \Psi$ is *not semisaturated,* so run the procedure recursively on each semisaturation of $\Gamma \vdash \Psi$, and

III.C.6.c.i. if every semisaturation is *derivable,* combine their derivations with appropriate applications of rules ($\wedge$L), ($\wedge$R), ($\vee$L), ($\vee$R), ($\Rightarrow$L), and ($\Rightarrow$RW) into a derivation of $\Gamma \vdash \Psi$,

————ii. otherwise some semisaturation has a countermodel, which is also a countermodel for $\Gamma \vdash \Psi$.

III.C.7. The derivation/countermodel search procedure must terminate, because every recursive call reduces the number of subformulae of members of Γ and Ψ that do *not* appear in the sequent (i.e., it takes us closer to saturation), so eventually every branch must end in a saturated sequent[20].

————8. For the propositional calculus, the termination of the procedure, and the finiteness of the countermodels that it constructs, yields a completely constructive formal proof that $\Gamma \models_C \Psi$ implies $\Gamma \vdash \Psi$. In fact, we prove constructively the stronger result that, for all finite Γ and Ψ, either $\Gamma \vdash \Psi$ or there is a finite model $\mathfrak{L}$ for which all formulae in Γ hold, and no formulae in Ψ.

————9. A more satisfying completeness proof would construct a uniform construction directly from an invariant realizer in some carefully chosen Läuchli model. We suspect that such a proof is not possible.

---

[19]The alternation of conjunctions with disjunctions in the constructive proof procedure leads the constructive satisfiability problem to be PSPACE-complete [41]. The corresponding classical proof search procedure has only conjunctive branching, leading to an NP-complete satisfiability problem [5, 13].

[20]Dyckhoff [8], Hudelmaier [21], and Lincoln, Scedrov, Shankar [32] have independently used a modification to ($\Rightarrow$L) that allows every sequent rule to reduce formulae in sequents to their subformulae, eliminating the larger formulae. This makes the termination condition particularly simple. We have not investigated the suitability of these alternative sequent rules for our model constructions.

Given Läuchli models

$$\mathfrak{L}_1 = \langle U_1, P_1, \Pi_1 \rangle \qquad \text{and} \qquad \mathfrak{L}_2 = \langle U_2, P_2, \Pi_2 \rangle$$

construct the combined model

$$\mathfrak{L} = \langle U, P, \Pi \rangle$$

where

Universe: $U = (\{0,1\} \times U_1) \cup (\{2\} \times U_1 \times U_2) \cup (\{3,4\} \times U_2) \cup \{g_1, g_2\}$

Evidence: P maps atomic propositional symbols to subsets of U as follows

- $P(\alpha) = (\{0,1\} \times P_1) \cup (\{2\} \times P_1 \times P_2) \cup (\{3,4\} \times P_2) \cup \{g_1, g_2\}$
  for each atomic propositional symbol α

Permutations: Π is the group generated by the permutations
$\{\varpi_{\pi_1,\pi_2} \mid \pi_1 \in \Pi_1 \text{ and } \pi_2 \in \Pi_2\} \cup \{\varsigma_1, \varsigma_2\}$, where

- $-\ \varpi_{\pi_1,\pi_2}(\langle i,a\rangle) = \langle i, \pi_1(a)\rangle$ for $i \in \{0,1\}$, $a \in U_1$
  $-\ \varpi_{\pi_1,\pi_2}(\langle 2,a,b\rangle) = \langle 2, \pi_1(a), \pi_2(b)\rangle$ for $a \in U_1$, $b \in U_2$
  $-\ \varpi_{\pi_1,\pi_2}(\langle j,b\rangle) = \langle j, \pi_2(b)\rangle$ for $j \in \{3,4\}$, $b \in U_2$
- $-\ \varsigma_1(\langle 0,a\rangle) = \langle 1,a\rangle$
  $-\ \varsigma_1(\langle 1,a\rangle) = \langle 0,a\rangle$
  $-\ \varsigma_1(g_1) = g_2$
  $-\ \varsigma_1(g_2) = g_1$
- $-\ \varsigma_2(\langle 3,b\rangle) = \langle 4,b\rangle$
  $-\ \varsigma_2(\langle 4,b\rangle) = \langle 3,b\rangle$
  $-\ \varsigma_2(g_1) = g_2$
  $-\ \varsigma_2(g_2) = g_1$

All permutations behave as the identity except where specified above.

Table 2: Läuchli countermodel for $\Gamma \vdash \Psi$, constructed from countermodels for its associates

III.D. Extension to Predicate Calculus.

III.D.1. Extend realizability semantics with dependent types.

————2. Extend the set of sequent rules with the natural rules (∀L), (∀R), (∃L), and (∃R).

————3. The derivation/countermodel search procedure above still works, but it does not always terminate. In case of nontermination, the infinite search tree constructs an infinite model.

————4. Unfortunately, while we can construct an infinite model above, we require the axiom of choice to prove that the model so constructed is a countermodel for $\Gamma \vdash \Psi$. For the most obvious formulation of completeness ($\Gamma \models_{\mathcal{L}} \Psi$ implies $\Gamma \vdash \Psi$) it is not even clear whether we can prove the double negation constructively.

————5. Some change in the set-theoretic representation of Läuchli models, analogous to Beth's variation on Kripke models, can probably yield a completely constructive result. The definition of such models is an open problem. It is possible that some other construction of Läuchli countermodels will provide a completely constructive proof, even with the current set-theoretic representation, but we conjecture that no such proof exists[21]

4   Acknowledgments

The ideas in this paper, as well as the presentation, benefitted from discussions with Frank Corley, William Tait, William Howard, and Gaisi Takeuti. James Lipton gave a very thorough critique of the final draft, and provided most of the crucial references to previous literature. The result is substantially better than it would have been without his help.

———

[21]Kreisel [27], van Dalen [47], and Leivant [31] have shown that truly constructive completeness theorems for the first-order predicate calculus, using models without fallible nodes, contradict an analogue of Church's thesis for realizers. See Beeson [1] for a discussion of the formal statements of Church's thesis.

References

[1] M. J. Beeson. *Foundations of Constructive Mathematics*. Springer-Verlag, 1980.

[2] E. W. Beth. *The Foundations of Mathematics, A Study in the Philosophy of Science*. Studies in Logic and the Foundations of Mathematics. North-Holland Publishing Company, Amsterdam, 1959.

[3] K. B. Bruce, A. R. Meyer, and J. C. Mitchell. The semantics of second-order lambda calculus. *Information and Computation*, 85(1):76–134, 1990. Reprinted in *Logical Foundations of Functional Programming*, ed. G. Huet, Addison-Wesley (1990) 213–273.

[4] A. Church. An unsolvable problem of elementary number theory. *American Journal of Mathematics*, 58:345–363, 1936.

[5] S. A. Cook. The complexity of theorem-proving procedures. In *3rd Annual ACM Symposium on Theory of Computing*, pages 151–158. Association for Computing Machinery, 1971.

[6] H. B. Curry and R. Feys. *Combinatory Logic Volume I*. Studies in Logic and the Foundations of Mathematics. North-Holland Publishing Company, Amsterdam, 1958.

[7] M. A. E. Dummett. *Elements of Intuitionism*. Oxford University Press, 1977.

[8] R. Dyckhoff. Contraction-free sequent calculi for intuitionistic logic. *Journal of Symbolic Logic*, 1991. To appear.

[9] S. Feferman. Constructive theories of functions and classes. In M. Boffa, D. van Dalen, and K. McAloon, editors, *Logic Colloquium 78*, pages 159–224. North-Holland, 1979. Proceedings of the colloquium held in Mons, August 1978.

[10] M. Felleisen, D. Friedman, E. Kohlbecker, and B. Duba. Reasoning with continuations. In *Proceedings of the First Annual Symposium on Logic in Computer Science*, pages 131–141, 1986.

[11] J. E. Fenstad, editor. *Selected Works in Logic*. Universitetsforlaget, 1970.

[12] M. C. Fitting. *Intuitionistic Logic, Model Theory, and Forcing*. Studies in Logic and the Foundations of Mathematics. North-Holland Publishing Company, Amsterdam, London, 1969.

[13] M. R. Garey and D. S. Johnson. *Computers and Intractability*. W. H. Freeman, 1979.

[14] G. Gentzen. Beweisbarkeit und Unbeweisbarkeit von Anfangsfällen der transfiniten Induktion in der reinen Zahlentheorie. *Mathematische Annalen*, 119:140–161, 1943.

[15] J.-Y. Girard, Y. Lafont, and P. Taylor. *Proofs and Types*. Cambridge Tracts in Theoretical Computer Science. Cambridge University Press, 1989.

[16] Jean-Yves Girard. Une extension de l'interprétation de Gödel à l'analysis, et son application à l'élimination des coupures dans l'analysis et la théorie des types. In J. E. Fenstad, editor, *Proceedings 2nd Scandinavian Logic Symposium*. North-Holland, 1971.

[17] Leon Henkin. Completeness in the theory of types. *The Journal of Symbolic Logic*, 15:81–91, 1950.

[18] A. Heyting. Die formalen Regeln der intuitionistischen Logik. *Sitzungsberichte der Preussischen Academie der Wissenschaften, Physikalisch-Matematische Klasse*, pages 42–56, 1930.

[19] A. Heyting. *Intuitionism, an introduction*. North-Holland, 1971.

[20] W. A. Howard. The formulae-as-types notion of construction. In J. P. Seldin and J. R. Hindley, editors, *To H. B. Curry: Essays on Combinatory Logic, Lambda Calculus and Formalism*, pages 479–490. Academic Press, 1980.

[21] J. Hudelmaier. *Bounds for Cut Elimination in Intuitionistic Logic*. PhD thesis, Universität Tübingen, 1989.

[22] P. T. Johnstone. Conditions related to de Morgan's law. In M. P. Fourman, C. J. Mulvey, and D. S. Scott, editors, *Applications of Sheaves. Proceedings of the Research Symposium on Applications of Sheaf Theory to Logic, Algebra, and Analysis*, pages 479–491. Springer Verlag, 1979. Lecture Notes in Mathematics 753.

[23] S. C. Kleene. On the interpretation of intuitionistic number theory. *The Journal of Symbolic Logic*, 10(4):109–124, December 1945.

[24] S. C. Kleene. Realizability. In A. Heyting, editor, *Constructivity in Mathematics*, pages 285–289. North-Holland Publishing Company, Amsterdam, 1959. Proceedings of the Colloquium Held in Amsterdam, August 26–31, 1957.

[25] S. C. Kleene. *Introduction to Metamathematics*. North-Holland, 1971.

[26] S. C. Kleene and R. E. Vesley. *The Foundations of Intuitionistic Mathematics, Especially in Relation to Recursive Functions*. Studies in Logic and the Foundations of Mathematics. North-Holland Publishing Company, Amsterdam, London, 1965.

[27] Georg Kreisel. On weak completeness of intuitionistic logic. *Journal of Symbolic Logic*, 27, 1962.

[28] S. A. Kripke. Semantical analysis of modal logic I: Normal modal propositional calculi. *Zeitschrift für Mathematische Logik und Grundlagen der Mathematik*, 9:67–96, 1963.

[29] S. A. Kripke. Semantical analysis of intuitionistic logic, I. In J. N. Crossley and M. A. E. Dummett, editors, *Formal Systems and Recursive Functions*, pages 92–130. North-Holland Publishing Company, Amsterdam, 1965. Proceedings of the Eighth Logic Colloquium, Oxford, July 1963.

[30] H. Läuchli. An abstract notion of realizability for which intuitionistic predicate calculus is complete. In A. Kino, J. Myhill, and R. E. Vesley, editors, *Intuitionism and Proof Theory*, Studies in Logic and the Foundations of Mathematics, pages 277–234. North-Holland Publishing Company, Amsterdam, London, 1970. Proceedings of

the Conference on Intuitionism and Proof Theory, Buffalo, New York, August 1968.

[31] D. M. E. Leivant. Failure of completeness properties for intuitionistic predicate logic for constructive models. *Annales Scientifiques de l'Université de Clermont-Ferrand II, Section Mathematiques*, 13:93–107, 1976.

[32] P. Lincoln, A. Scedrov, and N. Shankaar. Linearizing intuitionistic implication. In *Logic in Computer Science*, July 1991.

[33] J. Lipton. Kripke semantics for dependent type theory and realizabity interpretations. In *these proceedings*, 1991.

[34] M. Machtey and P. Young. *An Introduction to the General Theory of Algorithms*. North-Holland, 1978.

[35] J. C. Mitchell. Type systems for programming languages. In J. van Leeuwen, editor, *Handbook of Theoretical Computer Science, Volume B*, pages 365–458. North-Holland, Amsterdam, 1990.

[36] C. S. Peirce. On the algebra of logic: A contribution to the philosophy of notation. *American Journal of Mathematics*, 8:180–202, 1885.

[37] V. R. Pratt. Semantical considerations on Floyd-Hoare logic. In *17th Annual Symposium on Foundations of Computer Science*, pages 109–121. IEEE, October 1976.

[38] D. Prawitz. *Natural Deduction*. Almqvist & Wiksell, Stockholm, 1965.

[39] G. F. Rose. Propositional calculus and realizability. *Transactions of the American Mathematical Society*, 75:1–19, July–September 1953.

[40] T. Skolem. Über die Nicht-charakterisierbarkeit der Zahlenreihe mittels endlich oder abzählbar unendlich vieler Aussagen mit ausschließlich Zahlenvariablen. *Fundamenta Mathematicae*, 23:150–161, 1934. Reprinted in [11].

[41] R. Statman. Intuitionistic propositional logic is polynomial-space complete. *Theoretical Computer Science*, 9(1):67–72, 1979.

[42] R. Statman. Logical relations and the typed lambda calculus. *Information and Control*, 65:85–97, 1985.

[43] Sören Stenlund. *Combinators, λ-terms, and Proof Theory*. D. Riedel Publishing Company, Dordrecht-Holland, 1972.

[44] Alfred Tarski. Pojęcie prawdy w językach nauk dedukcyjnch. *Prace Towarzystwa Naukowego Warzawskiego*, 1933. English translation in [45].

[45] Alfred Tarski. *Logic, Semantics, and Metamathematics*. Oxford University Press, 1956.

[46] A. S. Troelstra. *Mathematical Investigation of Intuitionistic Arithmetic and Analysis*, volume 344 of *Lecture Notes in Mathematics*. Springer-Verlag, Berlin, 1973.

[47] A. S. Troelstra and D. van Dalen. *Constructivism in Mathematics: an Introduction, volume II*. Studies in Logic and the Foundations of Mathematics. North-Holland, 1988.

[48] A.S. Troelstra. *Choice Sequences: a chapter of intuitionistic mathematics*. Oxford Logic Guides. Clarendon Press, Oxford, 1977.

[49] A. M. Turing. On computable numbers, with an application to the Entscheidungsproblem. *Proceedings of the London Mathematical Society, Second Series*, 42:230–265, 1937.

[50] Johan van Benthem. Correspondence theory. In *Handbook of Philosophical Logic, Volume 2: Extensions of Classical Logic*, pages 167–247. D. Reidel, 1984.

[51] L. Wittgenstein. Tractatus logico-philosophicus. *Annalen der Natur-philosophie*, 1921. English translation in [52].

[52] L. Wittgenstein. *Tractatus Logico-Philosophicus*. Routledge and Kegan Paul, 1961.

Kripke Semantics for Dependent Type Theory and Realizability Interpretations

EXTENDED ABSTRACT

by

James Lipton

Department of Mathematics

University of Pennsylvania

Philadelphia, PA 19104

Abstract

Constructive reasoning has played an increasingly important role in the development of provably correct software. Both typed and type-free frameworks stemming from ideas of Heyting, Kleene, and Curry have been developed for extracting computations from constructive specifications. These include Realizability, and Theories based on the Curry-Howard isomorphism. Realizability – in its various typed and type-free formulations – brings out the algorithmic content of theories and proofs and supplies models of the "recursive universe". Formal systems based on the propositions-as-types paradigm, such as Martin-Löf's dependent type theories, incorporate term extraction into the logic itself.

Another, major tradition in constructive semantics originated in the model theory developed by Gödel, Herbrand and Tarski, resulting in the interpretations developed by Kripke and Beth, and in subsequent categorical generalizations. They provide a *complete* semantics for constructive logic. These models are a powerful tool for building counterexamples and establishing independence and conservativity results, but they are often less constructive and less computationally oriented.

It is highly desirable to combine the power of these approaches to constructive semantics, and to elucidate some connections between them. We define modified Kripke and Beth models for syntactic Realizability and Dependent Type theory, in particular for the one-universe Intensional Martin-Löf Theory ML_0^i and prove a completeness theorem for the latter theory. These models provide a new framework for reasoning about computational evidence and the process of term-extraction. They are defined over a *constructive* type-free metatheory based on the Feferman-Beeson theories of abstract applicative structure.

Our models have a feature which is shared by all published constructive completeness theorems for intuitionistic logic, known in the literature as "fallibility": there may be worlds in which some sentences are both false and true, a phenomenon which corresponds to the presence of empty types in various type disciplines. We also identify a natural lattice of truth values associated with type theory and realizability: the *degrees of inhabitation*.

1 Introduction

Kripke models were developed in 1963 by Saul Kripke. Similar interpretations were implicit in earlier, topological models of Tarski from the 1940's and in Beth's work in the 1950's. They were subsequently generalized by a number of researchers who strengthened some of the algebraic and topological features of the semantics. These models have proven to be a powerful tool for studying the metamathematics of intuitionistic formal systems.

Much of this power lies precisely in the fact that they bring to bear on the study of formal systems an exceptionally powerful and versatile arsenal of algebraic, topological and categorical tools, yielding new consistency, conservativity and independence results. Perhaps the greatest strength of the semantics as developed in Saul Kripke's original paper, is that it supplies the means for effective systematic construction of counterexamples forced by specific requirements, in the spirit of Cohen's independence results for set theory. Often a simple diagram suffices to give intuitionistic counterexamples. These models have also been used in computer science to provide interpretations of computations in terms of state transitions, or properties invariant under certain transformations ([28, 16, 29])

Our objective here is to develop a similar tool for type-theory, and other formal systems for carrying out term extraction from constructive reasoning ([5]), extending the line of research initiated by Mitchell and Moggi for simply typed λ-Calculus in [24], and for realizability-style interpretations ([3, 31, 13]). The models developed in this paper present a number of characteristics of interest: they require the use of covers, or *delayed satisfaction of disjunctive and existential formulas*, as well as the inhabitation of void in the semantics: *we allow inconsistent, or "fallible" nodes*. This has proven to be a critical component in all known intuitionistically valid completeness theorems (see [17, 18]). As with the realizability-Kripke models to be discussed below, and in the Effective Topos (see Hyland's [12]), the truth-value structure imposed by type theory is that of the *degrees of inhabitation* of definable sets [1]. That is to say, the Heyting Algebra formed by taking, for all definable (almost-negative) sets in HA or some applicative theory like APP , equivalence classes of their existential closures under provable equivalence, constitutes a natural domain of truth values of type theory, in a sense that will be made precise below. [2]

This paper is an extended abstract of a talk given at the CINCS conference. Many proofs are omitted or briefly sketched. Complete details can be found in [18, 19].

A Thumbnail Sketch of Kripke and Realizability semantics

The kind of Kripke model $\mathcal{K}$ we will be concerned with here is, in fact, a variant of the standard one in the literature. Strictly speaking, it is a *fallible Beth model*, because

- (fallibility) nodes are allowed to be inconsistent, i.e. some p may force *every* φ, provided not every node does so. We require $p \Vdash \bot \rightarrow p \Vdash \varphi$, a node forcing an inconsistency must force everything.

- We relax the definition of forcing of disjunction and existential quantification. Our variant, an instance of *forcing with covers over a site* (see Troelstra and van Dalen's [31], or Grayson's [10]), can be thought of as an analogue of Beth's forcing with *bars*. [3]

Definition 1.1 *A fallible Kripke/Beth structure* $\mathcal{B} = \langle B, \leq_B, \mathbf{D}, \Vdash_B, Cov \rangle$ *over a language* $\mathcal{L}$ *(containing e.g., relation symbols, R_i, and constant symbols c_i [4]) as follows:*
$\langle B, \leq_B \rangle$ *is a pre-order. Members of B are called nodes.* $\mathbf{D}$ *assigns a domain of individuals to nodes in a monotone way: $p \leq q \rightarrow \mathbf{D}(p) \subseteq \mathbf{D}(q)$.* $\Vdash$ *is a monotone binary relation between nodes and atomic sentences over $\mathcal{L}$:*

$$p \leq q \quad \& \quad p \Vdash R(a) \quad \Rightarrow q \Vdash R(a).$$

satisfying the covering property *(if every member of a cover of p forces a sentence, then so does p):*

$$Cov(p, \mathbf{S}) \,\&\, (\forall q \in \mathbf{S})(q \Vdash \varphi) \Rightarrow p \Vdash \varphi$$

where: Cov is a binary relation between nodes p and *sets* of nodes $\mathbf{S} \subseteq B$, satisfying a series of *cover axioms*. Rather than give a general formulation of cover axioms here, we will limit ourselves to giving the one required for our argument, at the beginning of section 2 below. [5]

The forcing relation is extended to all sentences φ over the language $\mathcal{L}$ as follows:

1. $p \Vdash \varphi \& \psi$ iff $p \Vdash \varphi$ and $p \Vdash \psi$

2. $p \Vdash \varphi \vee \psi$ iff $(\exists \mathbf{S}) \, Cov(p, \mathbf{S})$ and $(\forall q \in \mathbf{S}) \, q \Vdash \varphi$ or $q \Vdash \psi$

---

[1] or, in some cases, just the almost-negative ones

[2] Our work is, in some respects, a syntactic analogue of Hyland's *Effective Topos*. But the syntactic nature of the realizability makes a cruicial difference here, as can be seen from the main arguments. Our model is not a Topos, but rather, a locally cartesian category, with first order semantics. Our "object of truth values" is the set of formulas in one free variable in APP , or, in one case a subquofient we have called the inhabitation degrees.

[3] As a first approximation, we can think of forcing with covers as relaxing the condition that a node p of a Kripke model force a series of formulas by allowing instead that some set of nodes associated with p do the forcing. Such an associated set could be interpreted as a set of future states after state p: in which case we are tolerating some delay in the confirmation that p really does force something. See the references cited for other, more topological insights into this idea.

[4] since APP can be formulated relationally by taking App as a ternary predicate, this is all we need. Adding function symbols to the formalism is easy: the n-ary function symbol $\bar{f}$ is interpreted by the function f at node p if for every q above p $f : \mathbf{D}(q) \rightarrow \mathbf{D}(q)^n$ and the graph of f restricted to $\mathbf{D}(p)$ is contained in the graph of f at all higher nodes.

[5] See Grayson's [10] for a quite general formulation, or Troelstra and van Dalen, *op.cit.* for the original definition of forcing over a site, due to Joyal, and based on earlier ideas of Grothendieck.

3. $p \Vdash \varphi \to \psi$ iff $(\forall q \geq p)\, q \Vdash \varphi \Rightarrow q \Vdash \psi$

4. $p \Vdash \exists x \varphi(x)$ iff $(\exists S)\, Cov(p, \mathbf{S})$ and $(\forall q \in \mathbf{S})(\exists a \in \mathbf{D}(q))\, q \Vdash \varphi(a)$

5. $p \Vdash \forall x \varphi(x)$ iff $\quad (\forall q \geq p)(\forall a \in \mathbf{D}(q))\, q \Vdash \varphi(a)$.

We give a formulation of realizability due originally to Feferman, which has been extensively used in the literature, and modified by e.g. Troelstra, Van Dalen, Diller, Beeson and others (see [31]). It is defined over an abstract applicative theory, APP, of *partial application* using the logic of existence or *partial terms*. We refer the reader to [3] or [31] for details.

Abstract realizability and interpreting theories into APP In the presentation below, we may take the approach described in detail in [18] to incorporate other theories into our formal system: interpret an arbitrary first order theory S into APP by adding a universe or domain predicate U whose extension is the target of the interpretation. Then we leave "up to the reader" how to define the atomic realizability $|A|(x)$ of formulas A from the object language $\mathcal{L}(\mathbf{S})$ by terms x from the *realizing metalanguage* **APP**. We will only require that the atomic realizability be faithfully copied by the atomic forcing assignments. The essential point is that we wish to consider almost any syntactic realizability over any theory. The role of APP is only to supply abstract realizers. We will not explicitly develop this approach here, however, since the notation is cumbersome and the details not especially enlightening.

Definition 1.2 *Let A, B be sentences over the language of APP (APP$\underline{C}$). Then we define inductively the realizability formulas $|A|$ in one free variable as follows:*

$$\text{If } A \text{ is prime } |A|(x) \text{ is } A \,\&\, x \downarrow$$

$$|A \,\&\, B|(x) \equiv (|A| \times |B|)(x) \stackrel{\text{def}}{=} |A|(\pi_0 x) \,\&\, |B|(\pi_1 x)$$

$$|A \vee B|(x) \equiv (|A| + |B|)(x) \stackrel{\text{def}}{=} N(\pi_0 x) \,\&\, (\pi_0 x = 0 \to |A|(\pi_1 x)) \,\&$$
$$(\pi_0 x \neq 0 \to |B|(\pi_1 x))$$

$$|A \to B|(x) \equiv (|A| \Rightarrow |B|)(x) \stackrel{\text{def}}{=} \forall y[|A|(y) \to xy \downarrow \,\&\, |B|(xy)]$$

$$|\exists y A(y)|(x) \equiv (\textstyle\sum |A|)(x) \stackrel{\text{def}}{=} |A(\pi_0 x)|(\pi_1 x)$$

$$|\forall y A(y)|(x) \equiv (\textstyle\prod |A|)(x) \stackrel{\text{def}}{=} \forall y[(xy) \downarrow \,\&\, |A(y)|(xy)]$$

$|A|(x)$ is usually written $x \underset{\tilde{}}{r} A$. Note that if A is a formula in n variables over APP (or APP$\underline{C}$) then the above clauses defined an associated realizability formula in $n + 1$ variables. Note that if A is *prime*, A is logically equivalent to $|A|(x)$ for any variable x.

2 A Kripke Model for abstract realizability

Let $\underline{C} = \{c_i | i \in \omega\}$ be a denumerable set of fresh constants. APP$\underline{C}$ is the theory APP, together with the constants in $\underline{C}$, the axioms $c \downarrow$ for each $c \in \underline{C}$ and all schemas extended to the new language in the way specified in e.g. [3]. Let $\mathcal{L}$ be the language of APP$\underline{C}$, and Δ the set of all formal closed (variable-free) terms of $\mathcal{L}$.

We are now in a position to define the Kripke model $\mathcal{K}$ we want. The **nodes** of $\mathcal{K}$ are formulas of APP$\underline{C}$ in one free variable, with ordering given by

$$A \geq B \text{ iff } (\exists t \in \Delta) \text{ APP}\underline{C} \vdash \forall x[A(x) \to tx \downarrow \,\&\, B(tx)]$$

(We will sometimes denote this state of affairs by $A \overset{t}{\geq} B$, or $t : A \to B$). The **domain** of the Kripke Model is the constant domain Δ. Our notion of **covers** (see section 1) will be as follows: For any node A and set of nodes S we have $Cov(A, \mathbf{S})$ iff

1. $\forall B \in \mathbf{S}(\exists t \in \Delta)\, (t : B \to A)$

2. if whenever, for some object D $\forall B \in S \exists t_B \in \Delta$ $t_B : B \to D$ then $(\exists t \in \Delta)$ $t : A \to D$, i.e., $A = \inf(S)$ in $\langle |\mathcal{K}|, \leq \rangle$.

Finally, our atomic forcing **assignment** is given by:

$$A \Vdash \theta \quad \text{iff} \quad (\exists t \in \Delta)\ t : A \to |\theta| \tag{1}$$

for *atomic* [6] θ. In particular for $\theta \in \mathcal{L}(\mathbf{APP})$ this means true ground θ are forced by every node, and false ground θ only by provably uninhabited ones. Our main result is that the equivalence (1) holds for all formulas, not just the atomic ones.

Theorem 2.1 *For every node A and sentence φ*

$$A \Vdash \varphi \quad \text{iff} \quad (\exists t \in \Delta)\ t : A \to |\varphi|$$

(proof: see [19])

Now, we immediately have the desired result, to wit, that the model just constructed is elementarily equivalent to abstract realizability over APP.

Corollary 2.2 *Let φ be a sentence in the language of APP, $\mathcal{K}_{APP}$ the Kripke model described above. Then*

$$\mathcal{K}_{APP} \models \varphi \quad \text{iff} \quad \mathbf{APP} \vdash \exists x (x \underset{\sim}{r} \varphi)$$

(where $x \underset{\sim}{r} \varphi$ is the traditional notation for $|\varphi|(x)$).

On the "Degrees of Inhabitation"

In the preceding section, we constructed a Kripke model elementarily equivalent to abstract realizability over **APP**. In fact, via the modifications outlined in the remarks preceding definition (1.2) we have defined a uniform class of Kripke models for a quite general notion of realizability, in which the coding of the object theory and its atomic realizability are free to vary.

In particular, the model constructed above satisfies Church's thesis and the strong computational properties found in realizability semantics. But what is the structure of such models, and what light do they shed on the truth-value structure implicit in realizability? In this section we will briefly address this issue. To begin with, we note that the models have unusual properties. They are *closed under finite suprema and infima*: The cartesian product is the supremum, and $+$ the infimum, in fact they are weakly cartesian closed (we omit the proof in this abstract).

Also, in these models, every set of nodes has an upper bound, since we have "fallible" or inconsistent nodes at the top. Such fallible nodes were first discussed in the papers of Läuchli [16], Veldman [32] and de Swart [30], and seem to play a fundamental role in Kripke models associated with realizability, as well as in the intuitionistic completeness theorem of Friedman, Veldman, de Swart and Troelstra (see the discussion in [31]). This suggests that one should think of Kripke models of the type studied here as being developed in an intuitionistic metatheory, where consistency of nodes is not necessarily decidable. Our proofs are fully constructive and can be seen as a kind of syntactic counterpart to a constructive completeness theorem for Kripke semantics (this is brought out in more detail in [18]). These models are perhaps best conceived as *internal Kripke models*, that is to say, as models developed within a Topos or Kripke model. Another way of looking at the existence of inconsistent nodes is in terms of tableaux proofs in the style of Nerode, Fitting or Odifreddi (see e.g., [25, 26, 8]) in which we are not always constructively able to recognize when branches are infinite and consistent or finite and closed off.

One may regard the lattice-theoretic structure of these models as a syntactic analogue of reducibility orderings in recursion theory. Nodes ordered by

$$A \geq B \iff \exists t\ \mathbf{APP} \vdash \forall x (A(x) \to tx \downarrow \ \& \ B(tx))$$

constitute a structure we might call syntactic *degrees of inhabitation*, not unlike provable M-degrees (which are a stronger ordering: functions must preserve complements). A slight modification of our Kripke structures gives us a sharper picture, however. We will need a few definitions to make this precise.

---

[6]As remarked above, this atomic assignment can be replaced with little modification of the main arguments, by any formal interpreted realizability $|(\theta)^*|(t)$, where $|(\theta)^*|$ is the interpretation of a formula θ from any *object language* $\mathcal{L}'$ into the applicative metalanguage **APP**.

Definition 2.3 *A formula over a language L extending HA or* **APP** *is called* **negative** *if it contains no disjunctions or existential quantifiers, and* **almost negative** *if it contains no disjunctions and existential quantifiers are present only next to atomic subformulas.*

Negative and almost negative formulas play a role in realizability interpretations similar to that of absolute formulas in set theory: they are equivalent to their own realizability in a uniform way.

Definition 2.4 *A formula A is called* **self-realizing** *if there is a term* $\mathcal{J}_A$ *of* **APP** *such that, provably in* **APP**,

(i) $A(x) \rightarrow \mathcal{J}_A(x) \underset{\sim}{r} A$

(ii) $(q \underset{\sim}{r} A) \rightarrow A$.

The following properties of negative and almost-negative formulas are well-known. See Beeson *op.cit.* and [31] for proofs.

Lemma 2.5 *Every negative formula is self-realizing. If A is almost negative then A is equivalent to some negative formula B, provably in* **APP**. *Furthermore, every realizability formula* $|A|(x)$, *or* $x \underset{\sim}{r} A$ *is almost negative.*

Theorem 2.6 *If B is negative and C is any formula in two free variables, then*

$$\mathbf{APP} \vdash \forall x (B(x) \rightarrow \exists z C(x, z))$$

implies that for some closed term f

$$\mathbf{APP} \vdash \forall x (B(x) \rightarrow fx \downarrow \ \& \ C(x, fx)).$$

Inspection of the proofs of theorem 3.6, above, and, e.g., theorems 2.5 and 3.9 in [17] shows that the Kripke model $\mathcal{K}$, $\mathcal{K}_{APP}$, ($\mathcal{K}_{HA}$ in [17]) , are elementarily equivalent to their *almost-negative* reducts $\mathcal{K}^{an}$, $\mathcal{K}^{an}_{APP}$, $\mathcal{K}^{an}_{HA}$ given by restricting the corresponding partial orders $\langle K, \leq \rangle$ to

$$\{A : A \simeq \text{ an almost-negative formula }\}.$$

What is the significance of cutting the models down to the **a-n** reducts? Theorem 2.6 provides the key. For almost negative formulas, the partial order in $\mathcal{K}$ can be easily characterized: $A \geq B$ is equivalent to the existence of a proof

$$APP \vdash (\exists x A(x)) \rightarrow (\exists y B(y)).$$

In short, our model is simply the *Lindenbaum algebra of existential closures of almost negative formulas* with the order reversed, a structure we have dubbed the *provable degrees of inhabitation*. In fact, the result is not entirely surprising. Realizability semantics means *inhabiting* statements with computational *evidence*. Therefore a natural algebraic interpretation of realizability is obtained by taking as truth-values the different degrees of consistency of provable inhabitation. To be realized is to be forced by the provably inhabited formulas. Other models are obtained by relativizing to any degree that is consistent and independent over HA or APP. This also points the way towards the sort of converse studied in Läuchli's [16] and the author's [18]. If we are to construe arbitrary Kripke models as abstract realizability interpretations we must arrange to imbed the Heyting algebra of truth-values generated by the Kripke model into an algebra of degrees of inhabitation.

3 Interpreting Dependent Types

We now develop Kripke models for type theory in the same vein. We will take as our basic theory the formulation $\mathbf{ML}^i_0$ of "one-universe" Martin-Löf type theory presented in [31]. A similar presentation can be found in Beeson *op.cit.*. In order to harness the framework developed above to model type theories, we need to add one twist to the definitions just given. Using a syntax close to that of conventional first-order logic, we will be interpreting *dependent type expressions* of the form, e.g., $\prod x : D \cdot A(x)$. At first sight, this might seem like a natural "correlate" to the first-order predicate $(\forall x \in D)(A(x))$. However, in the type-free first-order language of arithmetic (or of **APP**) the latter formula is usually understood to be an abbreviation for $(\forall x)(D(x) \to A(x))$. This clearly violates the spirit of the Martin-Löf theory, in identifying type membership $x : D$, with what –for want of a better word – we call *parametricity*, $D(x)$. In our type-free *semantics* we will remedy this by using *realizability* to distinguish between the two notions: $x : D$ will become $|(D)^r|(x)$ where the bars denote realizability, and the r- superscript denotes a translation to be defined below. However, in the *syntax* we will require an extension of the definition of well-formed formula to include *formalized bounded quantification* in order to model dependent Π and Σ types. We call such new formulas *special* or *extended formulas* in this section.

Definition 3.1 *If D is a formula and $\theta(x)$ is a formula with x free, then $(\forall_{x \in D})\theta(x)$ and $(\exists_{x \in D})\theta(x)$ are formulas, with free variables $FV(\theta) \cup FV(D) \setminus \{x\}$. If D is a formula then τ_D is an atomic formula with one more variable free than D. $\tau_D(a)$ is simply a formalization of $a : D$ into our "extended first-order syntax." We also write $a : D$ for $\tau_D(a)$.*[7]

All schemas of first-order logic and the axioms of **APP** are to be extended to the new formulae. Their *meaning* and role in our work should be made clear by the way they are treated in the semantics.

Definition 3.2 *An extended Kripke model for $\mathbf{APP}\underline{C}$ is a Kripke model which satisfies the usual definition of covers and forcing for standard formulas of $\mathbf{APP}\underline{C}$, as well as the following conditions for each node p:*

$$p \Vdash (\exists_{x \in D})\theta(x) \overset{\text{def}}{\equiv} (\exists S)(Cov(p, S))(\forall q \in S)(\exists a \in D(q)) p \Vdash a : D \ \& \ \theta(a)$$

$$p \Vdash (\forall_{x \in D})\theta(x) \overset{\text{def}}{\equiv} (\forall q \geq p)(\forall a \in D(q)) \quad q \Vdash a : D \to \theta(a)$$

We must also define forcing of proofs *as well as formulas. For a a constant in $D(p)$,*

$$p \Vdash a : D \quad (\text{ or } \quad p \Vdash \tau_D(a))$$

is defined as follows:

$$p \Vdash a : C \vee D \overset{\text{def}}{\equiv} (\exists S)(Cov(p, S))(\forall q \in S)(q \Vdash \mathbf{p}_0 a = 0 \ and \ q \Vdash \mathbf{p}_1 a : C) \ or$$
$$(q \Vdash \mathbf{p}_0 a \neq 0 \ and \ q \Vdash \mathbf{p}_1 a : D)$$

$$p \Vdash a : (\exists_{x \in D})\theta(x) \overset{\text{def}}{\equiv} p \Vdash \mathbf{p}_0 a : D \ and \ p \Vdash \mathbf{p}_1 a : \theta(\mathbf{p}_0 a)$$

$$p \Vdash a : (\forall_{x \in D})\theta(x) \overset{\text{def}}{\equiv} (\forall q \geq p)(\forall a \in D(q))(q \Vdash u : D \Rightarrow q \Vdash au : \theta(u))$$

For *atomic* D, the $p \Vdash a : D$ must also be specified by the atomic forcing assignment for the model in question. Every occurrence of an application au above is *strict*: it is understood that $au \downarrow$. Also, every occurrence of the disjunctive flag condition $\mathbf{p}_0 a = 0$ or $\mathbf{p}_0 a \neq 0$ is preceded by a tacit $N(\mathbf{p}_0 a)$: the condition is decidable for natural numbers. We now need only modify our definition of realizability slightly, to take the new special formulas defined in (3.1) into account.

Now we define (extended) realizability for special formulas as follows:

Definition 3.3

$$|a : D|(x) \overset{\text{def}}{\equiv} |D|(a)$$

$$|\exists_{y \in D} A(y)| \overset{\text{def}}{\equiv} |D|(\mathbf{p}_0 z) \ \& \ |A(\mathbf{p}_0 z)|(\mathbf{p}_1 z)$$

$$|\forall_{y \in D} A(y)| \overset{\text{def}}{\equiv} \forall y[|D|(y) \to zy \downarrow \ \& \ |A(y)|(zy)].$$

We now make use of a well-known result in the semantics of Martin-Löf type theory: there is a natural translation of the theory into **APP** via abstract realizability, stemming from Martin-Löf's own informal semantics, and developed (in roughly similar ways) in the work of Troelstra and van Dalen, Allen, Beeson, and Diller, amongst others (see [31, 2]). The details of the translation of contexts Γ and judgements θ into formulas $[\Gamma]$ and $[\theta]$ of **APP** are and omitted in this abstract. Troelstra and van Dalen obtain the following soundness result for their translation. A similar result is to be found in Beeson, *op.cit.*

---

[7]The context will make it clear whether we are referring to the syntax of the type theory or the extended first-order formulae.

Theorem 3.4 *If* $ML_0^i \vdash \Gamma \gg \theta$, *then* $\mathbf{APP} \vdash [\Gamma] \to [\theta]$. *In particular, suppose the type A is provably inhabited in* ML_0^i, *that is to say, there is a proof in* ML_0^i *ending with the sequent*

$$\gg t : A$$

for some term t. Then $\mathbf{APP} \vdash [A](t^*)$

In order to mediate between type theory and the first-order language of partial application, we define a "reverse" r-translation $(A)^r$ of ML_0^i types into extended **APP** formulae. The key property of the translation is:

Lemma 3.5 (Reverse-Translation lemma) *Let A be a type in* ML_0^i. *Then the following equivalence is provable in* **APP** :

$$[A](e) \equiv |(A)^r|(e)$$

Now it is straightforward to define a Kripke model, $\mathcal{K}$, for ML_0^i along the lines of the preceding section, with adjustments for the special formulas defined above, which is sound in the sense that for any *extended* formula A:

$$A \text{ is provably inhabited in } ML_0^i \Rightarrow \mathcal{K} \models (A)^r$$

A straightforward modification of the definitions and arguments in section 2 (with some attention paid to bounded quantification) provides is with a Kripke model $\mathcal{K}$, for *extended* formulas, as defined in 3.2 which models extended realizability.

Theorem 3.6 *Let* φ *be any (possibly nonatomic) extended (i.e. special) sentence over the language of* $\mathbf{APP}_{\mathcal{C}}$. *Then*

$$A \Vdash \varphi \iff A \geq |\varphi|.$$

sketch of proof: The atomic case follows from the specification of our Kripke model. For any formulas free of *special subformulas* $a : D$, $(\exists_{x \in D})\theta(x)$, or $(\forall_{x \in D})\theta(x)$ this is the same result as in [17]. So we need consider only the disjunctive and special formulas (which are the only ones required to interpret ML_0^i). To give an idea of the arguments involved we will prove following lemma, which essentially asserts that for the trivial special case of a formula which *already supplies the inhabiting witness* our Kripke model $\mathcal{K}$ satisfies the conclusion of the preceding theorem in the obvious required way: it respects the choice of witness. The lemma is an immediate consequence of the definition of forcing of special formulas $a : D$.

Lemma 3.7 *For any node A, any term* $a \in \Delta$ *and any formula D, we have:*

$$A \Vdash a : D \iff A \overset{\lambda x \cdot a}{\geq} |a : D|,$$

that is to say:

$$\mathbf{APP}_{\mathcal{C}} \vdash \forall x (A(x) \to |D|(a)).$$

proof: The atomic case follows by definition.

Disjunction: suppose $A \Vdash a : C \vee D$. Then

$$A \Vdash p_0 a = 0 \ \& \ p_1 a : C \quad \text{or} \quad A \Vdash p_0 a \neq 0 \ \& \ p_1 a : D.$$

Recall that by definition of forcing of ground atomic formulae,

$$A \Vdash p_0 a = 0 \quad \equiv \quad \mathbf{APP}_{\mathcal{C}} \vdash \forall x (A(x) \to p_0 a = 0).$$

Therefore, by inductive hypothesis,

$$\forall x (A(x) \to p_0 a = 0 \ \& \ |C|(p_1 a)) \text{ or}$$
$$\forall x (A(x) \to p_0 a \neq 0 \ \& \ |D|(p_1 a)),$$

provably in **APP$\underline{C}$** . But in either case, we have $\forall x(A(x) \rightarrow |C \lor D|(a)$

Suppose $A \parallel\!\!-a : (\exists_{x \in D})\theta(x)$. Then

$$A \parallel\!\!-p_0 a : D \text{ and } A \parallel\!\!-p_1 a : \theta(p_0 a).$$

By inductive hypothesis, we have:

$$\textbf{APP}\underline{C} \vdash \forall x(A(x) \rightarrow |D(p_0 a)|$$

and

$$\textbf{APP}\underline{C} \vdash \forall x(A(x) \rightarrow |\theta(p_0 a)|(p_1 a))$$

from which

$$\textbf{APP}\underline{C} \vdash \forall x(A(x) \rightarrow |D(p_0 a)| \& |\theta(p_0 a)|(p_1 a))$$

which is the conclusion $A \overset{\lambda x \cdot a}{\geq} |a : (\exists_{x \in D})\theta(x)|$. The converse is straightforward and left to the reader.

Now, suppose $A \parallel\!\!-a : (\forall_{x \in D})\theta(x)$, i.e. for every term u in Δ and every $B \geq A$.

$$B \parallel\!\!-u : D \quad \Rightarrow \quad B \parallel\!\!-au : \theta(u). \tag{2}$$

Then we have: $A \times |u : D| \geq A$ and $A \times |u : D| \geq |u : D|$ by the canonical projections, so, by the inductive hypothesis $A \times |u : D| \parallel\!\!-u : D$ hence, by (2) and the inductive hypothesis again,

$$\textbf{APP}\underline{C} \vdash \forall x \forall y(A(x) \& |u : D|(y) \rightarrow |\theta(u)|(au)), \tag{3}$$

where $|u : D|(y)$ means $|D|(u)$. Now, since u may be taken to be *any* term in Δ, we pick a constant c in $\underline{C}$ not present elsewhere in (3). We now argue that, since c is a "fresh" constant, we can quantify over it. This is the whole point of adding the fresh constants $\underline{C}$ (as in Henkin's proof of the completeness theorem for first-order logic). But we cannot quite yet do so: c *may occur in the premisses* **APP**$\underline{C}$. Nonetheless, the occurrence is only in finitely many axioms $\Gamma_0(x)$ from the the *logical part* of **APP**$\underline{C}$, in the axiom schemas for equality, t substitutivity, etc. Picking a specific formulation of **APP**, we easily show, (see [19]) that we can effectively quantify over c obtaining:

$$\textbf{APP}\underline{C} \vdash \forall x(A(x) \rightarrow \forall z(|D|(z) \rightarrow |\theta(z)|(az))). \tag{4}$$

which is the conclusion of the lemma.

Conversely, suppose

$$\textbf{APP}\underline{C} \vdash \forall x(A(x) \rightarrow \forall u(|D|(u) \rightarrow |\theta(u)|(au))). \tag{5}$$

Pick u in Δ and assume that some function f witnesses $B \geq A$, and that $B \parallel\!\!-u : D$. Then, by inductive hypothesis,

$$\textbf{APP}\underline{C} \vdash \forall x(B(x) \rightarrow |D|(u))$$

and

$$\textbf{APP}\underline{C} \vdash \forall x(B(x) \rightarrow fx \downarrow \& A(fx))$$

so by (5) we must have

$$\textbf{APP}\underline{C} \vdash \forall x(B(x) \rightarrow |\theta(u)|(au))$$

so, by, inductive hypothesis again, $B \parallel\!\!-au : \theta(u)$ so $A \parallel\!\!-(\forall_{x \in D})\theta(x)$.

The remaining cases for the lemma are straightforward.

proof of theorem 3.6: We begin with disjunction. Suppose $A \Vdash \varphi \vee \theta$. Then, for some cover S of A we have

$$(\forall B \in S) B \Vdash \varphi \text{ or } B \Vdash \theta.$$

By the induction hypothesis, for each $B \in S$ there is a term t_B in Δ such that

$$\mathbf{APP\underline{C}} \vdash \forall x (B(x) \to t_B x \downarrow \ \& \ \varphi(t_B x)) \tag{6}$$

or

$$\mathbf{APP\underline{C}} \vdash \forall x (B(x) \to t_B x \downarrow \ \& \ \theta(t_B x)). \tag{7}$$

Suppose (6) holds. Then $\lambda x \cdot \langle 0, t_B x \rangle : B \to |\varphi \vee \theta|$. If (7) holds, then $\lambda x \cdot \langle 1, t_B x \rangle : B \to |\varphi \vee \theta|$. So, in all cases, $B \geq |\varphi \vee \theta|$. Now, by definition of cover in this model, we have $A \geq |\varphi \vee \theta|$, as we wanted to show.

Conversely, if for some term f in Δ we have, provably in $\mathbf{APP\underline{C}}$, $f : A \to |\varphi \vee \theta|$, then, letting

$$A_0(x) \ \overset{\text{def}}{\equiv} \ A(x) \ \& \ fx = 0$$
$$A_1(x) \ \overset{\text{def}}{\equiv} \ A(x) \ \& \ fx = 1$$

We have f witnessing both $A_0 \geq |\varphi|$ and $A_1 \geq |\theta|$. Now all that remains to be shown is that the pair of formulas $\{A_0, A_1\}$ is a cover for A. This means showing that, for some Δ - terms s and t, and some node D, if $A_0 \overset{s}{\geq} D$ and $A_1 \overset{t}{\geq} D$ then there is an h in Δ with $A \overset{h}{\geq} D$. Clearly we need only set

$$h \overset{\text{def}}{\equiv} \lambda x \cdot \text{if } fx = 0 \text{ then } sx \text{ else } tx.$$

Details of the proof for the existential and universal case can be found in [19]

From this we can immediately conclude:

Corollary 3.8 *Special formulas φ are true in the Kripke/Beth model $\mathcal{K} \Longleftrightarrow$ they are provably realizable in* **APP**.

Finally, we have

Corollary 3.9 (Soundness of the interpretation) *Let A be a type in $\mathbf{ML}_0^i$, provably inhabited in that theory. In other words, for some term u, the sequent $\gg u : A$ is provable in $\mathbf{ML}_0^i$. Then $\mathcal{K} \models (A)^r$*

4 Extensions and Constructive Completeness

Our construction provides a countable collection of non-equivalent models for $\mathbf{ML}_0^i$ since the model $\mathcal{K}_A$ taken by restricting attention to all nodes above a given node A is itself a Kripke model. It is not hard to see that $\mathcal{K}$ has countably many nonequivalent nodes. Pick any node A, and let B be a *sentence* over the language of $\mathbf{APP\underline{C}}$ independent of $\mathbf{APP\underline{C}} \cup \{\exists x A(x)\}$ We cannot have $A \geq B$, since this means that for some term f

$$\mathbf{APP} \vdash \forall x (A(x) \to fx \downarrow \ \& \ B).$$

But then, by existential elimination and arrow introduction, we have

$$\mathbf{APP} \vdash (\exists x A(x)) \to B,$$

contradicting independence. (Similarly we cannot have $A \geq \neg B$). We can, of course iterate this (tacitly using Gödel's incompleteness theorem to supply new sentences), obtaining a B_1 not below A or B, and then a B_2, etc. In effect our Kripke model *lifts independence results in* **APP** *to independence results in* $\mathbf{ML}_0^i$. If the realizability of $(\theta)^r$ is independent of **APP** then θ cannot be forced by the root node of $\mathcal{K}$, nor can its negation, hence neither is provably inhabited in $\mathbf{ML}_0^i$.

Fallible Kripke/Beth models first appeared (to the author's knowledge) in a 1970 paper by Läuchli ([16]) of which this paper (as well as many others in the bibliography) is a descendant. These models continued to appear in a seminal series of papers on *constructive proofs of completeness* due to Veldman, Friedman and others (see [31] for history and references). Along lines similar to [18] we can adapt the Veldman-Friedman-Troelstra-Van Dalen proof to establish, constructively, completeness of *extended fallible Kripke/Beth models* for extended **APP** formulas and for "one-universe" dependent type theory.

Theorem 4.1 (Completeness) *Let Γ be a context (premiss) in $\mathbf{ML}_0^i$. Then there is an extended, fallible Kripke/Beth model $\mathcal{K}$ such that for each sequent $\Gamma >> \theta$*

$$\mathbf{ML}_0^i \vdash \Gamma >> \theta \qquad \text{iff} \qquad \mathcal{K} \models \theta.$$

After enriching the sytax of $\mathbf{ML}_0^i$ with fresh constants, we use it to build a model along the lines of the preceding sections. We are able to define forcing of judgements directly (without the translation used before). The arguments are straightforward but somewhat lengthy. They are given in [19].

4.1 Conclusion

What are the main directions for continuing this work, what are the open problems? The results just shown provide a framework for developing a tableau-based refutation method for type theory (similar to Nuprl [4], but with Kripke counter-models) as well as for new conservativity and independence results. A natural question here is: How do we extend these results to second and higher order theories. Two directions suggest themselves: developing the second-order model theory of the subject (as initiated in [31], and extended in [6]), or formalizing these arguments in Constructive Set Theory. We discuss both approaches in [19]. Many questions remain open here, along with the matter of subrecursive realizability models, to be taken up in [20].

References

[1] Aczel, P. [1977] The Strength of Martin-Löf's Type Theory with One Universe, in: Mietissen, S. and Väänänen, J., (eds), *The proceedings of the symposiums on Mathematical Logic Helsinki 1975*, University of Helsinki, 1977.

[2] Allen,S. [1988], Ph. D. Dissertation, Cornell University, Ithaca, N.Y.

[3] Beeson, M. J. [1985a], *Foundations of Constructive Mathematics*, Springer-Verlag, Berlin.

[4] Constable, R. L., et al [1986], *Implementing Mathematics with the NUPRL Development System*, Prentice-Hall, N.J.

[5] Coquand, T. [1990], "On the analogy between propositions and types", in: *Logic Foundations of Functional Programming*, Huet, G. ed., Addison-Wesley, Reading, MA.

[6] Dragalin, A. G., [1988], "A Completeness Theorem for Higher Order Intuitionistic Logic: An Intuitionistic Proof", in *Mathematical Logic and its Applications*, D. Skordev, ed., Plenum.

[7] Feferman, S. [1975], "A language and axioms for explicit mathematics", in: *Algebra and Logic*, Lecture Notes in Mathematics No. 450, pp. 87-139, Springer, Berlin.

[8] Fitting, M. [1983], *Proof Methods for Modal and Intuitionistic Logics*, D. Reidel, Dordrecht, The Netherlands.

[9] Fourman, M. P. and D. S. Scott [1979], "Sheaves and logic", in: Fourman, Mulvey and Scott, (eds.), *Applications of Sheaves*, Mathematical Lecture Notes 753, pp.302-401, Springer-Verlag, Berlin.

[10] Grayson, R. J. [1983], "Forcing in intuitionistic systems without power set", Journal of Symbolic Logic 48, 670-682.

[11] Hyland, J. M. E., P. T. Johnstone and A. M. Pitts [1980], "Tripos Theory", Math. Proceedings of the Cambridge Phil. Society 88, 205-252.

[12] Hyland, J. M. E. [1982], "The effective topos". in: Troelstra, A. S. and D. S. van Dalen (eds.), *L.E.J. Brouwer Centenary Symposium*, North-Holland, Amsterdam.

[13] Kleene, S. C. [1952], *Introduction to Metamathematics*, North-Holland (1971 edition), Amsterdam.

[14] Kripke, S. [1965], "Semantical analysis of intuitionistic logic I", in: Crossley, J. N. and M. Dummett (eds.), *Formal Systems and Recursive Functions*, Proceedings of the Eighth Logic Colloquium, Oxford, 1963, North-Holland, Amsterdam, 92-130.

[15] Lambek, J. and P. J. Scott [1986], *Introduction to higher order categorical logic*, Cambridge Studies in Advanced Mathematics 7, Cambridge.

[16] Läuchli, H. [1970], "An abstract notion of realizability for which predicate calculus is complete", in: Myhill, J., A. Kino, and R. E. Vesley (eds.), *Intuitionism and Proof Theory*, North-Holland, Amsterdam, 227-234.

[17] Lipton, J. [1990], "Realizability and Kripke Forcing", in the *Annals of Mathematics and Artificial Intelligence*, Vol.4, North-Holland, Amsterdam. An expanded version appeared as technical report 90-1163, Dept. of Computer Science, Cornell University, Ithaca, NY.

[18] Lipton, J. [1990], "Constructive Kripke Semantics and Realizability", to appear in the proceedings of the *Logic for Computer Science* conference held at the Math. Sci. Research Institute, Berkeley, Nov. 1989.

[19] Lipton, J. [1992] "Type theory, Kripke models and degrees of Inhabitation." Submitted to the *Annals of Pure and Applied Logic*.

[20] Lipton, J. [1992], "Kripke Models for Subrecursive computation", to appear.

[21] Martin-Löf, P. [1982], "Constructive Mathematics and Computer Programming", in Logic, Methodology and Philosophy of Science IV, North Holland, Amsterdam.

[22] Martin-Löf, P. [1984], *Intuitionistic Type Theory*, Studies in Proof Theory Lecture Notes, BIBLIOPOLIS, Napoli, Italy.

[23] McCarty, D. C. [1986], "Realizability and recursive set theory", Annals of Pure and Applied Logic 32, 11-194.

[24] Mitchell, J. and E. Moggi [1987], "Kripke-style models for typed lambda calculus", *Proceedings from Symposium on Logic in Computer Science*, Cornell University, June 1987, IEEE, Washington, D.C..

[25] Nerode, A. [1989b], "Some lecures on Intuitionistic Logic I", Proceedings on Summer School on Logical Foundations of Computer Science, CIME, Montecatini, 1988, Lecture Notes in Mathematics, Springer-Verlag, Berlin.

[26] Nerode, A., and P. Odifreddi, [1990], *Lambda Calculi and Constructive Logics*, MSI Tech. Report '90-55

[27] Odifreddi, P. [1989], *Classical Recursion Theory*, North-Holland, Amsterdam.

[28] Plotkin, G., [1980], "Lambda definability in the full type hierarchy", in : Seldin, J.P. and J. R. Hindley (eds.), *To H.B. Curry: Essays in Combinatory Logic, Lambda Calculus and Formalism*, Academic Press, New York.

[29] Statman, R., [1982],"Logical relations and the typed Lambda Calculus", Information and Control.

[30] Swart, H. C. M. de [1976], "Another intuitionistic completeness proof", Journal of Symbolic Logic 41, 644-662.

[31] Troelstra, A. S. and D. van Dalen [1988], *Constructivism in Mathematics: An Introduction*, Vol. II, Studies in Logic and the Foundations of Mathematics, Vol. 123, North-Holland, Amsterdam.

[32] Veldman, W. [1976], "An intuitionistic completeness theorem for intuitionistic predicate logic", Journal of Symbolic Logic 41, 159-166.

Reflective Semantics of Constructive Type Theory
(Preliminary Report)

Scott F. Smith

The Johns Hopkins University[*]

Abstract

It is well-known that the proof theory of many sufficiently powerful logics may be represented internally by Gödelization. Here we show that for Constructive Type Theory, it is furthermore possible to define a semantics of the types in the type theory itself, and that this procedure results in new reasoning principles for type theory. Paradoxes are avoided by stratifying the definition in layers.

1 Introduction

Given a sufficiently powerful logical theory L such as Peano Arithmethic, it is well-known that the proof theory of L may be expressed internally via Gödelization. This is accomplished by defining a metafunction $\lceil A \rceil$ that encodes formulas A as data, and a predicate $Provable_L(\lceil A \rceil)$ which is true just when formula A is provable. This gives L knowledge of its own proof theory, but it doesn't know that it knows it: the embedded proof theory could just as well be for some different logic. What is needed then are principles of self-knowledge that connect the provability predicate with the actual proof theory. Feferman [Fef62] for one has studied adding such principles; the principle most relevant to this work is

> Given a theory L_k, extend it to give a theory L_{k+1} with added reflection axiom
> $\vdash_{L_{k+1}} Provable_{L_k}(\lceil A \rceil) \Rightarrow A$.

In other words, if we prove that we can prove it, we can prove it. Note that this is not a true self-knowledge principle, because L_k does not have self-knowledge of the reflection axiom. Given a theory L_1, this principle induces a hierarchy of theories $L_1, L_2, L_3, \ldots$, which we call the reflected proof hierarchy.

Logicians have studied this hierarchy to understand its proof-theoreric strength; here we are not directly interested in this issue. We are interested in how computer scientists have found it applicable for automated theorem proving systems. Suppose there was an assumption A from which we wished to prove B. If the theorem-proving system knew of a way of always proving B from A, not by having assumption $A \Rightarrow B$ but by observing some syntactic property of A, the system would still have to construct each step of the proof. Given a reflected proof system, this construction may be avoided. Suppose we were to prove the theorem

> $\forall \lceil A \rceil, \lceil B \rceil.$ "$\lceil A \rceil$ is of appropriate syntactic form" & $Provable_L(\lceil A \rceil) \Rightarrow Provable_L(\lceil B \rceil)$

---

[*]email scott@cs.jhu.edu, phone (410) 516-5299, fax (410) 516-6134.

Now, if a proof of B is desired and a proof of A is given, if A is of the appropriate form the above principle allows us to immediately conclude $Provable_L(\lceil B \rceil)$, whence B by the reflection axiom. Some systems incorporating this general paradigm are [ACHA90, BM81, DS79].

Cornell researchers [ACHA90] have formulated such a reflected proof hierarchy for the Nuprl type theory [CAB+86]. One advantage of studing reflected proof in such a context is there is a programming language built in to Nuprl, so it is possible to have the condition "$\lceil A \rceil$ is of appropriate syntactic form" be a decidible property that may actually be computed inside the Nuprl computation system, meaning the proof of that property also need not be constructed. Since formal proofs can get very large, this is an important method for shortening proof length.

Our goals are similar, but the method is different. Instead of reflecting proof, we reflect the *semantics*. Nuprl and other constructive type theories (CTT's) [Mar82] may be interpreted by inductively defining the types and their members [All87]. We show here that this inductive definition is actually expressible inside CTT. As in proof reflection there is a hierarchy of reflective semantic interpretations. The stratification is by the universe levels of the CTT. It is then possible to give an internal interpretation of proofs, leading to the *internal k-soundness* theorem. The hierarchy of internal k-soundness proofs then can be taken as an internal proof of soundness and thus consistency of the entire theory, contradicting Gödel's incompleteness theorem all but in name.

An axiom of self-awareness also may be added, and addition of the axiom does not lead to a hierarchy of proof theories, for the proof theory is not being reflected; the only hierarchy is based on universe levels.

2 Type Theory

This section is a short survey of constructive type theory (CTT) necessary for basic comprehension of this paper; for fuller descriptions and examples, see [CAB+86, Tho91]. The CTT used herein, called simply "CTT," is a variant of standard CTTs like Nuprl and Martin Löf's theories in that types do not come with equality $a = b \in A$; instead we just have $a \in A$, and equivalences are expressed via type-free equality $a \sim b$. This is type theory more in the spirit of Feferman class theory [Fef75].

We also wish to extend CTT to include a mechanism for recursive type definition. Recursive types for Nuprl have been defined by Mendler [CAB+86], and Dybjer has defined these types for Martin-Löf's theories [Dyb87]. CTT with recursive types added will be denoted CTTR. The internal presentation of the semantics of CTT is not possible without recursive types.

A type theory has as its language a collection of untyped *terms*. Some of these terms represent types, others computations; the two, perhaps surprisingly, are not separated. The terms of CTT include numbers and basic operations $0, 1, 2, \ldots$, $pred(a)$, $succ(a)$, $if\_zero(a; b; c)$; pairing and projection $\langle a, b \rangle$, $a.1$, $a.2$; injection and decision terms $inl(a)$, $inr(b)$, $decide(a; b; c)$; and functions and application $\lambda x.a$, $a(b)$. This language is thus a small functional programming language. On untyped terms we write $a \sim b$ meaning a is operationally equivalent to b, a notion we will not elaborate on here; see for instance [Plo75, Smi91a].

Letters a–d, A–D will range over terms, and v–z, V–Z will range over variables. Although terms and types are formally of the same sort, we use capitol letters to denote types and small letters to denote terms. Notions of *bound* and *free* variables, *open* and *closed* terms and captureless substitution of b for x in a $a[b/x]$ are standard.

2.1 Types

The expressiveness of type theory is largely due to the diversity of types that are definable. Here we list the types of CTT and CTTR.

(i) N is a type of natural numbers.

(ii) $a \sim b$ is a type that is inhabited (by placeholder 0) just when a and b are indistinguishable as computations.

(iii) a *in* A is a type that internalizes the assertion $a \in A$: $0 \in (a \ in \ A)$ just when $a \in A$.

(iv) $x{:}A \times B(x)$ is a dependent product type (the *type* $B(x)$ depends on the *member* of the type A) whose inhabitants are pairs $\langle a, b \rangle$, with $a \in A$ and $b \in B(a)$.

(v) $x{:}A \to B(x)$ is a dependent function type whose inhabitants are functions $\lambda x.b$, where for all $a \in A$, $(\lambda x.b)(a) \in B(a)$.

(vi) $U_1, U_2, \ldots, U_k, \ldots$ are universes or large types, types that have types as members. The levels are constructed in sequence: U_1 has as members all types closed under the type forming operators; U_2 has as members all types closed under the same operators, plus the type U_1; *et cetera*.

(vii) CTTR has parameterized recursive types in addition to all the types mentioned up to now. $rec(X; x{:}A.B(X)(x))(a)$ is a parameterized recursive type. In simplified non-parametric form $rec(X; B(X))$, it denotes the solution to a recursive type equation of the form $X = B(X)$. A parameter x is added to give solutions to more general type equations $X(x) = B(X)(x)$, with initial parameter $a \in A$.

2.2 Formulas as types

One of the defining features of constructive type theory is it is a theory of realizability. Formulas are viewed as types for which the members are the realizers of the formula. To prove the formula, show that when it is viewed as a type, the type is inhabited, i.e. has some member. Formulas are defined in terms of types as follows.

$$A \Rightarrow B \stackrel{\text{def}}{=} A \to B$$
$$A \ \& \ B \stackrel{\text{def}}{=} A \times B$$
$$A \vee B \stackrel{\text{def}}{=} A + B \stackrel{\text{def}}{=} x{:}N \times if\_zero(x; A; B)$$
$$\neg A \stackrel{\text{def}}{=} A \to false$$
$$false \stackrel{\text{def}}{=} 0 \sim 1$$
$$\forall x{:}A. \ B \stackrel{\text{def}}{=} x{:}A \to B$$
$$\exists x{:}A. \ B \stackrel{\text{def}}{=} x{:}A \times B$$

The type universes U_k also serve as universes of formulas, and the type $A \to U_k$ has as members predicates on the type A.

2.3   Refinement proof and extraction

An assumption list Γ is of the form $x_1 \in A_1, x_2 \in A_2, \ldots x_n \in A_n$. One form of assertion may be made:

$$\Gamma \vdash a \in A$$

which asserts under assumptions Γ, term a inhabits type A.

The rules of CTT may be presented in goal-directed or *refinement-style* fashion: a rule is applied to a goal, and this gives subgoals which when proven realize the goal. Proofs are thus trees with nodes being goals and children of a node being its subgoals. The leaves of the tree are goals with no subgoals. A sample rule, the introduction rule for (non-dependent) products, is

$$\vdash \langle a, b \rangle \in A \times B$$
$$\vdash a \in A$$
$$\vdash b \in B$$

where the first line is the goal and the indented lines are the subgoals. A full set of rules for CTT is given in appendix A.

3   Semantics of CTT

CTT and CTTR may be proven sound by giving an inductive definition of the types and their inhabitants, using the method of Allen [All87]. Types are defined by induction on their structure. In the presence of dependent types it is impossible to first define the types and then define membership predicates for the types, because for a term $x{:}A \rightarrow B(x)$ to be a type, $B(a)$ must be a type for all members a of type A. This means the *members* of A must be defined before the *type* $x{:}A \rightarrow B(x)$ is considered well-formed. The solution Allen developed is to simultaneously define types and their inhabitants. This inductive definition can be viewed as a term model of CTT.

The main technical result of this paper is to show that Allen's method of inductive definition may be expressed inside CTTR itself, giving an internal definition of truth. We first give the inductive definition of CTT and show it has reasonable properties. The definition is then expressed in CTTR. Inductive definition of recursive types requires considerable extra notation, so we will settle for a summary at the end of the paper of the steps necessary to inductively define CTTR. Mendler [Men87] has developed techniques for giving an inductive definition of recursive types.

The development of the semantics of CTT now proceeds in three phases: untyped equality on terms is defined, the inductive definition of types is given, and the theory is proven sound with respect to this definition.

First, an evaluator for untyped closed terms and equivalence thereupon is defined. See [Smi91b] for complete definitions.

DEFINITION 3.1 For closed CTT terms a and b, $a \mapsto b$ iff a computes under call-by-name syntactic reduction to b.

Equality $a \sim b$ is defined to mean a and b are indistinguishable when placed in any context. $c[-]$ is defined to be a context, i.e. a term with a distinguished hole "$-$" in which another term may be placed. $c[a]$ denotes $c[-]$ with a placed in the hole, possibly capturing free variables of a.

DEFINITION 3.2 $a \sim b$ iff for all closing contexts $c[-]$. $c[a] \mapsto a'$ iff $c[b] \mapsto b'$.

A collection of properties may be proven about this equivalence that in turn allows the $\sim$ rules of CTT to be justified; see [Smi91b].

Given the collection of terms with equivalence relation, the types and their inhabitants may be simultaneously defined. Open terms are considered only when interpreting hypothetical assertions, so until that point all terms may be implicitly taken to be closed.

DEFINITION 3.3 A *type interpretation* is a two-place predicate $\tau(A, \epsilon)$ where A is a term, and ϵ is a one-place predicate on terms.

For $\tau(A, \epsilon)$ to be true means in type interpretation τ, A is a type with its members specified by ϵ. The following definitions make this explicit.

DEFINITION 3.4   (i) A Type$_\tau$ iff $\tau(A, \epsilon)$ for some ϵ

(ii) $a \in_\tau A$ iff $\tau(A, \epsilon)$ for some ϵ and $\epsilon(a)$.

A Type$_\tau$ thus means A is a type in interpretation τ, and $a \in_\tau A$ means a is a member of the type A.

Recall we have both small types and a hierarchy of large types U_k. The universes are predicative, so $U_k \in U_k$ fails. This is important because it means larger universes may be defined in terms of the (small) type constructors and the smaller universes. The definition of the full theory thus proceeds a universe at a time: ν_k is the type interpretation for universe U_k.

To give a well-formed inductive type definition, a monotonic operator is defined, and the least fixed point is then taken. The proof of monotonicity of the operator will be omitted.

DEFINITION 3.5 ν_k is the least fixed-point of the following monotonic operator Ψ_k on type interpretations:

$$\Psi_k(\tau) \stackrel{\text{def}}{=} \tau', \text{ where } \tau'(T, \epsilon) \text{ is true if and only if}$$

EITHER $T \sim N$, in which case
        for all t, $\epsilon(t)$ iff $t \sim n$, where n is $0, 1, 2, \ldots$
OR $T \sim (a \sim b)$, in which case
        for all t, $\epsilon(t)$ iff $t \sim 0$ and $a \sim b$
OR $T \sim a$ *in* A, in which case A Type$_\tau$ and
        for all t, $\epsilon(t)$ iff $t \sim 0$ and $a \in_\tau A$
OR $T \sim x{:}A \rightarrow B(x)$, in which case
        A Type$_\tau$ and for all $a \in_\tau A$, $B(a)$ Type$_\tau$, and
        for all t, $\epsilon(t)$ iff $t \sim \lambda x.b$ and for all $a \in_\tau A$, $t(a) \in_\tau B(a)$
OR $T \sim x{:}A \times B(x)$, in which case
        A Type$_\tau$ and for all $a \in_\tau A$, $B(a)$ Type$_\tau$, and
        for all t, $\epsilon(t)$ iff $t \sim \langle a, b \rangle$ and $a \in_\tau A$ and $b \in_\tau B(a)$
OR $T \sim U_{k'}$, in which case
        $k' < k$ and for all t, $\epsilon(t)$ iff $\nu_{k'}(t, \epsilon')$ for some ϵ'.

This completes the definition is each universe level, and thus the entire CTT. Some abbreviations are now made: $a \in_k A \stackrel{\text{def}}{=} a \in_{\nu_k} A$, $a \in A \stackrel{\text{def}}{=} a \in_k A$ for some k, and A Type$_k \stackrel{\text{def}}{=} A$ Type$_{\nu_k}$, A Type $\stackrel{\text{def}}{=} A$ Type$_k$ for some k.

Some straightforward lemmas about typehood and membership are now proven. They can be taken as defining what it means for the different forms of term to be types, and for what it means to be an inhabitant of the different types.

LEMMA 3.6 Type formation is characterized by the following properties.

(i) N Type, $a \sim b$ Type.

(ii) a in A Type iff A Type.

(iii) $x{:}A \to B(x)$ Type iff A Type and for all $a \in A$, $B(a)$ Type.

(iv) $x{:}A \times B(x)$ Type iff A Type and for all $a \in A$, $B(a)$ Type.

(v) if $A \in U_k$ then A Type.

(vi) if A Type then $A \in U_k$ for some k.

(vii) if $a \in A$ then A Type.

LEMMA 3.7 Type membership is characterized by the following properties.

(i) $c \in a \sim b$ iff $c \sim 0$ and $a \sim b$.

(ii) $c \in N$ iff $c \sim 0, 1, 2, \ldots$.

(iii) $c \in (a$ in $A)$ iff $(a$ in $A)$ Type and $c \sim 0$ and $a \in A$.

(iv) $c \in x{:}A \to B(x)$ iff $x{:}A \to B(x)$ Type and $c \sim \lambda x.c(x)$ and for all $a \in A$, $c(a) \in B(c)$.

(v) $c \in x{:}A \times B(x)$ iff $x{:}A \times B(x)$ Type and $c \sim \langle c.1, c.2 \rangle$ and $c.1 \in A$ and $c.2 \in B(c.1)$.

These two collections of lemmas closely correspond to the type formation and introduction/elimination rules, respectively. From the lemmas, it is then direct to show

THEOREM 3.8 (SOUNDNESS) If $\vdash a \in A$ in CTT, then $a \in A$, meaning the interpretation is sound.

COROLLARY 3.9 (INTUITIONISTIC CONSISTENCY) There is no proof of $\vdash 0 \in (0 \sim 1)$ in CTT.

4 Reflective semantics of CTT

The previous section gave a semantics of CTT; in this section we show how this semantics may be phrased inside CTTR. The key insights are that parameterized recursive types are powerful enough that general inductive definitions may be expressed, and that the type definitions ν_k may be expressed in U_{k+1}, meaning each universe is defined inside the next one.

A *Gödelization* of the CTT terms consists of a type *Term* encoding all terms of the theory, together with operations to construct, destruct, and compare terms. To convert terms to and from Gödelized form, we use metafunctions $\lceil a \rceil$ and $\lfloor a \rfloor$, respectively. We also will use the convention that a metavariable with a hat ($\hat{a}$) means the variable is a Gödelized term, so $\hat{a} \in$ *Term*. The proof theory (the rules of appendix A) may also be Gödelized as follows.

(i) *Sequent* is a type of sequents, lists of pairs of terms.

(ii) *Ptree* is a type of primitive proof trees.

(iii) *Proof* is a type of proof trees that are well-formed, i.e. the provability of the children of a node imply the provability of the node by one of the rules.

(iv) *Provable*($\lceil \Gamma \vdash a \in A \rceil$) iff $\lceil \Gamma \vdash a \in A \rceil$ is the top node of a *Proof* tree.

A rigorous Gödelization of the terms and proofs of CTT in CTT may be found in [ACHA90]; details will not be given here.

Since here we are here concerned with truth and not just proof, the evaluator and equivalence defined in the previous section also need to be expressed in CTT. Define an n-step evaluator $Eval \in Term \times \mathrm{N} \to Term$, and an untyped term equality predicate $Equal \in Term \times Term \to \mathrm{U}_1$. For this paper, we take these as given: they amount to expressing the definitions of $\mapsto$ and $\sim$ inside CTT. Since the proofs in [Smi91a] are constructive, this internalization should be routine.

The internalized equality should properly reflect the external equality.

LEMMA 4.1 (i) If for some t, $t \in Equal(\lceil a \rceil, \lceil b \rceil)$, then $a \sim b$.

(ii) If for some t, $t \in \neg Equal(\lceil a \rceil, \lceil b \rceil)$, then $a \not\sim b$.

4.1 The inductive definition

We will define a family of predicates $CTT_k \in Term \times (Term \to \mathrm{U}_k) \to \mathrm{U}_{k+1}$ for $k = 1, 2, 3, \ldots$. Each CTT_k is intended to be an internal expression of the type definition ν_k. Then, it is possible to prove a family of theorems in CTT that collectively express its own soundness and thus consistency.

The predicates CTT_k are now defined inductively using parameterized recursive types. The definition below thus is only expressible in CTTR; see section 5 for a sketch of how predicates $CTTR_k$ may be defined.

DEFINITION 4.2

$CTT_k \stackrel{\text{def}}{=} \lambda \langle \hat{A}, M_{\hat{A}} \rangle . rec(X; \langle \hat{A}, M_{\hat{A}} \rangle : Term \times Term \to \mathrm{U}_k.$

$\qquad Equal(\hat{A}, \lceil \mathrm{N} \rceil) \,\&\, \forall \hat{a} : Term.\, M_{\hat{A}}(\hat{a}) \iff \exists \hat{n} : Term.\, Equal(\hat{a}, \hat{n}) \,\&\, Natnum(\hat{n})$
$\qquad \lor$
$\qquad \exists \hat{b}, \hat{c} : Term.\, Equal(\hat{A}, \lceil \lfloor \hat{b} \rfloor \sim \lfloor \hat{c} \rfloor \rceil) \,\&\,$
$\qquad\qquad \forall \hat{a} : Term.\, M_{\hat{A}}(\hat{a}) \iff Equal(\hat{a}, \lceil 0 \rceil) \,\&\, Equal(\hat{b}, \hat{c})$
$\qquad \lor$
$\qquad \exists \hat{b}, \hat{B} : Term.\, Equal(\hat{A}, \lceil \lfloor \hat{b} \rfloor \text{ in } \lfloor \hat{B} \rfloor \rceil) \,\&\, \exists M_{\hat{B}} : Term \to \mathrm{U}_k.\, X(\langle \hat{B}, M_{\hat{B}} \rangle) \,\&\,$
$\qquad\qquad \forall \hat{a} : Term.\, M_{\hat{A}}(\hat{a}) \iff Equal(\hat{a}, \lceil 0 \rceil) \,\&\, M_{\hat{B}}(\hat{b})$
$\qquad \lor$
$\qquad \exists \hat{B} : Term.\, \exists \hat{C} : Term \to Term.\, Equal(\hat{A}, \lceil x : \lfloor \hat{B} \rfloor \to \lfloor \hat{C} \rfloor(x) \rceil) \,\&\, \exists M_{\hat{B}} : Term \to \mathrm{U}_k.\, X(\langle \hat{B}, M_{\hat{B}} \rangle) \,\&\,$
$\qquad\qquad \exists M_{\hat{C}} : Term \to Term \to \mathrm{U}_k.\, \forall \hat{b} : Term.\, M_{\hat{B}}(\hat{b}) \Rightarrow X(\langle \hat{C}(\hat{b}), M_{\hat{C}}(\hat{b}) \rangle) \,\&\,$
$\qquad\qquad \forall \hat{a} : Term.\, M_{\hat{A}}(\hat{a}) \iff Equal(\hat{a}, \lceil \lambda x. \lfloor \hat{a} \rfloor(x) \rceil) \,\&\, \forall \hat{b} : Term.\, M_{\hat{B}}(\hat{b}) \Rightarrow M_{\hat{C}}(\hat{b})(\lceil \lfloor \hat{a} \rfloor(\lfloor \hat{b} \rfloor) \rceil)$
$\qquad \lor$
$\qquad \exists \hat{B} : Term.\, \exists \hat{C} : Term \to Term.\, Equal(\hat{A}, \lceil x : \lfloor \hat{B} \rfloor \times \lfloor \hat{C} \rfloor(x) \rceil) \,\&\, \exists M_{\hat{B}} : Term \to \mathrm{U}_k.\, X(\langle \hat{B}, M_{\hat{B}} \rangle) \,\&\,$
$\qquad\qquad \exists M_{\hat{C}} : Term \to Term \to \mathrm{U}_k.\, \forall \hat{b} : Term.\, M_{\hat{B}}(\hat{b}) \Rightarrow X(\langle \hat{C}(\hat{b}), M_{\hat{C}}(\hat{b}) \rangle) \,\&\,$
$\qquad\qquad \forall \hat{a} : Term.\, M_{\hat{A}}(\hat{a}) \iff Equal(\hat{a}, \lceil \langle \lfloor \hat{a} \rfloor.1, \lfloor \hat{a} \rfloor.2 \rangle \rceil) \,\&\, M_{\hat{B}}(\lceil \lfloor \hat{a} \rfloor.1 \rceil) \,\&\, M_{\hat{C}}(\lceil \lfloor \hat{a} \rfloor.1 \rceil)(\lceil \lfloor \hat{a} \rfloor.2 \rceil)$
$\qquad \lor$
$\qquad Equal(\hat{A}, \lceil \mathrm{U}_{k-1} \rceil) \,\&\, \forall \hat{B} : Term.\, M_{\hat{A}}(\hat{B}) \iff \exists M_{\hat{B}} : Term \to \mathrm{U}_{k-1}.\, CTT_{k-1}(\hat{B}, M_{\hat{B}})$
$\qquad \lor$
$\qquad \vdots$
$\qquad \lor$
$\qquad Equal(\hat{A}, \lceil \mathrm{U}_1 \rceil) \,\&\, \forall \hat{B} : Term.\, M_{\hat{A}}(\hat{B}) \iff \exists M_{\hat{B}} : Term \to \mathrm{U}_1.\, CTT_1(\hat{B}, M_{\hat{B}}))(\langle \hat{A}, M_{\hat{A}} \rangle)$

LEMMA 4.3 $CTT_k \in U_{k+1}$.

This completes the definition; now we establish its reasonableness. A first step is the proof of soundness of the previous section may be internalized. First, we define the notion of a proof occurring entirely within some maximum universe.

DEFINITION 4.4 For arbitrary k, $Provable_k(t)$ iff $Provable(t)$ and all universes appearing in the proof are of level at most k.

Since all proofs have some maximum universe level, if $Provable(t)$ then $Provable_k(t)$ for some k.
    All k-provable statements can then internally be shown sound.

THEOREM 4.5 (INTERNAL k-SOUNDNESS) For any natural number k, the following theorem may be proved in CTTR: $\vdash \forall \hat{i}, \hat{T} : Term. Provable_k(\lceil \vdash \lfloor \hat{i} \rfloor \in \lfloor \hat{T} \rfloor \rceil) \Rightarrow \exists M : Term \to U_k. CTT_k(\hat{T}, M) \,\&\, M(\hat{i})$.

PROOF. This is just a matter of repeating the arguments of the previous section inside type theory. The most interesting facet of that procedure is proofs by induction on the definition of ν_k now proceed by the (Rec induction) rule on CTT_k definitions.
QED.

COROLLARY 4.6 (INTERNAL k-CONSISTENCY) The consistency of proofs at any particular level k may then be proven, i.e. $\vdash \neg Provable_k(\lceil 0 \in (0 \sim 1) \rceil)$.

The interpretation of $CTT_k(\lceil A \rceil, M)$ in the external semantics ν_k may be connected to the external semantics itself, and this proves useful.

THEOREM 4.7 (INTERNAL-EXTERNAL) If $t \in_{k+1} \exists M : Term \to U_k. CTT_k(\lceil A \rceil, M) \,\&\, M(\lceil a \rceil)$ then $\nu_k(A, \epsilon)$ and $\epsilon(a)$.

This means internal truth implies external truth. Rules may be added to the theory to reflect this fact.
    CTT is extended to give the theory CTTI by adding the rule (In), which we present for the simplified case that the hypothesis list is empty.

(In) $\qquad\qquad \vdash a \in A$
$\qquad\qquad\qquad \vdash t \in \exists M : Term \to U_k. CTT_k(\lceil A \rceil, M) \,\&\, M(\lceil a \rceil)$

With the (In) rule, proving a term is in a type is accomplished by showing the term to be in the membership predicate for the type in the internal semantics. This rule is justified to be sound by theorem 4.7.

5   Semantics of Recursive Types

Recursive types present more difficulties. For $rec(X; x:A.B(X)(x))(a)$ to be a type, the formation rule states the body $B(X)(x)$ must be a type for all types X, but this means a new type is defined by quantifying over all types, obviously circular. The solution is to use Girard's candidat de réductibilité method [Gir71] and interpret X as an arbitrary predicate, not a type. This breaks the circularity, but at the expense of needing environments to bind all such X to some predicate. Mendler [Men87] gives a detailed semantics for CTT with parameterized recursive types. The reflective semantics for the full CTTR type theory then involves encoding Mendler's method inside CTTR.

5.1 Uses of reflective semantics

We sketch some application of these ideas through a simple example. Many types are obviously well-formed, but in an automated implementation of a type theory such as Nuprl, the complete proofs of well-formedness must be given. Since a significant amount of processing time is spent proving types are well-formed, it would be useful to be able to immediately recognize some simple types and avoid constructing their proofs of well-formedness.

Suppose the function $SimpleType(\hat{A})$ was defined such that $SimpleType(\hat{A}) \sim true$ just when $\hat{A}$ meets some criteria of being a simple well-formed type. Then, we may prove a theorem

$$\vdash \forall \hat{A}: Term.\ SimpleType(\hat{A}) \sim true \Rightarrow \exists M: Term \rightarrow \mathrm{U}_1.\ CTTR_2(\lceil \lfloor \hat{A} \rfloor \in \mathrm{U}_1 \rceil, M)$$

meaning all such simple types may be shown, by the internal semantics, to be types. Then, any time it is necessary to show a type well-formed, $\vdash A \in \mathrm{U}_1$, we may try the following procedure:

(i) Compute $SimpleType(\lceil A \rceil)$; fail if the result is not $true$.

(ii) apply the (In) rule to give the subgoal $\vdash CTTR_2(\lceil A \in \mathrm{U}_1 \rceil, M)$.

(iii) using the above theorem and modus ponens, the proof is complete.

A The rules

The rules will be presented refinement-style, giving a goal sequent, followed by subgoal sequents which if proven imply the truth of the goal.

We confine various technical conventions for the presentation of the rules to this paragraph. The hypothesis list is always increasing going from goal to subgoals, even though subgoals do not list goal hypotheses. In the hypothesis list, x_i may occur free in any A_{i+j} for positive j. The free variables in the conclusion are no more than the x_i. α-conversion is an unmentioned rule. To improve readability, the assertion $0 \in a \sim b$ will be abbreviated $a \sim b$, likewise for $0 \in (a\ in\ A)$. Since there is at most one inhabiting object, 0, it need not be mentioned. Also, in hypothesis lists, $x \in a \sim b$ will be abbreviated $a \sim b$, since x is known to be 0. Abbreviate $a \sim b \rightarrow 0 \sim 1$ as $a \not\sim b$.

A.1 Computation

The type $a \sim b$ asserts a and b are equivalent computations. In accordance with the principle of propositions-as-types, this type is inhabited (by the placeholder 0) just when it is true.

(Computation) $\Gamma \vdash t \sim t'$

   where t and t' are one of the following pairs:

t	t'
$(\lambda x.b)(a)$	$b[a/x]$
$\langle a, b \rangle.1$	a
$\langle a, b \rangle.2$	b
$succ(n)$	n', where n' is one larger than n
$pred(n)$	n', where n' is one smaller than n or 0 if $n = 0$
$if\_zero(0; a; b)$	$a.$

(Computation) $\Gamma \vdash if\_zero(a; b; c) \sim c$
            $\vdash a \not\sim 0$

(Contradict) $\Gamma, a \sim b \vdash c \in C$

    where a and b are different values by inspection.

(Sim refl) $\Gamma \vdash a \sim a$

(Sim sym) $\Gamma, a \sim b \vdash b \sim a$

(Sim trans) $\Gamma, a \sim b, b \sim c \vdash a \sim c$

A.2   Membership

The expression a *in* A reifies the assertion $a \in A$ as a type; $0 \in a$ *in* A just when $a \in A$. The inhabitant 0 is implicit in the rules below.

(Member intro) $\Gamma \vdash a$ *in* A
                      $\vdash a \in A$

(Member elim) $\Gamma \vdash a \in A$
                      $\vdash a$ *in* A

A.3   Natural number

N is a type of natural numbers.

(N intro) $\Gamma \vdash a \in N$

    Where a is one of 0, 1, 2,. ...

(N succ) $\Gamma \vdash succ(a) \in N$
                    $\vdash a \in N$

    There is a symmetric rule (N pred).

(N induction) $\Gamma, a \in N \vdash b(a) \in B(a)$
                    $a \sim 0 \vdash b(a) \in B(a)$
                    $a \not\sim 0, x \in N, x \not\sim 0, b(pred(x)) \in B(pred(x)) \vdash b(x) \in B(x)$

A.4   Dependent function

Dependent functions $x{:}A \to B(x)$ are also commonly notated $\Pi x{:}A.B(x)$. We let $A \to B$ abbreviate $x{:}A \to B$ where x is not free in B.

(Func intro) $\Gamma \vdash \lambda x.b \in x{:}A \to B(x)$
                    $x \in A \vdash b \in B(x)$
                    $\vdash A \in U_k$

(Func elim) $\Gamma \vdash c \in C$
                    $b \sim \lambda x.b(x), b(a) \in B(a) \vdash c \in C$
                    $\vdash a \in A$
                    $\vdash b \in x{:}A \to B(x)$

    where x is not free in b.

A.5 Dependent product

Dependent products $x{:}A \times B(x)$ are also commonly notated $\Sigma x{:}A.B(x)$, but "dependent product" is a more apt computational description. These are types of pairs for which the type of the second component of the pair depends on the value of the first component. Let $A \times B$ abbreviate $x{:}A \times B$ where x is not free in B.

(Prod intro) $\Gamma \vdash \langle a, b \rangle \in x{:}A \times B(x)$
           $\vdash a \in A$
           $\vdash b \in B(a)$
           $x \in A \vdash B(x) \in U_k$

The third subgoal assures that $x{:}A \times B(x)$ is a sensible type.

(Prod elim) $\Gamma \vdash c \in C$
           $a \sim \langle a.1, a.2 \rangle, a.1 \in A, a.2 \in B(a.1) \vdash c \in C$
           $\vdash a \in x{:}A \times B(x)$

A.6 Universe

(U form) $\Gamma \vdash U_k \in U_{k+1}$

(U cumulativity) $\Gamma \vdash A \in U_{k+1}$
           $\vdash A \in U_k$

($\sim$ form) $\Gamma \vdash a \sim b \in U_k$

(Member form) $\Gamma \vdash (a \text{ in } A) \in U_k$
           $\vdash A \in U_k$

(N form) $\Gamma \vdash N \in U_k$

(Func form) $\Gamma \vdash x{:}A \rightarrow B(x) \in U_k$
           $\vdash A \in U_k$
           $x \in A \vdash B \in U_k$

(Prod form) $\Gamma \vdash x{:}A \times B(x) \in U_k$
           $\vdash A \in U_k$
           $x \in A \vdash B(x) \in U_k$

A.7 Miscellaneous

(Cut) $\Gamma \vdash a \in A$
           $\vdash b \in B$
           $x \in B \vdash a \in A$

(Hypothesis) $\Gamma \vdash x \in A$
     where $x \in A$ occurs in Γ.

(Subst) $\Gamma \vdash c[a] \in C[a]$
           $\vdash a \sim b$
           $\vdash c[b] \in C[b]$

(Prop member) $\Gamma, x \in A \vdash x \sim 0$
     where A is b in B, or $a \sim b$.

A.8 Recursive Types and CTTR

CTTR is CTT extended to have parameterized recursive types. $rec(X; x{:}A.B(X)(x))(a)$ denotes solutions to parameterized recursive type equations $X(x) \overset{\text{def}}{=} B(X)(x)$, where x is a parameter and has initial value a. Parameterized recursive types for CTT are presented in [CAB+86], chapter 12, and another version appears in [Men87]. Define $R_1 \subseteq_A R_2$ for $R_{1/2} \in A \to U_k$ as $\forall x{:}A. \forall y{:}R_1(x).y \in R_2(x)$.

Rec intro) $\qquad \Gamma \vdash b \in rec(X; x{:}A.B(X)(x))(a)$
$\qquad\qquad \vdash b \in B(rec(X; x{:}A.B(X)(x)))(a)$
$\qquad\qquad \vdash rec(X; x{:}A.B(X)(x))(a) \in U_k$

Rec elim) $\qquad \Gamma, y \in rec(X; x{:}A.B(X)(x))(a) \vdash c \in C$
$\qquad\qquad y \in B(rec(X; x{:}A.B(X)(x)))(a) \vdash c \in C$

Rec induction) $\qquad \Gamma, x \in rec(X; x{:}A.B(X)(x)))(a) \vdash c(x)(a) \in C(x)(a)$
$\qquad\qquad Z \in A \to U_k, y \in A, z \in (\forall y{:}A.\, x \in Z(y) \to c(x)(y)\ in\ C(x)(y)), x \in B(Z)(y)$
$\qquad\qquad \vdash c(x)(y)\ in\ C(x)(y)$

Rec form) $\qquad \Gamma \vdash rec(X; x{:}A.B(X)(x)) \in A \to U_k$
$\qquad\qquad X \in A \to U_k, x \in A \vdash B(X)(x) \in U_k$
$\qquad\qquad X_1 \in A \to U_k, X_2 \in A \to U_k, y \in X_1 \subseteq_A X_2 \vdash e \in B(X_1) \subseteq_A B(X_2)$

The second subgoal assures the body $B(X)(x)$ to be monotonic, making the least fixed point sensible.

References

[ACHA90] S. F. Allen, R. L. Constable, D. Howe, and W. Aitken. The semantics of reflected proof. In *Proceedings of the Fifth Annual Symposium on Logic in Computer Science*, pages 95–105, 1990.

[All87] S. F. Allen. A non-type-theoretic semantics for type-theoretic language. Technical Report 87-866, Department of Computer Science, Cornell University, September 1987. Ph.D. Thesis.

[BM81] R. S. Boyer and J S. Moore. Metafunctions: Proving them correct and using them efficiently as new proof procedures. In R. S. Boyer and J Strother Moore, editors, *The Correctness Problem in Computer Science*, chapter 3. Academic Press, 1981.

[CAB+86] R. L. Constable, S. F. Allen, H. Bromley, W. R. Cleveland, J. Cremer, R. Harper, D. Howe, T. Knoblock, N. P. Mendler, P. Panangaden, J. Sasaki, and S. F. Smith. *Implementing Mathematics with the Nuprl Proof Development System*. Prentice-Hall, Englewood Cliffs, New Jersey, 1986.

[DS79] M. Davis and J. T. Schwartz. Metamathematical extensibility for theorem verifiers and proof-checkers. *Computers and Mathematics with Applications*, 5:217–230, 1979.

[Dyb87] P. Dybjer. Inductively defined sets in Martin-Löf's set theory. Technical report, Department of Computer Science, University of Göteburg/Chalmers, 1987.

[Fef62]   S. Feferman. Transfinite recursive progressions of axiomatic theories. *Journal of Symbolic Logic*, 27:259–316, 1962.

[Fef75]   S. Feferman. A language and axioms for explicit mathematics. In J. N. Crossley, editor, *Algebra and Logic*, volume 450 of *Lecture notes in Mathematics*, pages 87–139. Springer-Verlag, 1975.

[Gir71]   J.-Y. Girard. Une extension de l'interprétation de Gödel à l'Analyse, et son application à l'Élimination des coupures dans l'Analyse et la Théorie des types. In J. E. Fenstad, editor, *Second Scandinavian Logic Symposium*, pages 63–92, Amsterdam, 1971. North-Holland.

[Mar82]   P. Martin-Löf. Constructive mathematics and computer programming. In *Sixth International Congress for Logic, Methodology, and Philosophy of Science*, pages 153–175, Amsterdam, 1982. North Holland.

[Men87]   P. F. Mendler. Inductive definition in type theory. Technical Report 87-870, Department of Computer Science, Cornell University, September 1987. Ph.D. Thesis.

[Plo75]   G. Plotkin. Call-by-name, call-by-value, and the λ-calculus. *Theoretical Computer Science*, pages 125–159, 1975.

[Smi91a]   S. F. Smith. From operational to denotational semantics. In *MFPS 1991*, Lecture notes in Computer Science, 1991. (To appear).

[Smi91b]   S. F. Smith. Partial computations in constructive type theory. Submitted to *Journal of Logic and Computation*, 1991.

[Tho91]   S. Thompson. *Type Theory and Functional Programming*. Addison-Wesley, 1991.

Are subsets necessary in Martin-Löf type theory?

Simon Thompson

Computing Laboratory, University of Kent at Canterbury

Canterbury, CT2 7NF, U.K.

e-mail: sjt@ukc.ac.uk

Introduction

Martin-Löf's theory of types, expounded in [Martin-Löf, 1975, Martin-Löf, 1985] and discussed at greater length in [Nordström *et al.*, 1990, Thompson, 1991] is a theory of types and functions or alternatively of propositions and proofs which has attracted much recent interest in the computing science community. There seems to have emerged a consensus that the system provides a good *foundation* for integrated program development and proof, but that for the system to be usable in practical projects a number of additions need to be made to it. Pre-eminent among these is the proposal to add a *subset* construction to the system, so that members of the type

$$\{ x : A \mid B \}$$

are those members of A with the property B. This is in contrast to the representation of such a type by

$$(\exists x : A) . B$$

whose members consist of pairs (a, p) with $a : A$ and $p : B[a/x]$ a *proof* of or *witness* to the fact that the property B is true of a. This latter representation is faithful to the principle of *complete presentation* which requires that it should be evident from any object that it has the type asserted of it. The witness is a proof of this fact.

In this paper after presenting an overview of the two important subset constructions we examine the reasons given for the addition of the subset type and argue that we can achieve the desired results *without* complicating the system by such an augmentation.

The first reason for adding a subset type is that it allows for the separation of the computational information in an object from the proof theoretic information it might contain – this we examine in sections 3 and 4 where we argue that this separation is better achieved by *naming* the appropriate portions of an object, using the axiom of choice where necessary to identify these portions.

Note that we take the axiom of choice as valid – this is the case for all Martin-Löf's systems, and is a simple consequence of the *strong* elimination rule for the existential quantifier, which allows the second projection from a pair to have dependent type. We should also observe that for most constructivists the axiom of choice is unexceptionable – given the interpretation of the quantifiers, a choice function can be read off in the obvious way.

The other reason advanced for the subset type is that it can contribute to the efficiency of evaluation, since by suppressing the proof-theoretic portion of an expression, any evaluation in this portion will no longer be necessary. We argue in section 5 that exactly the same effect is achieved if we use *lazy* evaluation to implement the system.

It is therefore evident that we can achieve the effects required *without* adding to and therefore complicating the system of type theory, as expounded in Martin-Löf's original papers, as long as we

---

Parts of this article are based on material originally appearing in *Type Theory and Functional Programming* ©Addison Wesley Publishing Company, 1991, and are included with the publisher's permission.

are prepared to work in a lazy implementation of the theory. Lazy implementations used to have the reputation of being slow, but recent work (see, for example, [Peyton Jones, 1987]) has shown that this need not be the case.

1   Type theory

In this section we provide a short review of those aspects of constructive type theory relevant to the discussion which follows.

The basic intuition underlying the system of type theory is that

> to prove is to construct.

In particular,

- a proof of $A \wedge B$ is a pair of proofs of A and B;

- a proof of $A \Rightarrow B$ is a transformation of proofs of A into proofs of B;

- a proof of $A \vee B$ is either a proof of A or a proof of B;

- a proof of $(\forall x : A) . B(x)$ takes a in A to a proof of $B(a)$, and

- a proof of $(\exists x : A) . B(x)$ is a witness a in A together with a proof of $B(a)$.

This intuitive presentation can be formalised in a collection of deduction rules which mention the *judgement*

> $a : A$

which is thought of as expressing

> a is a proof of the formula A

Four rules are presented for each connective. A *formation* rule describes how the formula is formed, it is a rule of *syntax* in other words

Formation Rule for $\wedge$

$$\frac{A \ is \ a \ formula \quad B \ is \ a \ formula}{(A \wedge B) \ is \ a \ formula}(\wedge F)$$

A second rule gives circumstances under which a proof of the formula can be found – this is the *introduction* rule. In the case of conjunction, a proof can be formed from proofs of the component formulas.

Introduction Rule for $\wedge$

$$\frac{p : A \quad q : B}{(p, q) : (A \wedge B)}(\wedge I)$$

The *elimination* rule or rules embody the fact that proofs can only be constructed according to the introduction rule(s): any proof of a conjunction can be decomposed to yield proofs of the individual components.

Elimination Rules for $\wedge$

$$\frac{r : (A \wedge B)}{fst \ r : A}(\wedge E_1) \qquad \frac{r : (A \wedge B)}{snd \ r : B}(\wedge E_2)$$

$$\frac{A \text{ is a type} \quad B \text{ is a type}}{(A \Rightarrow B) \text{ is a type}} (\Rightarrow F) \qquad \frac{\begin{array}{c}[x:A]\\ \vdots\\ e:B\end{array}}{(\lambda x:A).e \ : \ (A \Rightarrow B)} (\Rightarrow I)$$

$$\frac{q:(A \Rightarrow B) \quad a:A}{(q\,a) \ : \ B} (\Rightarrow E) \qquad ((\lambda x:A).e)\,a \ \rightarrow \ e[a/x]$$

Figure 1: Rules for function space

Moreover, if we form a proof (a, b) of $A \wedge B$ from proofs of A and B and extract the proofs of A and B from (a, b) using $(\wedge E)$ we derive the proofs we started with – this is described by the *computation* rules.

Computation Rules for $\wedge$

$$\text{fst } (p, q) \ \rightarrow \ p \qquad \text{snd } (p, q) \ \rightarrow \ q$$

It is striking that these rules can also be thought of as rules for a typed functional programming language, if we replace '... *is a formula*' by '... *is a type*'. The rules for the conjunction are those for the product type.

In a similar way, implication can be thought of as forming the *function space*, disjunction a *sum* type and the absurd proposition an *empty* type. Rules for the function type are given in Figure 1. The connective $\Rightarrow$ is introduced by means of a λ-abstraction: the assumption of the object x of A is discharged in the process of forming $(\lambda x:A).e$, as the variable x has become *bound*. The discharge is indicated by the surrounding brackets [...].

The formal system of type theory contains as well as the propositional connectives we have discussed, both basic types such as the natural numbers and lists and the *quantifiers* – we turn to these now. Further details of the full rules for systems of type theory can be found in [Thompson, 1991] and elsewhere.

The universal quantifier can be thought of as defining a *generalised* function space in which the type of the result of a function depends upon the value of the argument(s).

Formation Rule for $\forall$

$$\frac{A \text{ is a formula} \quad P \text{ is a formula}}{(\forall x:A).P \text{ is a formula}} \begin{array}{c}[x:A]\\ \vdots\end{array} (\forall F)$$

The dependence can be seen here from the fact that P can contain the variable x free, and so depend upon a value of type A. The universal quantifier is introduced by a λ-abstraction, where it is assumed that x occurs free in no other assumption than $x:A$.

Introduction Rule for $\forall$

$$\frac{\begin{array}{c}[x:A]\\ \vdots\\ p:P\end{array}}{(\lambda x:A).p \ : \ (\forall x:A).P} (\forall I)$$

The elimination and computation rules generalise those for implication.

Elimination Rule for ∀

$$\frac{a:A \quad f:(\forall x:A).P}{f\,a \;:\; P[a/x]}(\forall E)$$

Computation Rule for ∀

$$((\lambda x:A).p)\,a \;\rightarrow\; p[a/x]$$

A constructive interpretation of the *existential* quantifier takes objects of existential type to be *pairs*, the first half of which gives the witnessing element, and the second half the *proof* that this element has the property required.

Formation Rule for ∃

$$[x:A]$$
$$\vdots$$
$$\frac{A \text{ is a formula} \quad P \text{ is a formula}}{(\exists x:A).P \text{ is a formula}}(\exists F)$$

Introduction Rule for ∃

$$\frac{a:A \quad p:P[a/x]}{(a,p) \;:\; (\exists x:A).P}(\exists I)$$

The elimination and computation rules show the decomposition of an existential proof object into its component parts.

Elimination Rules for ∃

$$\frac{p \;:\; (\exists x:A).P}{Fst\,p \;:\; A}(\exists E_1') \qquad \frac{p \;:\; (\exists x:A).P}{Snd\,p \;:\; P[Fst\,p/x]}(\exists E_2')$$

Computation Rules for ∃

$$Fst\,(p,q) \;\rightarrow\; p \qquad Snd\,(p,q) \;\rightarrow\; q$$

The existential quantifier can be thought of as a type constructor in a number of different ways. It forms an infinitary sum of the types $B(a)$ as a varies over the tag type A; it can be thought of as forming modules, when the type A is a universe, and most importantly here, it forms a *subset* of A, consisting of those elements of A with the property B. In keeping with a constructivist approach, the element a carries with it the proof that it belongs to the subset – otherwise how can it be said to reside there?

The elimination rule $(\exists E_2')$ is unusual in that its conclusion contains the proof object p on the right-hand side of a judgement; this is in contrast to the other rules of the system. These other rules reduce to the rules of first-order intuitionistic predicate calculus if the proof objects are omitted; this cannot be done with $(\exists E_2')$ since the proof object appears in the formula part of the judgement. The rules presented are equivalent to the *strong* elimination rule:

$$[x:A; y:B]$$
$$\vdots$$
$$\frac{p:(\exists x:A).B \quad c:C[(x,y)/z]}{Cases_{x,y}\,p\,c \;:\; C[p/z]}(\exists E)$$

In turn, this rule has been shown in [Swaen, 1989] to be equivalent to the axiom of choice plus a weaker rule of elimination which corresponds to the usual rule of elimination in first-order logic. We show the derivability of the axiom of choice from the strong rule now.

The axiom of choice has the statement

$$(\forall x\!:\!A).\,(\exists y\!:\!B).\,C(x,y) \Rightarrow (\exists g\!:\!A \Rightarrow B).\,(\forall x\!:\!A).\,C(x,(g\,x))$$

Suppose that $f : (\forall x\!:\!A).\,(\exists y\!:\!B).\,C(x,y)$ then

$$Fst\,(f\,x) \,:\, B$$

and

$$Snd\,(f\,x) \,:\, C(x, Fst(f\,x))$$

Therefore

$$\lambda x_A.\,(Fst\,(f\,x)) \,:\, (A \Rightarrow B)$$

and we write g for this function. Also,

$$\lambda x_A.\,(Snd\,(f\,x)) \,:\, (\forall x\!:\!A).\,C(x, Fst(f\,x))$$

giving

$$\lambda x_A.\,(Snd\,(f\,x)) \,:\, (\forall x\!:\!A).\,C(x,(g\,x))$$

We thus have an object

$$(\lambda x_A.\,(Fst\,(f\,x))\,,\,\lambda x_A.\,(Snd\,(f\,x)))$$

of type

$$(\exists g\!:\!A \Rightarrow B).\,(\forall x\!:\!A).\,C(x,(g\,x))$$

Abstracting over f gives the proof of the axiom of choice.

2 The subset type

What are the formal rules for the subset type? The rules we give now were first proposed in [Nordström and Petersson, 1983], and used in [Constable and others, 1986], page 167, and [Backhouse et al., 1989], section 3.4.2. Formation is completely straightforward

Formation Rule for *Set*

$$[x\!:\!A]$$
$$\vdots$$
$$\frac{A\ is\ a\ type \quad B\ is\ a\ type}{\{\,x\!:\!A \mid B\,\}\ is\ a\ type}(SetF)$$

as is the introduction rule,

Introduction Rule for *Set*

$$\frac{a\!:\!A \quad p\!:\!B[a/x]}{a \,:\, \{\,x\!:\!A \mid B\,\}}(SetI)$$

How should a set be eliminated? If we know that $a\!:\!\{\,x\!:\!A \mid B\,\}$ then we certainly know that $a\!:\!A$, but also that $B[a/x]$. What we don't have is a specific proof that $B[a/x]$, so how could we encapsulate this? We can modify the existential elimination rule ($\exists E$) so that the hypothetical judgement $c : C$ is derived assuming some $y\!:\!B[a/x]$, but that c and C cannot depend upon this y. We use the fact that $B[a/x]$ is provable, but we cannot depend on the proof y itself:

Elimination Rule for *Set*

$$[x:A; y:B]$$
$$\vdots$$
$$\frac{a:\{x:A\mid B\}\quad c(x):C(x)}{c(a)\;:\;C(a)}(SetE)$$

where y is not free in c or C. Since no new operator is added by the elimination rule, there is no computation rule for the subset type. We should note that this makes these rules different from others in type theory. This is also evident from the fact that they fail to satisfy the *inversion principle* of [Schroeder-Heister, 1989]

[Salvesen and Smith, 1989] shows that these rules are weaker than might at first be thought, especially if we adopt an *intensional* version of type theory. In fact, we cannot in a consistent manner derive the formula

$$(\forall x:\{z:A\mid P(z)\}). P(x) \tag{1}$$

for most formulas P. This has the consequence that we cannot derive functions to take the head and tail of a non-empty list, if we choose to represent the type of non-empty lists by a subset type,

$$\{l:[A]\mid nonempty\ l\}$$

where the predicate *nonempty* is defined by a recursion over a universe thus:

$$nonempty\ [\,]\qquad \equiv_{df}\quad \bot$$
$$nonempty\ (a :: x)\quad \equiv_{df}\quad \top$$

The situation in the extensional theory is better, but there are still cases of the formula (1) which are not derivable consistently. Because of these weaknesses, Martin-Löf proposed a new *subset* theory which incorporates the judgement P *is true* into the system.

If the representation of the judgement is to be an improvement on TT, as far as subsets are concerned, it is desirable that the system validates the rule

$$\frac{a\;:\;\{x:A\mid P\}}{P(a)\;is\;true} \tag{2}$$

This can be done if we move to a system in which propositions and types are distinct. In [Nordström *et al.*, 1990] can be found the **subset theory** in which the new judgements

$$P\ prop\qquad \text{and}\qquad P\ is\ true$$

are added to the system, together with a set of *logical* connectives, distinct from the type forming operations introduced in their extensional version of type theory. This system does allow the derivation of (2) but at the cost of losing the isomorphism between propositions and types and making the system more complex.

3 What is a specification?

The judgement $a:A$ can be thought of as expressing 'a proves the proposition A' and 'a is an object of type A', but it has also been proposed, in [Martin-Löf, 1985, Petersson and Smith, 1985] for example, that it be read as saying

$$a \text{ is a program which meets the specification } A \tag{†}$$

It is misleading to apply this interpretation to every judgement $a : A$. Take for instance the case of a function f which sorts lists; this has type $[A] \Rightarrow [A]$, and so,

$$f : [A] \Rightarrow [A]$$

Should we therefore say that it meets the specification $[A] \Rightarrow [A]$? It does, but then so do the identity and the reverse functions! The type of a function is but one aspect of its specification, which should describe the relation between its input and output. This characterisation takes the form

The result $(f\, l)$ is ordered and a permutation of the list l

for which we will write $S(f)$. To assert that the specification can be met by some implementation, we write

$$(\exists f : [A] \Rightarrow [A]) . S(f)$$

What form do objects of this type take? They are pairs (f, p) with $f : [A] \Rightarrow [A]$ and p a proof that f has the property $S(f)$. The confusion in (†) is thus that the object a consists not of a program meeting the specification, but of such a program together with a *proof* that it meets that specification.

In the light of the discussion above, it seems sensible to suggest that we conceive of specifications as statements $(\exists o : T) . P$, and that the formal assertion

$$(o, p) : (\exists o : T) . P$$

be interpreted as saying

The object o, of type T, is shown to meet the specification P by the proof object p.

an interpretation which combines the logical and programming interpretations of the language in an elegant way. This would be obvious to a constructivist, who would argue that we can only assert (†) if we have the appropriate evidence, namely the proof object.

In developing a proof of the formula $(\exists o : T) . P$ we construct a pair consisting of an object of type T and a proof that the object has the property P. Such a pair keeps separate the computational and logical aspects of the development, so that we can extract directly the computational part simply by choosing the first element of the pair.

There is a variation on this theme, mentioned in [Nordström *et al.*, 1990] for instance, which suggests that a specification of a function should be of the form

$$(\forall x : A) . (\exists y : B) . P(x, y) \qquad (3)$$

Elements of this type are functions F so that for all $x : A$,

$$F\, x : (\exists y : B) . P(x, y)$$

and *each* of these values will be a pair (y_x, p_x) with

$$y_x : B \quad \text{and} \quad p_x : P(x, y)$$

The pair consists of value and proof information, showing that under this approach the program and its verification are inextricably mixed. It has been argued that the only way to achieve this separation is to replace the inner existential type with a *subset* type, which removes the proof information p_x. This can be done, but the intermingling can be avoided *without* augmenting the system – we simply have to give the intended function a *name*. That such a naming can be achieved in general is a simple consequence of the *axiom of choice*, which states that

$$(\forall x : A) . (\exists y : B) . P(x, y) \Rightarrow (\exists f : A \Rightarrow B) . (\forall x : A) . P(x, f\, x)$$

Applying *modus ponens* to this and (3) we deduce the specification

$$(\exists f : A \Rightarrow B).(\forall x : A).P(x, f\,x) \tag{4}$$

Note that the converse implication to that of the axiom of choice is easily derivable, making the two forms of the specification logically equivalent.

It is worth noting that some functions are not specified simply by their input/output relation, one example being a hashing function[1]. This means that specifications will necessarily have the $(\exists o : T).P$ form in general.

This analysis of specifications makes it clear that when we seek a program to meet a specification, we look for the *first* component of a member of an existential type; the second proves that the program meets the constraint part of the specification. As long as we realise this, it seems irrelevant whether or not our system includes a type of first components, which is what the subset type consists of. There are other arguments for the introduction of a subset type, which we turn to now.

4   Subsets in specifications

We have seen that the intermingling of computation and verification which appears to result from an interpretation of specifications as propositions can be avoided by the expedient of using the axiom of choice in the obvious way.

In this section we look at other uses of the subset type within specifications and show that in many of these we can again avoid the subset type by separating from a complex specification exactly the part which is computationally relevant in some sense. This is to be done by *naming* in an appropriate manner the operations and objects sought, as we did in the previous section when we changed the $\forall \exists$ specification into an $\exists \forall$ form. This reversal of quantifiers which arises by naming the function is known to logicians as **Skolemizing** the quantifiers. We believe the alternative is superior for two reasons:

- it is a solution which requires no addition to the system of type theory, and

- it allows for more delicate distinctions between proof and computation.

The method of Skolemizing can be used in more complex situations, as we now see.

Take as an example a simplification of the specification of the Dutch (or Polish) national flag problem as given in [Nordström *et al.*, 1990]. We now show how it may be written without the subset type. The original specification has the form

$$(\forall x : A).\{\,y : \{\,y' : B \mid C(y')\,\} \mid P(x, y)\,\}$$

with the intention that for each a we find b in the subset $\{\,y' : B \mid C(y')\,\}$ of B with the property $P(a, b)$. If we replace the subsets by existential types, we have

$$(\forall x : A).(\exists y : (\exists y' : B).C(y')).P(x, y)$$

This is logically equivalent to

$$(\forall x : A).(\exists y : B).(\,C(y) \wedge P(x, y)\,) \tag{5}$$

and by the axiom of choice to

$$(\exists f : A \Rightarrow B).(\forall x : A).(\,C(f\,x) \wedge P(x, (f\,x))\,)$$

which is inhabited by functions *together with proofs of their correctness*. It can be argued that this expresses in a clear way what was rather more implicit in the specification based on sets – the

---

[1] I am grateful to Michael O'Donnell for this observation.

formation of an existential type bundles together data and proof, the transformation to (5) makes explicit the unbundling process.

As a second example, consider a problem in which we are asked to produce for each a in A with the property $D(a)$ some b with the property $P(a, b)$. There is an important question of whether the b depends just upon the a, or upon both the a and the proof that it has the property $D(a)$. In the latter case we could write the specification thus:

$$(\forall x : (\exists x' : A) . D(x')) . (\exists y : B) . P(x, y)$$

and Skolemize to give

$$(\exists f : (\exists x' : A) . D(x') \Rightarrow B) . (\forall x : (\exists x' : A) . D(x')) . P(x, (f\, x))$$

If we use the equivalence between the types

$$((\exists x : X) . P) \Rightarrow Q \qquad (\forall x : X) . (P \Rightarrow Q)$$

(which is the logical version of the isomorphism between 'curried' and 'uncurried' versions of binary functions) we have

$$(\exists f : (\forall z : A) . (D(z) \Rightarrow B)) . (\forall x' : A) . (\forall p : D(x')) . P((x', p), (f\, x'\, p))$$

which makes manifest the functional dependence required. Observe that we could indeed have written this formal specification directly on the basis of the informal version from which we started.

If we do *not* wish the object sought to depend upon the proof of the property D, we can write the following specification:

$$(\exists f : A \Rightarrow B) . (\forall x' : A) . (\forall p : D(x')) . P((x', p), (f\, x')) \tag{6}$$

in which it is plain that the object $(f\, x')$ in B is not dependent on the proof object $p : D(x')$. Observe that there *is* still dependence of the property P on the proof p; if we were to use a subset type to express the specification, thus, we would have something of the form

$$(\forall x' : \{\, x' : A \mid D(x') \,\}) . (\exists y : B) . P'(x', y)$$

where the property $P'(x, y)$ relates $x' : A$ and $y : B$. This is equivalent to the specification

$$(\exists f : A \Rightarrow B) . (\forall x' : A) . (\forall p : D(x')) . P'(x', (f\, x'))$$

in which the property P' must not mention the proof object p, so that with our more explicit approach we have been able to express the specification (6) which cannot be expressed under the naïve subset discipline.

5 Computational Irrelevance; Lazy Evaluation

The natural definition of the 'head' function on lists is over the type of non- empty lists, given thus:

$$(nelist\, A) \equiv_{df} (\exists l : [A]) . (nonempty\, l)$$

where the predicate *nonempty* was defined above. The head function itself, hd, is given by

$$
\begin{array}{lll}
hd & : & (nelist\, A) \Rightarrow A \\
hd\, ([\,], p) & \equiv_{df} & abort_A\, p \\
hd\, ((a :: x), p) & \equiv_{df} & a
\end{array}
$$

which is formalised in type theory by a primitive recursion over the list component of the pair.

Given an application

$$hd\, ((2 :: \ldots), \ldots)$$

computation of the result to 2 can proceed in the absence of any information about the elided portions. In particular, the *proof* information is not necessary for the process of computation to proceed in such a case. Nonetheless, the proof information is crucial in showing that the application is properly typed; we cannot apply the function to a bare list, as that list might be empty. There is thus a tension between what are usually thought of as the *dynamic* and *static* parts of the language. In particular it has been argued that if no separation is achieved, then the efficiency of programs will be impaired by the welter of irrelevant information which they carry around – see section 3.4 of [Backhouse *et al.*, 1989] and section 10.3 of [Constable and others, 1986].

Any conclusion about the efficiency of an object or program is predicated on the evaluation mechanism for the system under consideration, and we now argue that a *lazy* or outermost first strategy has the advantage of not evaluating the computationally irrelevant.

If we work in an intensional system of type theory, then using the results of [Martin-Löf, 1975] the system is both strongly normalising and has the Church Rosser property. This means that *every* sequence of reductions will lead us to the same result. Similar results are valid if we evaluate to weak head normal form in the extensional case.

We can therefore choose how expressions are to be evaluated. There are two obvious choices. **Strict** evaluation is the norm for imperative languages and many functional languages (Standard ML, [Harper, 1986], is an example). Under this discipline, in an application like

$$f\ a_1\ \ldots\ a_n$$

the arguments a_i are evaluated fully before the whole expression is evaluated. In such a situation, if an argument a_k is computationally irrelevant, then its evaluation will degrade the efficiency of the program. The alternative, of **normal order** evaluation is to begin evaluation of the whole expression, prior to argument evaluation: if the value of an argument is unnecessary, then it is not evaluated.

To be formal about the definition, we say that evaluation in which we always choose the leftmost outermost redex is **normal order** evaluation. If in addition we ensure that no redex is evaluated more than once we call the evaluation **lazy**. (For more details on evaluation strategies for functional languages, see [Peyton Jones, 1987], for example).

In a language with structured data such as pairs and lists, there is a further clause to the definition: when an argument is evaluated it need not be evaluated to normal form; it is only evaluated to the extent that is necessary for computation to proceed. This will usually imply that it is evaluated to weak head normal form (see [Peyton Jones, 1987]). This means that, for example, an argument of the product type $A \wedge B$ will be reduced to a pair (a, b), with the sub-expressions a and b as yet unevaluated. These may or may not be evaluated in subsequent computation.

Under lazy evaluation computationally irrelevant objects or components of structured objects will simply be *ignored*, and so no additional computational overhead is imposed. Indeed, it can be argued that the proper definition of computational relevance would be that which chose just that portion of an expression which is used in calculating a result under a lazy evaluation discipline.

Another example is given by the following example of the *quicksort* function over lists. Quicksort is defined by

$$qsort\ l \equiv_{df} qsort'\ l\ (\#l)\ p$$

where p is the canonical proof that $(\#l) \leq (\#l)$. The auxiliary function is given by

$$qsort'\ :\ (\forall n : N).(\forall l : [N]).((\#l \leq n) \Rightarrow [N])$$

$$
\begin{aligned}
&qsort'\ n\ [\ ]\ p & &\equiv_{df}\ [\ ] \\
&qsort'\ 0\ (a :: x)\ p & &\equiv_{df}\ abort_{[N]}p' \\
&qsort'\ (n+1)\ (a :: x)\ p \\
&\quad \equiv_{df}\quad qsort'\ n\ (filter\ (lesseq\ a)\ x)\ p_1 \\
&\quad\quad\quad +\!\!+\ [a]\ +\!\!+ \\
&\quad\quad\quad qsort'\ n\ (filter\ (greater\ a)\ x)\ p_2
\end{aligned}
$$

The function has three parameters: a list (l), a natural number (n), and a proof that the length of l is less that or equal to n. In the second clause of the definition we use the proof p to construct a proof p' that 0 is smaller than itself, a contradiction. In the recursive calls to the function, we construct proofs p_1 and p_2 which witness the facts that $(filter(lesseq\,a)\,x)$ and $(filter(greater\,a)\,x)$ have length at most n if x has.

We have built an implementation of a system of type theory without universes by means of a translation of it into Miranda which is implemented in a lazy fashion. The quicksort function above will sort a list without calculating any of the terms p_i in any of the invocations of the function – proof of their computational irrelevance.

There is one drawback to the lazy implementation – no irrelevant terms are evaluated, but there are cases in which tuples are formed and destroyed, as were the tuples in the quicksort example. It seems too high a price to pay for the programmer to have to include for her- or himself an indication of how a program may be optimised, especially as this kind of use analysis can be performed most effectively by the techniques of abstract interpretation, as discussed in [Abramsky and Hankin, 1987] for instance. Linked to this is the syntactic characterisation of computational relevance, which involves an examination of the different forms that types (*i.e.* propositions) can take – to be found in section 3.4 of [Backhouse *et al.*, 1989]. It is not hard to see that under lazy evaluation the objects deemed to be irrelevant will not contribute to the final result, and will remain unevaluated.

6 Conclusion

To summarise, there are two responses to the use of subsets in type theory. Their use in separating the computational from the proof theoretic can be achieved using the appropriate names for functions whose existence is assured by the validity of the axiom of choice in type theory.

If proof theoretic information remains in an expression, we contend that if it is indeed irrelevant to the computational behaviour of a function, it will not be evaluated under a lazy evaluation strategy, and so we advocate this as an implementation technique which avoids the unnecessary evaluation which is a consequence of a strict evaluation scheme. As we mentioned earlier, there will be some cases in which structures are formed needlessly – we see their elimination as the role of the implementation of the system, and would view abstract interpretation as an ideal tool for this purpose.

Using the subset type to represent a subset brings problems; as we saw in the previous section, it is not possible in general to recover the witnessing information from a subset type, especially in an intensional system like TT, and so in these cases, the existential type should be used, retaining the witnessing information. Even in cases where such information can be recovered, we gain this only at the cost of having to work in a more complex system, especially in the case where the addition of the judgement P *is true* will give a confusion between similar results in the type and proposition modes.

References

[Abramsky and Hankin, 1987] Samson Abramsky and Chris Hankin, editors. *Abstract Interpretation of Declarative Languages*. Ellis-Horwood, 1987.

[Backhouse *et al.*, 1989] Roland Backhouse, Paul Chisholm, Grant Malcolm, and Erik Saaman. Do-it-yourself type theory. *Formal Aspects of Computing*, 1, 1989.

[Constable and others, 1986] Robert L. Constable et al. *Implementing Mathematics with the Nuprl Proof Development System*. Prentice-Hall Inc., 1986.

[Harper, 1986] Robert Harper. Introduction to Standard ML. Technical Report ECS-LFCS-86-14, Laboratory for Foundations of Computer Science, Department of Computer Science, University of Edinburgh, November 1986.

[Martin-Löf, 1975] Per Martin-Löf. An intuitionistic theory of types: Predicative part. In H. Rose and J. C. Shepherdson, editors, *Logic Colloquium 1973*. North-Holland, 1975.

[Martin-Löf, 1985] Per Martin-Löf. Constructive mathematics and computer programming. In C. A. R. Hoare, editor, *Mathematical Logic and Programming Languages*. Prentice-Hall, 1985.

[Nordström and Petersson, 1983] Bengt Nordström and Kent Petersson. Types and specifications. In *IFIP'83*. Elsevier, 1983.

[Nordström et al., 1990] Bengt Nordström, Kent Petersson, and Jan M. Smith. *Programming in Martin-Löf's Type Theory — An Introduction*, volume 7 of *International Series of Monographs on Computer Science*. Oxford University Press, 1990.

[Petersson and Smith, 1985] Kent Petersson and Jan Smith. Program derivation in type theory: The Polish flag problem. In Peter Dybjer et al., editors, *Proceedings of the Workshop on Specification and Derivation of Programs*. Programming Methodology Group, University of Goteborg and Chalmers University of Technology, 1985. Technical Report, number 18.

[Peyton Jones, 1987] Simon Peyton Jones. *The Implementation of Functional Programming Languages*. Prentice Hall International, 1987.

[Salvesen and Smith, 1989] Anne Salvesen and Jan Smith. The strength of the subset type in Martin-Löf's type theory. In *Proceedings of the Third Annual Symposium on Logic in Computer Science*. IEEE Computer Society Press, 1989.

[Schroeder-Heister, 1989] Peter Schroeder-Heister. Judgements of higher levels and standardized rules for logical constants in Martin-Löf's theory of logic. In Peter Dybjer et al., editors, *Proceedings of the Workshop on Programming Logic*. Programming Methodology Group, University of Goteborg and Chalmers University of Technology, 1989. Technical Report, number 54. This paper was written in 1985.

[Swaen, 1989] Marco D. G. Swaen. *Weak and Strong Sum-Elimination in Intuitionistic Type Theory*. PhD thesis, University of Amsterdam, 1989.

[Thompson, 1991] Simon Thompson. *Type Theory and Functional Programming*. Addison Wesley, 1991.

Development Transformation Based on Higher Order Type Theory

Jianguo Lu, Jiafu Xu

Institute of Computer Software, Nanjing University,

Nanjing 210008, P.R.China

1.Introduction

A lot of difficulties in program development result from two conflicting goals, i.e., the clarity and efficiency of the program that is being developed. A solution to this problem is offered by transformational approach to program development[2]. One issue in program transformation system is that complete automation is almost impossible. Even the transformation of a very simple program needs a lot of human interference[5]. Although human guidance in isolated program development is almost indispensable, past program development experience should be utilized to aid the development of similar programs. This motivation results in the research on analogical programming[6]. In most cases, analogical programming is a kind of modification/adaptation of program development process that is embodied in one form or another. Hence we propose program development by transforming development of one program to another similar one.

One crucial issue in development transformation is the representation of program derivation process and the correctness of the program of the transformed development with respect to its new specification. Usually, specification, program and the development from the former to the latter are denoted in different notations, which hamper the manipulation of program developments and the verification of the program of the transformed development. Much work has been devoted to building systems for checking and building formal proofs in various logical systems[4][7]. However, existing work is largely oriented towards proof construction and checking in a general setting. Little work is directed at program development in a logic framework, even less attention has been paid to the analogical derivation of the formally represented program development.

Our approach is to represent specification, program, and program development in a single framework of λ calculus based on intuitionistic type theory, i.e., specifications and programs are the types of the calculus, while program development is a λ term in the calculus. A direct benefit of this approach is that proving correctness of the development is reduced to the problem of type checking, and the correctness of the program of the transformed development with respect to its specification is guaranteed by that the type representing the specification is inhabited by an appropriate element. Based on this notion of correctness of program development, we present a development transformation procedure, which transforms a development for one problem to a new development that solves another problem not identical but analogous to the original one, avoiding constructing the new development from scratch. The correctness preserving property of the development transformation procedure is proved. The approach is examplified in the Fold/Unfold program transformation paradigm[2].

The paper is organized as follows. In section 2 we will briefly introduce a calculus similar to Λ calculus[13], and use this calculus to represent program development. Section 3 will discuss development transformation problem, its solution, and the development transformation procedure. In section 4, a very simple example is given to illustrate the development transformation method. Finally, some related work is discussed.

2.Representation of development

There is an unexpected analogy between terms of typed λ calculus and natural deductions. This correspondence relates types with propositions and terms with natural deduction proofs[7][8][11]. Since program construction can be looked as a constructive theorem proving process, it is natural to conjecture that there is also a correspondence between terms of typed λ calculus and program developments. The following tries to build this correspondence, which is largely based on the work of [13] [14][15].

2.1. The calculus

Following AUTOMATH[1], there is a point of view different from that in ordinary typed λ calculus. We make no distinction between terms and types, i.e., types are first class objects. The syntax of the calculus can be defined as follows:

```
term::= empty
      | variable | constant
      | λ variable:term.term
      | (term(term)).
```

$\lambda x:A, (\lambda x:A)B$ are examples of terms. Subterm and substitution are defined as usual.

The typing function Typ is defined as in [13]. The Calculus has the reduction relations, i.e., α, β, η reductions. Conversion is the equivalence relation generated by the closure of these reductions, which is denoted as $=_{PD}$ in the following discussion. The calculus has the usual properties such as Strong normalization and Church-Rosser property. The validity condition is similar to the strong functionality and the closedness in [13]. A term t is consistent with another term s (denoted as Con(t,s)) if there exist terms y,t1,s1, such that

$$\mathrm{Typ}(t) =_{PD} [y:t1]s1, \text{ and}$$
$$\mathrm{Typ}(s) =_{PD} t1.$$

For ease of reading, we adopt some notational abbreviations. Terms such as (t(s1))(s2) will be written as t(s1,s2). Sometimes we use infix notation, e.g., we write a+ b instead of plus(a,b).

2.2. Expressing program, specification and development

By specification we mean a pair of pre- and post- conditions, or a concise but inefficient program as in [2]. Programs are functional ones with recursive equations. Program developments are restricted to the Fold/Unfold program transformation method. We first introduce some conventional language constructs expressed in the calculus, then programs, specifications and developments.

- Propositions and objects

Suppose we have two primitive types 1 and o, which represent the type of propositions and the type of objects, respectively. Then the fact that P is a proposition can be represented as P:1. The universal quantification $V_{x \in A} P$ can be represented by $\lambda x:A.P$.

- Declarations and assumptions

The declaration that x is of type A can be represented by the term λx:A. The assumption that proposition P holds can be represented by λx:P, which asserts that P has x as its proof. For readability, we simply write x:A for declaration, and P for assumption when the proof of P is out of concern. To approximate the constructs in programming languages, the concatenation of declarations(assumptions), say (a:A)(b:B), will be written as (a:A; b:B).

- Definitions

    The abbreviation that x (the type of x is B) denotes A in the context of C can be expressed by (λx:B.C)A, where the type of A is convertible to B. It can be justified by the β reduction rule that every occurrence of x in C is replaced by A. Hereafter we simply use the notation (x:= A;C).

- Deductions and functions

    That the expressibility of functions by λ terms is well known. When P and Q are propositions, the fact that Q is true under the assumption P can be represented by λx:P.Q, which has the intuitive meaning that for every proof of P we can get a proof of Q. The x in λx:P above is not always relevant. Sometimes for simplicity we wish to omit x. To enable this, and to reflect the intuitive meaning of the deductive connotation of λ abstraction, we write [P]Q or [x:P]Q instead of λx:P.Q.

- Polymorphism

    It is very desirable to allow the definition of polymorphic functions. We provide this facility by using type variables, i.e., the declaration t:TYPE states that t range over types.

- Specifications and programs

    The representation of functions and first order predicate formulae has been discussed above. Here is an example. The functional program to compute the factorial function can be written as:

$$
\begin{aligned}
\text{FAC} := [\quad & \text{fac: [nat]nat;} \\
& \text{f1 : fac(1)= 1;} \\
& \text{f2 : [n:nat] fac(n+ 1)= fac(n) * (n+ 1)} \\
&],
\end{aligned}
$$

where nat is a predefined type, * is an operation of this type, '= ' is an equality symbol that is defined in an equality theory. Once fac is specified in this manner, equations, say fac(3)= 6, can be proved in this context. Its proof object is unfold(f1, unfold(f2,f2(2))), whose type is convertible to fac(3)= 6.

Here the transformation rule unfold is specified in the following discussion.

• Development

By development we mean a sequence of refinements by applications of transformation rules. Transformation rules can be represented as constants of the calculus, e.g., the unfold rule can be declared as :

$$\text{unfold} : [\; t,s: \text{Type}; \; x,y{:}t; \; z{:}s; \; F{:}[t]s]$$
$$[x=y] \; [z= F(x)] \; z= F(y).$$

The other rules, such as fold, laws, instantiation(·inst), and reflexivity(refl), can be defined in a similar way. For clarity, we write s't to denote the application t(s), and t..s to assert that t has type s, following the notations introduced by [13][15]. Thus, a segment of program transformation in [2] can be expressed as follows:

$$\text{Bd1} := [F{:}[\text{nat}]\text{nat}]$$
$$[\; \text{eu}: [n,\text{acc:nat}] \; F(n,\text{acc})= \text{fac}(n)*\text{acc}]$$
$$\text{Inst}(\text{eu},1) \; .. \; [\text{acc:nat}]F(\text{acc})= \text{fac}(1)*\text{acc}$$
$$\text{'unfold}(f1) \; .. \; [\text{acc:nat}]F(\text{acc})= 1*\text{acc}$$
$$\text{'law}(a*1= a) \; .. \; [\text{acc:nat}]F(\text{acc})= \text{acc}.$$

The validity of the above development is obvious.

3.Development transformation

Program transformation has proved very difficult to automate completely. One problem is that even though the system has had an experience with a program transformation problem, it is still completely at a loss when encountering a similar but not identical one. The new program still has to be developed from scratch. We propose a kind of analogical program derivation method that transfers program development experience from one program to another similar one. The issue of analogy recognition, i.e., finding an appropriate base problem, is not considered here.

3.1. Analogical correspondence

The analogical correspondence C between two terms t and s of the calculus is a subset of the Cartesian product of subterms of t and s, i.e.,

$$C \subseteq subterm(t) \times subterm(s).$$

The problem of finding analogical correspondence of two terms is discussed in [9]. Here we assume C is given beforehand.

Two terms t and s are analogous with respect to C (denoted as t ~ s wrt C) if they are the same in some abstract level. It is defined as

t ~ s wrt C if t and s can be generalized into a common term r that is valid and satisfies some additional conditions(for details see [9]),and there exist substitutions ξ,γ, such that $r\xi = t$, $r\gamma = s$, and merge $(\xi,\gamma) \subseteq C$.

Here merge is defined as

$$merge(\xi,\gamma) = \{(t,s) | \exists\ x \in VAR(\Sigma)\ (x\xi = t, x\gamma = s)\}.$$

3.2. Correctness of program with respect to its specification

Let ε be a sequence of recursive equations $Ei(i = 1,...,m)$ that define a function symbol f. ε can be looked as a functional, the function defined by this set of equations is the least fixed point of ε. Let ε' be derived by the rules of the fold/unfold program transformation method, g,f are the least fixed point of ε' and ε, respectively. It is known that $g \leq f$, that is, ε' is partially correct with respect to ε.

It is obvious that expressions without where abstraction in the Fold/Unfold method can be expressed as terms in the calculus. Suppose the expressions E,F...are represented as E^o, F^o, ..., etc.. The where abstraction, say E where u= F, can be expressed as $[u:= F^o]\ E^o$. Recursive equations, say E $\Leftarrow$ F, can be expressed as $E^o = F^o$. Hence each recursive equation Eq can be represented as a term in the calculus. Suppose its representation is Eq^o.

Suppose equations (E1,...,Em) can be transformed into (F1,..., Fn) in the Fold/Unfold method, then there are terms d1,..., dn, such that $[x1:E1^o;...; xm:Em^o]di$ is valid, and $Typ(di) =_{PD} Fi^o$, where i = 1,...,n. In this sense, function defined by $(F1^o; ...; Fn^o)$ is partially correct with respect to $(E1^o;...; Em^o)$, and (d1;...;dn) is the development from $(E1^o; ...;Em^o)$ to $(F1^o; ... ; Fn^o)$.

3.3. Development transformation problem

Suppose a base problem specification Bs, its program Bp, and the development Bd are given, along with the target problem specification Ts and the analogical correspondence C between the base and target problem.

The development transformation problem is a triple (Ts,(Bs,Bp,Bd), C). Here Ts, Bs, Bp, and Bd are terms, C is an analogical correspondence between Bs and Ts, $Typ(Bd) =_{PD} Bp$, and [Δ;Bs]Bd is valid, where Δ is a sequence of declarations and definitions.

The solution of the problem is a term Td such that

- [Δ;Ts]Td is valid,
- Typ(Bd) ~ Typ(Td) wrt C.

Once the solution Td is obtained, the program of the target specification Ts can be derived from Td by simple computation of the type Td, which is almost a mechanical procedure.

3.4. Development transformation

Now we offer the rules of development transformation.

Suppose Bs ~ Ts wrt C, [Δ;Bs]Bd is valid, $Typ(Bd) =_{PD} Bp$. The fact that Bd can be transformed into Td in the context of (Δ;Ts) (denoted as Bd → Td) can be defined by the following rules, which are classified into two classes:

a) Decomposition rules

The general form of the rules is

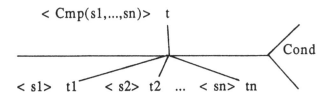

It describes how to decompose a base term to other terms, as well as how to compose analogical correspondence of the original term from existing ones. The meaning of the rule is that if ti → si, i= 1,...,n, and Cond is satisfied, then t → Cmp(s1,..., sn). Here ti, si, and t are terms, i= 1,...,n. Cmp(s1,...,sn) is the composition of s1,...., sn according to the rules applied.

b) Primitive rules

The general form is

    < s>   t   Cond

This kind of rules directly transform the base term to target term. The meaning of the rule is that if s and t satisfy the condition Cond, then t → s.

Some typical rules are as follows. For details see [9].

r1)   < s(s1)>   t(t1)

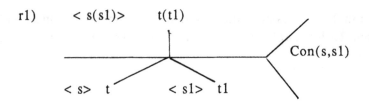

Con(s,s1)

     < s>   t   < s1>   t1

r2)   < t(s1)>   t(t1)

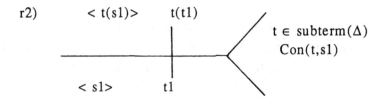

t ∈ subterm(Δ)
Con(t,s1)

     < s1>   t1

r3)   < [x:s1]s>   [x:t1]t

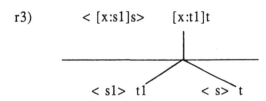

     < s1>   t1   < s>   t

r4)   < t(s1,s2)>   t(t1,t2)

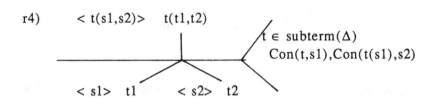

t ∈ subterm(Δ)
Con(t,s1),Con(t(s1),s2)

     < s1>   t1   < s2>   t2

r5)   < s>   t   if (t,s) ∈ C,
                                  or ∃ C' ⊆ C, such that t ~ s wrt C'

r6)   < x>   x   if $x \in$ VARΔ)

The development transformation process is accomplished by two separate stages, i.e., the top-down decomposition of the base term, and the bottom-up composition of the target term. In the first stage, the base term is repeatedly decomposed into other terms, and a derivation tree is generated. The leaf is the node that is generated by the application of the primitive rules. Once the analogical correspondence of the subnodes of term t is derived, the analogical correspondence of t can be composed from its subnodes according to the rules applied. The transformed development possesses the following property:

Theorem: Given development transformation problem (Ts,(Bs,Bp,Bd),C), if Bd $\rightarrow$ Td, then Td is a solution of the problem.

Proof: Follows from the properties of the analogical correspondence. For details see [9].

4. Example

To illustrate the problem, we introduce a naive example.

Given a concise but inefficient program computing the sum:

Ts:= [sum:[nat]nat;
       s1 : sum(0)= 0;
       s2 : [n:nat]sum(n+ 1)= sum(n)+ (n+ 1)
       ].

Suppose the base problem is the factorial program represented above, whose specification and complete development are:

Bs:= [fac:[nat]nat;
      f1 : fac(1)= 1;
      f2 : [n:nat]fac(n+ 1)= fac(n)*(n+ 1)
      ].

Bd1:= [F:[nat,nat]nat]
      [eu:[n,acc:nat]F(n,acc)= fac(n)*acc]
      Inst(eu,1) .. [acc:nat]F(1,acc)= fac(1)*acc
      'unfold(f1).. [acc:nat]F(1,acc)= 1*acc
      'law(a*1= a) .. [acc:nat]F(1,acc)= acc;

Bd2:= [F:[nat,nat]nat]
    [eu:[n,acc:nat]F(n,acc)= fac(n)*acc]
    Inst(eu,n+ 1) .. [acc:nat]F(n+ 1,acc)= fac(n+ 1)*acc
    'unfold(f2).. [acc:nat]F(n+ 1,acc)= (fac(n)*(n+ 1))*acc
    'law(a*(b*c)= (a*b)*c) .. [acc:nat]F(n+ 1,acc)= fac(n)*((n+ 1)*acc)
    'fold(eu) ..[acc:nat]F(n+ 1,acc)= F(n,(n+ 1)*acc);

Bd3:= [F:[nat,nat]nat]
    [eu:[n,acc:nat]F(n,acc)= fac(n)*acc]
    f2 ..[n:nat]fac(n+ 1)= fac(n)*(n+ 1)
    'fold(eu)..[n:nat]fac(n+ 1)= F(n,n+ 1).

From this development it is obvious that fac(n+ 1) is equal to F(n,n+ 1), which achieves better efficiency than fac(n+ 1) through the use of accumulator.

Suppose the initial analogical correspondence between the base and target problem is {(fac,sum),(f1,s1),(f2,s2), (*,+), (1,0)}, and there is a correspondence between the commutativity laws of the multiply and plus operators. Applying the development transformation rules above we can obtain a similar development as follows, which is valid according to the theorem in section 3.

Td1:= [F:[nat,nat]nat]
    [eu:[n,acc:nat]F(n,acc)= sum(n)+ acc]
    Inst(eu,0) .. [acc:nat]F(0,acc)= sum(0)+ acc
    'unfold(f1).. [acc:nat]F(0,acc)= 0+ acc
    'law(a+ 0= a) .. [acc:nat]F(0,acc)= acc;

Td2:= [F:[nat,nat]nat]
    [eu:[n,acc:nat]F(n,acc)= sum(n)+ acc]
    Inst(eu,n+ 1) .. [acc:nat]F(n+ 1,acc)= sum(n+ 1)+ acc
    'unfold(f2).. [acc:nat]F(n+ 1,acc)= (sum(n)+ (n+ 1))+ acc
    'law(a+ (b+ c)= (a+ b)+ c) .. [acc:nat]F(n+ 1,acc)= sum(n)+ ((n+ 1)+ acc)
    'fold(eu) ..[acc:nat]F(n+ 1,acc)= F(n,(n+ 1)+ acc);

Td3:= [F:[nat,nat]nat]
    [eu:[n,acc:nat]F(n,acc)= sum(n)+ acc]
    f2 ..[n:nat]sum(n+ 1)= sum(n)+ (n+ 1)
    'fold(eu)..[n:nat]sum(n+ 1)= F(n,n+ 1).

Hence, we obtain a tail recursive program computing the sum:

```
SP:= [ F   : [nat,nat]nat;
       sum  : [nat]nat;
       sp1  : [acc:nat] F(0,acc)= acc;
       sp2  : [n,acc:nat] F(n+ 1,acc)= F(n,n+ 1+ acc);
       sp3  : [n:nat] sum(n+ 1)= F(n,n+ 1)
       ].
```

The partial correctness of this program is guaranteed by the validity of the transformed development.

The difficulty in the above development transformation is the introduction of the new lemma, i.e., the commutativity of the plus operator. The development transformation is more complicated when the rules in the base problem are no longer applicable. In such cases, some rules in the original problem should be skipped, or new rules inserted into the rule sequence, which is also discussed in [12].

5.Discussion

Analogical programming needs to expand the analogical relations between specifications to their development, and then to their programs. As far as we know, the specifications, programs, and its developments are usually expressed in three different levels, which make it difficult to expand analogical relations, and especially, difficult to verify the correctness of the analogically derived consequence. As we can see from the above discussion, the higher order type theory provides a neat setting to express specifications, programs and developments in a single logic framework, which greatly facilitates the reuse of program development.

Peter Madden discussed the transformation of constructive program synthesis proofs to adapt programs to special situations, or improve program efficiency by transforming one proof to another form[10]. Our work differs from it in that we concern the transformation of developments(proofs) between different problems, while Madden transforms a proof of one problem to another proof of the same problem or of the specialized case. Nederpelt described how to represent mathematical proof in the Λ calculus [13], Sintzoff and Weber presented a more realistic design calculus to unify the concepts in program development[14][15].

Acknowledgement

The authors are grateful to Michael J.O'Donnell and an anonymous reviewer for their helpful comments, and to Yi Bo for beneficial discussions with him.

References

[1] N.G.de Bruijn, *A Survey of the Project AUTOMATH*, in: J.P.Seldin and J.R.Hindley (editors), To H.B.Curry: Essays on Combinatory Logic, Lambda Calculus, and Formalism, Prentice Hall, 1980.

[2] R.M.Burstall, *Transformational System for Developing Recursive Programs*, JACM 24(1),pp.44-67.

[3] J.Carbonell, *Derivational Analogy: A Theory of Reconstructive Problem Solving and Expertise Acquisition*, in R.S. Michalski et al(ed.), Machine Intelligence: An Artificial Intelligence Approach, Morgan Kaufmann,1986.

[4] R. Constable, et al, *Implementing Mathematics with NuPRL Proof Development System*, Prentice Hall, 1985.

[5] John Darlington, *An Experimental Program transformation and Synthesis System*, Artificial Intelligence 16(1981), pp. 1-46.

[6] Nachum Dershowitz, *Programming By Analogy*, in R.S.Michalski et al (eds.), Machine Learning II: An Artificial Intelligence Approach,Morgan Kaufmann, 1986.

[7] R. Harper, F.A.Honsell, G.Plotkin, *A Framework for Defining Logics*, Proceedings of the second symposium on Logic in Computer Science, PP. 194-204, IEEE, 1986.

[8] W.A.Howard, *The Formula as Types Notion of Construction*, in: J.P.Seldin and J.R.Hindley (editors), To H.B.Curry: Essays on Combinatory Logic, Lambda Calculus, and Formalism, Prentice Hall, 1980.

[9] Jianguo Lu, *Research on Analogical Program Derivation Based on Type Theory*, Ph.D. Dissertation, Nanjing University, 1991.

[10] P.Madden, *Automatic Programming Optimization via the Transformation of NuPRL Synthesis Proofs*, UK IT 88 Conference Publication, Swansea, UK, 4-7 July, 1988.

[11] Per Martin-Lof, *Constructive Mathematics and Computer Programming*, in Logic, Methodology and Philosophy of Science, pp.153-175, North Holland, 1982.

[12] J. Mostow, *Design by Derivational Analogy: Issues in the Automated Replay of Design Plans*, Artificial Intelligence 40(1989), pp.119-184.

[13] R.P.Nederpelt, *An approach to theorem proving on the basis of typed lambda calculus*, 5th Conference on Automated Deduction, LNCS 87, pp.181-190, 1980.

[14] M. Sintzoff, *Understanding and Expressing Software Construction*, in P.Pepper(ed.), Program Transformation and Programming Environment, Springer, 1984.

[15] M. Weber, *A Meta Calculus for Formal System Development*, Ph.D. Dissertation, der Universitat Karlsruhe, 1990.

[16] Jiafu Xu, *Reports on a Software Automation R&D Project*, Information Processing'89, G.X.Ritter(ed.), Elsevier Science Publisher, B.V.(North-Holland), Aug. 1989.

Classical Proofs as Programs:
How, What and Why

Chetan R. Murthy[*]

Cornell University

Abstract

We recapitulate Friedman's conservative extension result of (suitable) classical over constructive systems for Π_2^0 sentences, viewing it in two lights: as a translation of programs from an almost-functional language (with $\mathcal{C}$) back to its functional core, and as a translation of a constructive logic for a functional language to a classical logic for an almost-functional language. We investigate the computational properties of the translation and of classical proofs and characterize the classical proofs which give constructions in concrete, computational terms, rather than logical terms. We characterize different versions of Friedman's translation as translating slightly different almost-functional languages to a functional language, thus giving a general method for arriving at a sound reduction semantics for an almost-functional language with a mixture of eager and lazy constructors and destructors, as well as integers, pairs, unions, etc. Finally, we describe how to use classical reasoning in a disciplined manner in giving classical (yet constructivizable) proofs of sentences of greater complexity than Π_2^0. This direction offers the possibility of applying classical reasoning to more general programming problems.

1 Introduction

It is well-known that constructive type theories and logics can serve as reasoning systems for functional programming languages. In (13), we demonstrated that classical arithmetic can serve in a like fashion as a total-correctness reasoning system for programs written in an almost-functional programming language based on the integers, and incorporating the $\mathcal{C}$ (8) operator (pronounced "control", a relative of

---

[*]Supported in part by an NSF graduate fellowship and NSF grant CCR-8616552 and ONR grant N00014-88-K-0409

call/cc (2). This result follows from Friedman's conservative extension of (suitable) classical over constructive logics for Π_2^0 sentences, which shows that a classical proof of a Π_2^0 sentence can be translated to a constructive proof of the same sentence, and Griffin's (11) work showing a correspondence between the classical axiom and C. In (13) we used these results to show the soundness of a particular almost-call-by-name[1] evaluation semantics for proofs in classical arithmetic (regarded as programs), by the method of Friedman specialized to a modified Kolmogorov translation, hence showing that classical proofs of Π_2^0 sentences compute *evidence* (3) for those propositions in a direct and explicit manner. That is, we showed that classical proofs of Π_2^0 sentences could be construed as computing evidence in the same sense as constructive proofs, and, moreover, the programming language associated with such proofs is a functional language augmented with the nonlocal control operator C. Here we seek to extend this work, and address three problems:

- Friedman's work is restricted to Π_2^0 sentences. Why? And can we understand this computationally? Why does this restriction come about? We will find that the restriction to Π_2^0 is natural and can be accounted for in a manner which is (we feel) satisfying to the computer scientist, as well as to the logician. Moreover, we will give simple examples which demonstrate that, in general, classical proofs of Π_3^0 and Σ_2^0 sentences do *not* compute evidence for the propositions they stand as classical witnesses for.

- There are other translations besides the Kolmogorov translation for which Friedman's work applies, and which we can use in proving the same conservative extension results. We define a modified Kuroda translation, give the translations on programs which accompanies it, and assert the soundness of an almost-call-by-value evaluation semantics for proofs in classical arithmetic. We go on to show a "mixed-Kolmogorov" translation which allows us to prove the soundness of an evaluation semantics for a programming language with both by-name and by-value applications. This result extends to eager/lazy pairing and injections also.

- We use this same methodology and demonstrate how to use classical reasoning in a disciplined manner to give classical, but automatically constructivizable, proofs of sentences which are more complex than Π_2^0. We give a simple extension to the proof system of classical arithmetic which allows us to accomplish this goal.

---

[1]The "almost" prefix is due to the nature of the Kolmogorov translation, which prohibits CPS-translation from translating integer-typed expressions; hence, these expressions cannot be guaranteed to evaluate in a by-name manner. However, the type system guarantees that these expressions will evaluate to integer constants, and will never abort.

Our work is based on the principle that we can understand a nonfunctional language by understanding its translation into a functional language; equally, we can give a logic for a nonfunctional language by "pulling back" a logic for a functional language thru the appropriate translation.

2 Constructive Type Theories

Constructive type theories are based on the Curry-Howard isomorphism, under which propositions are identified with types, and their proofs with members of types. This isomorphism relies heavily on the notion of *evidence*. Evidence is informally defined as values in a programming language, and is assigned to propositions as follows:

- Evidence for $a = b$, where a, b contain no free variables, would be completely axiomatic, consisting in the computation of a, b down to numerals (they could be, for instance, $2 * 5 = 5 + 5$, which would need to be computed down to $10 = 10$).

- Evidence for $A \wedge B$ would be evidence for A and for B. This could be given as a pair, $\langle u, v \rangle$, where u is evidence for A, and v is evidence for B.

- Evidence for $A \Rightarrow B$ would be a function which, when given evidence for A, would compute evidence for B.

- Evidence for $A \vee B$ would be evidence for A, or for B, and a tag telling us which disjunct we were getting evidence for. This could be represented as $inl(u)$ (inject-left), where u is evidence for A, or $inr(v)$, where v is evidence for B.

- Evidence for $\exists x \in \mathrm{N}.R(x)$ would be an integer, n, and evidence for $R(n)$. This, again, could be represented as a pair.

- Evidence for $\forall x \in \mathrm{N}.R(x)$ would be a function which, given $n \in \mathrm{N}$, would compute evidence for $R(n)$.

This definition can be equally well viewed as assigning types to program values, and it can be extended to expressions in a confluent programming language in a straightforward way. An example of a constructive type theory is Heyting Arithmetic (HA) (4), which is essentially Peano Arithmetic (PA), or number theory, without the axiom of excluded middle. In such a type theory, a constructive proof of $\forall x. \exists y. R(x, y)$ is evidence for this proposition - i.e., a function which, will compute evidence for $\exists y. R(X, y)$. This in turn is a pair $\langle Y, Z \rangle$, such that Z is evidence for $R(X, Y)$.

We use a particular format for expressing the typing judgments of our logic (alternatively, for expressing the rules of our logic). We will write $\Gamma \vdash_T M : \phi$, where Γ is a (possibly empty) list of typing assumptions $x : A$, ϕ is a proposition, and M is a program phrase whose free variables are among those declared in Γ. The reading of this sequent is that, under the typing assumptions Γ, program phrase M has type ϕ. E.g.,

$$\Gamma \vdash M(N):B$$
```
BY modus ponens
 ⊢ M:A ⇒ B
 ⊢ N:A
```

is a sequent-calculus version of the typing rule for function application, as well as being the constructive rule of modus ponens.

3 "Denotational" Translations of Nonconstructive Logics

It is by now common to use denotational semantics to interpret nonfunctional languages into pure functional ones. These target languages enjoy beautiful properties, among them the availability of workable correctness logics - the constructive type theories. Given that every reasonable sequential programming language can be thus translated, one wonders what sort of (type) theory would result if one were to "pull back" a type system thru the semantic interpretation function. For example, if τ is the translation, and if $\tau(M)$ has type ϕ, is there some meaningful type we can assign to $\tau^{-1}(\phi)$? It is also usually true that $M \succ M'$ (where $\rightarrow$ is reduction) implies $\tau(M) = \tau(M')$, in some appropriate equational theory of functional languages. Does this also mean that $M : \tau^{-1}(\phi)$ implies $M' : \tau^{-1}(M')$? This would be a type-preservation theorem about the relation of our typing system to evaluation. Finally, for values, say, observable datatypes, does $M : \phi$ imply $\tau(M) : \phi$? The latter would tell us that the translation on values of datatypes was the identity.

Here we will show that the call-by-name CPS-translation, $\bullet$, a semantic translation of $\lambda + C$ (a functional language augmented with the "control" operator) into λ, induces a classical type sytem on $\lambda + C$. It will turn out that the classical type system can be viewed as arising first, as a type sytem for a non-Church-Rosser language, and that we can think of the CPS-translation as fixing which reductions are valid by stipulating that $M \succ M'$ only if $\underline{M} = \underline{M'}$.

4   CPS-Translation Basics

The continuation-passing-style (CPS) translation is a mapping from lambda-terms to lambda-terms which mimicks the operational semantics of evaluation of the original term by explicit reductions of the translated term. We will not go into a lengthy discussion of the various translations here; let it suffice to say that there are at least two well-known translations in the literature, the call-by-value translation and the call-by-name translation, each of which mimicks the respective operational semantics. For further information and tutorial, the reader is encouraged to consult (14, 9, 11). One central fact about CPS-translation is that, for call-by-value (resp. by-name) evaluation, a lambda-calculus program M converges to a value b iff its call-by-value (resp. by-name) CPS-translation, applied to the proper top-level continuation (usually $\lambda\,x.x$) converges to this same value.

5   How: Friedman's Translation

Friedman showed that one could translate a classical arithmetic (PA) proof of a Π^0_2 sentence into a constructive arithmetic (HA) proof of the same by a simple extension of the method of double-negation translation. We give the proof for the case where the double-negation translation used is the Kolmogorov translation (12).

Definition 1 (Kolmogorov Translation) Given a sentence ϕ in PA, define $\overline{\phi}$, the Kolmogorov double-negation translation of ϕ, as being the simultaneous double-negation of every propositional position in ϕ:

$$
\begin{aligned}
\overline{(A \vee B)} &\longmapsto & \neg\neg(\overline{A} \vee \overline{B}) \\
\overline{(A \wedge B)} &\longmapsto & \neg\neg(\overline{A} \wedge \overline{B}) \\
\overline{(\exists x \in A.B)} &\longmapsto & \neg\neg(\exists x \in A.\overline{B}) \\
\overline{(\forall x \in A.B)} &\longmapsto & \neg\neg(\forall x \in A.\overline{B}) \\
\overline{(A \Rightarrow B)} &\longmapsto & \neg\neg(\overline{A} \Rightarrow \overline{B}) \\
\overline{P} &\longmapsto & \neg\neg(P) \qquad \text{(P prime)}
\end{aligned}
$$

Theorem 1. (Double-Negation Embedding) If $\vdash_{PA} \phi$, then $\vdash_{HA} \overline{\phi}$.

This theorem tells us that the Kolmogorov translation converts a classically provable sentence into a constructively provable sentence, but it does *not* tell us what form the constructive proof will take. Friedman's discovery was that (if one gave the constructive proof properly) one could replace instances of falsehood ($\bot$) with an arbitrary proposition, in particular ϕ. Friedman showed:

Theorem 2 (Friedman's A-Translation) If $\vdash_{PA} \phi$, where ϕ is Σ_1^0, and is $\bot$-free, then

$$\vdash_{HA} \overline{\phi}[\phi/\bot].$$

In fact, for correctly-constructed translations, the A-translation does not disturb the constructive extracts. I.e., if $\vdash_{HA} M : \overline{\phi}$, then $\vdash_{HA} M : \overline{\phi}[\phi/\bot]$. Friedman then showed:

Theorem 3 (Conservative Extension) If we have a proof $\vdash_{PA} \phi$, where ϕ is Σ_1^0, then we can construct a proof $\vdash_{HA} \phi$.

Proof: Suppose $\phi \equiv \exists y \in \mathsf{N}.f(y) = 0$, and let $A \equiv \phi$. Let $\underset{\phi}{\neg}(T) \equiv T \Rightarrow \phi$. Then from a classical proof of ϕ, we obtain a constructive proof of

$$\underset{\phi\,\phi}{\neg\neg}(\exists y \in \mathsf{N}.\underset{\phi\,\phi}{\neg\neg}(f(y) = 0)).$$

To recover a proof of ϕ, it suffices to prove

$$\underset{\phi}{\neg}(\exists y \in \mathsf{N}.\underset{\phi\,\phi}{\neg\neg}(f(y) = 0)),$$

which we refer to as "Friedman's top-level trick." This is trivial, but a deep analysis of the ramifications of the choice of the "top-level trick" would shed light on the essentially non-lazy features of the interactive top-loop in lazy languages, and we omit it for reasons of space. ∎ The result for Π_2^0 sentences follows by considering free integer variables.

6 What: Classical Proofs as Programs

So given a classical proof π_K of a Π_2^0 sentence, we can obtain a proof π_J (corresponding to a functional program) by translation which stands as a constructive witness for that sentence, and computes evidence for it. But we can do more. By "guessing" algorithmic content for the classical rules and verifying them via translation, we can extract evidence from PA proofs directly. As an example, consider the following pair of (original,double-negation-translated) rules:

$$\Gamma \vdash \texttt{<what?>}:P \qquad \overline{\Gamma} \vdash \lambda\ k.\underline{M}(\lambda\ m.m(\lambda\ g.g(\lambda\ v, h.v(k)))\lambda\ x.x):\overline{P}$$
$$\text{BY}\ \neg\neg\text{-elim} \qquad \text{BY translated}\ \neg\neg\text{-elim}$$
$$\vdash\ M:\neg\neg(P) \qquad \vdash\ M:\overline{\neg\neg(P)}\ (=\neg\neg(\neg\neg(\overline{P} \Rightarrow \neg\neg(\bot)) \Rightarrow \neg\neg(\bot))$$

If one looks carefully, one sees that the translated extract is the *call-by-name CPS-translation of the term CM*! This operator, pronounced "control," and described in (8), is a nonlocal control operator, and its evaluation semantics are (informally) expressed as

$$E[\mathcal{C}M]\ \rightarrow_1\ M(\lambda\ x.\mathcal{A}(E[x]))$$
$$E[\mathcal{A}M]\ \rightarrow_1\ M.$$

where $\mathcal{A}$ is the "abort" operator, syntactic sugar for $\mathcal{C}(\lambda\ ().M)$, and intuitively, $\mathcal{C}$ applies its argument to an abstraction of the evaluation context ($E[]$) in which it is evaluated. Thus, $\mathcal{C}$ allows a term to "goto" another evaluation context, discarding the context in which it is currently evaluating. We can repeat this process for every rule of our classical logic. For each rule, we examine the translated rule fragment, and attempt to discover the proper "classical extract" which, when translated, will give the translated extract. This defines a programming language, $Prog_K$, and a translation $\bullet$ back to its functional core, $Prog_J$. Here are a few of the clauses of the translation:

$$\underline{x}\ \equiv\ x\ (x\ a\ variable)$$
$$\underline{MN}\ \equiv\ \lambda\ k.\underline{M}(\lambda\ m.m\underline{N}k)$$
$$\underline{\lambda\ x.M}\ \equiv\ \lambda\ k.k(\lambda\ x.\underline{M})$$
$$\underline{\mathcal{C}M}\ \equiv\ \lambda\ k.\underline{M}(\lambda\ m.m(\lambda\ g.g(\lambda\ v, h.v(k)))\tau_\phi).$$

But, having arrived at a candidate set of rules (complete with extractions), we still must show that these rules are sound. This comes in three stages:

- Show that reduction to a *weak-head-normal-form* (1) terminates. Apply "colon-translation", discovered by Plotkin (14), and extended by Griffin (11).

- Show that reduction is sound. We consider each reduction rule in turn, and show that $(M \rightarrow_1 M') \Rightarrow (\underline{M} = \underline{M'})$. The evaluation rule for $\mathcal{C}$ is difficult, but Felleisen's methods (8) suffice.

- Show that reduction preserves typing, i.e., $(\Gamma \vdash M : T) \wedge (M \rightarrow_1 N) \Rightarrow (\Gamma \vdash N : T)$.

The last obligation - to show that reduction preserves typing - is in some sense the real "meat" behind any typing system. While it is easy to show that all the original reduction rules of the functional core language preserve typing, we cannot show that if $E[\mathcal{C}M] : \phi$, where M has type $\neg\neg(T)$, then $M(\lambda \ x.\mathcal{A}E[x])$ has the same type ϕ. Let us demonstrate this:

$$
\begin{array}{rcl}
M & : & \neg\neg(T) \\
x & : & T \\
E[x] & : & \phi \\
\mathcal{A}E[x] \equiv \mathcal{C}\lambda \ ().M & : & untypable \\
\lambda \ x.\mathcal{A}E[x] & : & untypable
\end{array}
$$

Thus, we see that $M(\lambda \ x.\mathcal{A}E[x])$ is ill-typed. We can remedy this problem in one of two ways. Griffin (11) chose to assume that the type of the entire program was, instead of some arbitrary ϕ, the type $\bot$. Then, the typing above goes thru trivially. His method corresponded to wrapping a complete program M in a term $\mathcal{C}\lambda \ k.k(M)$.

Besides being on a somewhat logically insecure footing, this choice does not give meanings directly to the programs we wish to write – which do not have these "wrappers" around them. Moreover, it obscures the fact that in reality programs do not have type $\bot$, but rather concrete types. We choose instead to replace every $\bot$ in the entire proof (and in every rule) with ϕ (a pre-A-translation), without changing the type of the program. This can happen when ϕ is $\bot$-free (e.g. when ϕ is Σ_1^0). The reader may verify that this typing works. So what have we wrought with our modification? Suppose we start with a classical proof of a $\bot$-free Σ_1^0 sentence, ϕ. If we replace every instance of $\bot$ in every sequent and every rule (including the classical axiom) with ϕ, then the "classical extract" is unchanged, just as the constructive extract is unchanged by such a modification. Moreover, in this modified logic, called $PA(\phi)$, *reduction preserves typing*.

Theorem 4 (Reduction Preserves Typing) Given a proof $\vdash_{PA} M : \phi$, we can show $\vdash_{PA(\phi)} M : \phi$, and if $M \to M'$, then

$$\vdash_{PA(\phi)} M' : \phi.$$

As we noted before, we can prove that reduction to a weak-head-normal-form (1) will terminate (via a colon-translation argument). Consider now the case $\phi \equiv \exists x \in \mathbb{N}.f(x) =$ We know that the only *whnfs* of this type are pairs $\langle N, axiom \rangle$, where $N \in \mathbb{N}$ and *axiom* is a proof of $f(N) = 0$ (that is, $f(N) = 0$ is true). But such a value is *also* of type ϕ in PA, or even in HA. Thus, even though we had to move to $PA(\phi)$ to prove that reduction preserves typing, when we finish reduction, we are left with a well-typed value in PA.

Theorem 5 (Classical Evaluation Semantics) Given a proof $\vdash_{PA} M : \phi$, if ϕ is Σ_1^0, then $M \to V$, where V is a *value*, and $\vdash_{PA} V : \phi$.

So we have found an evaluation semantics for (nearly) classical proofs which is sound, and gives evidence for Σ_1^0 sentences. But what import does this have for real classical proofs?

6.1 Why Π_2^0?

So finally we come to the question: *Why* does Friedman's translation only work for Π_2^0 sentences? Let us look at the Σ_1^0 sentences. A Σ_1^0 type can always be written as $\exists x \in \mathbb{N}.f(x) = 0$. Members of this type are pairs $\langle X, Y \rangle$ where $X \in \mathbb{N}$, and Y is a witness of $f(X) = 0$ - that is, *axiom*. Σ_1^0 sentences have the special property that their *whnfs* are concrete data values, and specifically contain no embedded function closures, which also means they contain no unevaluated continuations. Quite simply, when evaluating a program of Σ_1^0 type ϕ, we always end up with a value devoid of function closures. And this value has type ϕ in PA exactly because it has that type in $PA(\phi)$.

Now a Π_2^0 type can be written $\forall x \in \mathbb{N}.\exists y \in \mathbb{N}.f(x,y) = 0$. An expression of this type is a function F with domain $\mathbb{N}$, such that $F(X)$ computes evidence of type

$\exists y \in \mathsf{N}.f(X, y) = 0$. W.l.o.g we can assume that F is of the form $\lambda\, x.M$. When F is applied to $X \in \mathsf{N}$, $F(X)$ has type $\exists y \in \mathsf{N}.f(X, y) = 0$ ($\in \Sigma_1^0$), and so it will compute evidence for this type. But the intuitive reason why F has type $\forall x \in \mathsf{N}.\exists y \in \mathsf{N}.f(x, y) = 0$ is that when fed a concrete datum, it produces a concrete datum, again, without any embedded function closures or instances of C.

Consider now a Σ_2^0 type, written $\phi \equiv \exists x \in \mathsf{N}.\forall y \in \mathsf{N}.f(x, y) = 0$. A proof of such a type is a program which computes to a pair $\langle X, G \rangle$, where $X \in \mathsf{N}$, and, in $PA(\phi)$, G has type $\forall y \in \mathsf{N}.f(X, y) = 0$. G is a function and when G is applied to a value Y, $G(Y)$ can reduce to evidence for $f(X, Y) = 0$. Alternatively, $G(Y)$ can "unwind" evaluation to a previous context, in which case $G(Y)$ is *not* evidence for $f(X, Y) = 0$. So if we had an oracle which could tell us whether, for any Y, $G(Y)$ would apply an unevaluated continuation, thus unwinding its evaluation context, we could decide whether G was evidence for $\forall y \in \mathsf{N}.f(X, y) = 0$. This is not decidable in general, so we cannot determine if G is evidence for $\forall y \in \mathsf{N}.f(X, y) = 0$.

For example, there is a simple classical proof of

$$\psi \equiv \exists n \in \mathsf{N}.\forall m \in \mathsf{N}.f(n) \leq f(m).$$

This sentence expresses the fact that every boolean function attains a minimum (where $0 < 1$). Suppose we had a constructive proof of this fact, from which we extracted a program. Intuitively, this cannot be the case because the constant function $f : n \longmapsto 1$ is indistinguishable within n steps from a function which becomes zero after $n + 1$ (or some suitably large number) of values. So if our program could examine the constant function $f : n \longmapsto 1$ and determine that it attained a minimum of 1 after N steps of computation, then we could "spoof" it by giving it as input a function which attained zero only after a suitably long time (say a stack of N 2's). And our program would report an incorrect minimum on this input. We give a classical proof of ψ below:

```
⊢ ψ  BY double-negation elim
 ¬(ψ) ⊢ ⊥  BY function-elim
  ⊢ ψ  BY intro 0
   ⊢ ∀m ∈ N.f(0) ≤  f(m)  BY function intro
    m : N ⊢  f(0) ≤  f(m)  BY cases on  f(0) ≤  f(m)
     f(0) ≤  f(m) ⊢  f(0) ≤  f(m)  BY hypothesis
     f(m)  <  f(0) ⊢  f(0) ≤  f(m)
      BY double-negation elim
        THEN function-elim
      f(m)  <  f(0) ⊢ ψ  BY intro m, etc
```

The computation in this proof is

$$\mathcal{C}\lambda k.k\langle 0,$$
$$\lambda m.if\ f(0) \le f(m)\ then\ axiom$$
$$else\ k\langle m,\ \lambda\ m'.axiom\rangle\rangle$$

Intuitively, the program given by the proof will make a "guess" that 0 is the desired n. Then, given m, it will check if $f(0) \le f(m)$. If so, then it will simply report success. If not, then $f(m) < f(0)$, which means that $f(m) = 0$. So the program will unwind the context back to before it chose 0, and instead choose m. As a result, our program does not really provide evidence for the truth of the proposition it purports to be a proof of, but rather, provides a program which, given a counterexample, will "throw" back to a place in the computation where it can change the "answer" to disqualify the counterexample.

There is one special case. If G contains no expressions $\mathcal{AM}$, then it is easy to show that G is evidence for $\forall y \in N.f(X, y) = 0$, because every unevaluated continuation is an expression $\mathcal{AM}$. Since G has type $\forall y \in N.f(X, y) = 0$ in $PA(\phi)$, and $\mathcal{AM}$ does not occur in G, then G has the same type in PA. Thus G has the same type in $PA(\forall y \in N.f(X, y) = 0)$. And from this, we can show that G computes evidence for $\forall y \in N.f(X, y) = 0$.

As one can see, the criterion for deciding that a proof of a Σ_2^0 (or stronger) sentence actually computes evidence for that sentence is nontrivial, and for any realistic implementation, intractable.

7 Other Translations as Variant Evaluation Semantics

There are other double-negation translations which, in concert with A-translation, suffice to translate classical proofs of Π_2^0 sentences into constructive proofs. One such translation is a simple modification of the Kuroda negative translation. This translation is very simple - to translate a term, double-negate the *body* of every universal quantification and r.h.s. of every implication, and the outside of the entire term [2]. We can again "discover" a $Prog_K$, and a corresponding translation $\bullet$. This time, it turns out that $Prog_K$ has call-by-value semantics. Here is a fragment of the translation:

$$
\begin{aligned}
\underline{x} &\equiv \lambda\, k.k(x)\ (x\ a\ variable) \\
\underline{MN} &\equiv \lambda\, k.\underline{M}(\lambda\, m.\underline{N}(\lambda\, n.m(n)k)) \\
\underline{\lambda\, x.M} &\equiv \lambda\, k.k(\lambda\, x.\underline{M}) \\
\underline{CM} &\equiv \lambda\, k.\underline{M}(\lambda\, m.m(\lambda\, v, h.k(v))(\lambda\, x.x))
\end{aligned}
$$

While the left-hand-sides look just like the call-by-name language already proposed, it is *not*, since the lambda-abstraction and application here, as well as the C, are call-by-value. In fact, if one consults a standard work on CPS-translation (14, 8), one finds that this translation is indeed a call-by-value translation. By a process analogous to that for the Kolmogorov translation, and using techniques essentially identical to those found in (14, 8), we can show that a by-value evaluation semantics is sound for this programming language and type system, again for Π_2^0 sentences only.

We can go further, though, and consider "mixed-mode" translations. Consider the following Kolmogorov translation of the rule of modus ponens:

$\Gamma \vdash \lambda\, k.\underline{M}(\lambda\, m.\underline{N}(\lambda\, n.m(\lambda\, k.k(n))k)):\overline{B}$
    BY translated modus ponens
    $\vdash \underline{M}:\overline{A \Rightarrow B}\ (= \neg\neg(\overline{A} \Rightarrow \overline{B}))$
    $\vdash \underline{N}:\overline{A}.$

We can add a second rule of modus ponens to our classical logic, "by-value modus ponens", which does the obvious thing - the application term extracted will function in a by-value manner. In a like manner, we can add eager pairing, or even half-eager pairing (which would evaluate, say, the first component of a pair when constructing the pair). This translation has the undesirable quality that eager features are not

---

[2]The original Kuroda translation did not double-negate the r.h.s. of implications

reflected in the translated types. It is a simple (though tedious) matter to give another translation which does reflect the eagerness of expressions in the forms of types.

7.1 An *A Priori* Classical Programming Language

The fact that we can invent new translations which enforce different evaluation orders is not surprising. In fact, we could imagine that, rather than starting with a classical logic, and discovering a programming language by observing that double-negation translation looked like CPS-translation, we had instead just discovered, from first principles, a programming language in classical proofs. We could have taken the lambda-calculus, and added the operator $\mathcal{C}$, to arrive at the programming language $\lambda + \mathcal{C}$. One usually writes the evaluation semantics for such a language as:

$$
\begin{aligned}
E[(\lambda\ x.b)N] &\quad \triangleright_1 \quad E[b[N/x]] \\
E[\mathcal{C}M] &\quad \triangleright_1 \quad M(\lambda\ x.AE[x]) \\
E[AM] &\quad \triangleright_1 \quad M.
\end{aligned}
$$

But this definition has the undesirable quality that we have not specified what, if anything, is meant by "$E[]$". In deterministic evaluation, $E[]$ specifies an *evaluation context*(7), which, loosely speaking, identifies unambiguously in a program the next redex for contraction. To remedy this fault, Felleisen (6) introduced the notion of local and global reduction rules. Local reduction rules were those which were valid in any program context, and global rules were those which were only valid on entire programs. Speaking in these terms, one can rephrase the evaluation rules for a programming language in two classes, using $\rightarrow_1$ for local rules, and $\triangleright_1$ for global rules:

$$
\begin{aligned}
(\lambda\ x.b)N &\quad \rightarrow_1 \quad b[N/x] &\quad (\beta) \\
(\mathcal{C}M)N &\quad \rightarrow_1 \quad \mathcal{C}\lambda\ k.M(\lambda\ f.k(fN)) &\quad (\mathcal{C}_L) \\
M(\mathcal{C}N) &\quad \rightarrow_1 \quad \mathcal{C}\lambda\ k.N(\lambda\ a.k(Ma)) &\quad (\mathcal{C}_R) \\
(AM)N &\quad \rightarrow_1 \quad M &\quad (A_L) \\
M(AN) &\quad \rightarrow_1 \quad N &\quad (A_R) \\
\mathcal{C}M &\quad \triangleright_1 \quad M(\lambda\ x.Ax) &\quad (\mathcal{C}_{\downarrow_1}) \\
AM &\quad \triangleright_1 \quad M &\quad (A_{\downarrow_1}).
\end{aligned}
$$

Again, the $\triangleright_1$ rules are only applicable to entire programs. We can show that all these rules can be given proper typings in the same manner as was done for the other, global rules. That is, we can show that each of the local rules can be given a typing in classical logic, but that the global rules require us to "pre-A-translate" the logic. However, as has been observed numerous times, the reduction system specified above is *not* Church-Rosser. To make it Church-Rosser, one must further restrict the applicability of some of the rules. For instance, when the redex is $(\mathcal{C}M)(\mathcal{C}N)$, we are faced with a choice of $\mathcal{C}_L$ or $\mathcal{C}_R$. Likewise, given a redex $(\lambda\ x.b)(\mathcal{C}N)$, we must choose whether to apply β or $\mathcal{C}_R$.

The choice of a translation, in a very real sense, is equivalent to deciding on the precedence of the various rules when they conflict. So, for instance, choosing the call-by-name CPS-translation, and the accompanying Kolmogorov-translation, is equivalent to choosing to disallow $\mathcal{C}_R$ completely. And choosing the call-by-value CPS-translation outlined previously is equivalent to choosing to allow $\mathcal{C}_L$ everywhere, call-by-value β (β_v), and $\mathcal{C}_R$ only when M is a value, i.e. a λ-abstraction.

To demonstrate this, the reader may amuse himself by CPS-translating, with both the call-by-name and call-by-value translations, each of β, β_v, $\mathcal{C}_L$, $\mathcal{C}_R$, and the restricted version of $\mathcal{C}_R$ (where M is a λ-abstraction), and verifying the facts stated above.

8 Extending up the Arithmetic Hierarchy

In a certain sense, the results we have presented are all that can be hoped for. It is a simple matter to give classically provable Σ_2^0 sentences $(\exists n \in \mathbb{N}.\forall m \in \mathbb{N}.f(n,m) = 0)$ from which we have no *hope* of extracting constructions. The reason, as we have discussed, lies in our use of classical reasoning to prove the proposition $\forall m \in \mathbb{N}.f(n,m) = 0$. If we could somehow prevent this, then the value G from our previous example would be purely functional, and we would have a proof of the Σ_2^0 sentence. To achieve this, we must selectively disallow classical reasoning upon particular propositions. We do this with a new term-forming construct, the $J(\bullet)$ [3] operator, and new rules which govern when we may introduce or eliminate the J operator. Syntactically, the J operator can be applied to any proposition, and results in a proposition. It introduces

---

[3] The name is chosen to remind the reader of "intuitionistic"

no new free variables, and binds no variables. The Kolmogorov-translation of $J(T)$ is $\neg\neg(T)$. Note that the type T is *not* internally translated.

There is only one rule for the J operator:

$$\overline{n} : \mathsf{N}, \ \overline{x} : J(\Gamma) \vdash_K (\lambda^v \ \overline{x}.M)(\overline{x}) : J(T)$$
$$\texttt{BY J-protect}$$
$$\overline{n} : \mathsf{N}, \ \overline{x} : \Gamma \vdash_J M : T.$$

The meaning of $\vdash_J$ (resp. $\vdash_K$) is that the proof from this point down (leaf-ward) in the proof tree must be given constructively (resp. classically). Intuitively, $J(\bullet)$ is a marker which delineates where classical reasoning may not be used. The remainder of the rules of PA treat J-protected terms as atomic propositions. The translation of the J-protect rule is basically the by-value translation of $(\lambda^v \ \overline{x}.M)(\overline{x})$, but we protect from translation the term M. For the case where $\overline{x} : J(\Gamma) \equiv x : J(\psi_1), y : J(\psi_2)$, the translated sequent would be:

$$\overline{n} : \mathsf{N}, \ x : \neg\neg(\psi_1), \ y : \neg\neg(\psi_2) \vdash_J \lambda \ k.x(\lambda \ x.y(\lambda \ y.k(M))) : \neg\neg(T)$$
$$\texttt{BY translated J-protect}$$
$$\overline{n} : \mathsf{N}, \ x : \psi_1, \ y : \psi_2 \vdash_J M : T.$$

and it is a relatively simple matter to show:

Theorem 6 (PA+J Conservative Extension)
If $\vdash_{PA+J} \forall x \in \mathsf{N}.J(\psi_1(x)) \Rightarrow \exists y \in \mathsf{N}.J(\psi_2(x, y))$, then $\vdash_{HA} \forall x \in \mathsf{N}.\psi_1(x) \Rightarrow \exists y \in \mathsf{N}.\psi_2($

Proof: Along the same lines as Friedman's conservative extension proof. The translation of the J-protect rule has already been shown. ∎

Moreover, the constructive extract we obtain from this proof is as outlined above, and we can give sound evidence semantics for the original classical proof as before. However, it is easy to show that we cannot, in this classical logic, prove

$$\exists n \in \mathsf{N}.J(\forall m \in \mathsf{N}.f(n) \leq f(m)).$$

One simply notes that the proof given, and in fact any proof, of this fact, re-uses the fact that $\neg(\exists n \in \mathsf{N}.J(\forall m \in \mathsf{N}.f(n) \leq f(m)))$ while proving $\forall m \in \mathsf{N}.f(n) \leq f(m)$.

This re-use is prohibited, because before proving this universal, one must invoke J-protect, which guarantees that the negative hypothesis is not available.

9 Conclusions and Related Work

This work is based on the principle that we can understand the algorithmic content of a nonconstructive reasoning system by translating proofs into a constructive system, and then using the constructive content there to infer what the direct algorithmic content should be for the nonconstructive system. Our work is based on the fundamental logical results of Friedman (10), who showed that classical arithmetic (and other classical theories) conservatively extends its constructive counterpart, and the work of Griffin (11), who discovered that one could give the control operator C as the algorithmic extract of the classical axiom. Griffin did not employ A-translation, though, and did not explore the implications of A-translation results for total-correctness of programs written using C. Our results elsewhere (13) showed that such programs correspond to classical proofs, and that their total-correctness corresponds to the constructivizability of these proofs. We have seen that, based upon the observation that Friedman's translation is a CPS-translation, we can "pull back" a reasoning system for an almost-functional language from a reasoning system for its functional core. We used this idea to show why certain classical proofs were always constructivizable, and why other classes could not in general be so.

We showed how different translations allowed us to pull back radically different reasoning systems for our programming language, corresponding to different operational semantics. We have seen that in fact we can start with a non-Church-Rosser "classical" programming language, with a classical type system, and, by specifying the translation into a constructive logic/functional programming language, we restrict reduction in the classical language to be confluent. Finally, we showed how to extract evidence systematically from proofs of non-Π_2^0 sentences.

In the future, we wish to explore more fully the "design space" of translations. There are many, many double-negation translations which suffice for Friedman's results, and it would be quite interesting to see what sorts of logics arise out of "pulling back" reasoning systems with these. We also would like to begin using classical theories with the J operator as a way of giving natural correctness proofs for nontrivial programs using C. The availability of J allows us to reason about nontrivial postconditions without "coding them up" as integer equalities, e.g. "σ is a most general unifier of $\langle X, Y \rangle$" is awkward to express as a predicate $f(\sigma, X, Y) = 0$. Finally, there is the task of integrating these results with the hierarchies of control operators proposed by, among others, Danvy and Filinski (5).

References

[1] H. P. Barendregt. *The Lambda Calculus: Its Syntax and Semantics*, volume 103 of *Studies in Logic and the Foundations of Mathematics*. North-Holland, Amsterdam, revised edition, 1984.

[2] W. Clinger and J. Rees. The *revised*³ report on the algorithmic language scheme. *SIGPLAN Notices*, 21(12):37–79, 1986.

[3] R. Constable. The semantics of evidence. Technical Report TR 85–684, Cornell University, Department of Computer Science, Ithaca, New York, May 1985.

[4] D. A. V. Dalen and A. S. Troelstra. *Constructivism in Mathematics*. North-Holland, 1989.

[5] O. Danvy and A. Filinski. Abstracting control. In *Proceedings of the 1990 ACM Conference on Lisp and Functional Programming*, pages 151–160, 1990.

[6] M. Felleisen. *The Calculi of $\lambda_v - CS$ conversion: A Syntactic Theory of Control and State in Imperative Higher-Order Programming Languages*. PhD thesis, Indiana University, 1987.

[7] M. Felleisen and D. Friedman. Control operators, the SECD machine and the λ-calculus. In *Formal Description of Programming Concepts III*, pages 131–141. North-Holland, 1986.

[8] M. Felleisen, D. Friedman, E.Kohlbecker, and B. Duba. Reasoning with continuations. In *Proceedings of the First Annual Symposium on Logic in Computer Science*, pages 131–141, 1986.

[9] M. J. Fischer. Lambda-calculus schemata. In *Proceedings of the ACM Conference on Proving Assertions about Programs*, volume 7 of *Sigplan Notices*, pages 104–109, 1972.

[10] H. Friedman. Classically and intuitionistically provably recursive functions. In Scott, D. S. and Muller, G. H., editor, *Higher Set Theory*, volume 699 of *Lecture Notes in Mathematics*, pages 21–28. Springer-Verlag, 1978.

[11] T. G. Griffin. A formulae-as-types notion of control. In *Conference Record of the Seventeenth Annual ACM Symposium on Principles of Programming Languages*, 1990.

[12] A. N. Kolmogorov. On the principle of the excluded middle. In J. van Heijenoort, editor, *From Frege to Gödel: A Source Book in Mathematical Logic, 1879–1931*, pages 414–437. Harvard University Press, Cambridge, Massachusetts, 1967.

[13] C. Murthy. An evaluation semantics for classical proofs. In *Proceedings of the Fifth Annual Symposium on Logic in Computer Science*, 1991.

[14] G. Plotkin. Call-by-name, call-by-value, and the λ-calculus. *Theoretical Computer Science*, pages 125–159, 1975.

CLASSICAL TYPE THEORY

Maria Napierala

Oregon Graduate Institute

19600 NW von Neumann Drive

Beaverton, OR 97006, USA

1 Introduction

Double-negation translations, e.g., of Gödel, Kolmogorov, encode classical formulas into intuitionistic formulas. Due to such translations, it was shown [Fri78] that classical proofs of Π_2^0 sentences (i.e., sentences of the form $\forall x \exists y . R(x,y)$ where R is a primitive recursive relation) can be translated into constructive proofs of the same sentences. The computational significance of this result was not exploited until recently [Mur90]. We made a similar observation with different "translation". It is well-known [Pra65] that the second-order propositional logic encodes propositional connectives into their intuitionistic semantics. The definition of the condition for existential results to be witnessed by the explicit solutions is also possible in this logic. Even though implicit in this interpretation, its computational content has not been investigated.

Another well-known result [Gir70] is that all functions provably total in the second-order Peano arithmetic PA^2 are representable in the system F [Gir70] or the second-order λ-calculus [Rey74], formal systems of typed terms which lie behind proofs in the second-order propositional logic. The power of the system F comes from the operation of abstraction on types. This *universal abstraction* makes the programs expressible in F decompose their arguments completely, i.e. the corresponding functions are computed "by-values" only. This is a defect of the system, a price paid for its power. For example, if a predecessor function

$$pred(0) = 0 \qquad pred(Sx) = x$$

(where S is a successor function) is programmed in F, the second equation will only be satisfied for x being a numeral $\bar{n}$, which means that the program decomposes the argument x completely to $SSS...S0$ (with n occurrences of S), then reconstructs it leaving out the last symbol S [GLT89]. Of course we would like to remove the first S instead. This would not change the result of computation and it would make it economical. If it was required that x always computes to a numeral, the second equation would be computed by subtracting 1.

As another example consider *factorial* which is defined by the following equations:

$$fact(0) = 1, \quad fact(Sx) = (Sx) * fact(x)$$

The program for this function in F decomposes x to a numeral $\bar{n}$, then computes a new integer by multiplying together all integers $\bar{i}$ (where $i = 1,..,n$) obtained during the reconstruction of $\bar{n}$, except that it takes 1 instead of 0. If it was required that x always computes to a numeral,

$$fact(0) = 1, \quad fact(x) = (x)*fact(x-1)$$

the second equation would be computed (*lazily*) by multiplying x (i.e., *placeholder* for the data) by the rest of computation (i.e., *continuation*).

The goal is to distinguish terms that are computable to *values*, i.e., to terms without unevaluated continuations. We shall also refer to values without unevaluated continuations as *explicit values* to stress the distinction from values as are understood in intuitionistic formal systems. We want to derive a system of typed terms and reduction rules, such that the reduction process always terminates in an explicit value. What we want to know is what are the types whose values can be "explicitly" computed, and what are the actual programs computing such values. Since we are not restricting the logic to be intuitionistic, the programs will represent the classical proofs of a class of sentences, which provide the evidence for the sentences in a constructive sense [Con85]. The class of such sentences will be determined by the types of programs which will be derived. We shall see that the explicit values are not only integers but all other usual data types in computer science like lists of values, pairs and sums of values, trees, etc. The notion of a value is extended to *explicit* functions, i.e., functions that given a value return a value.

2 Prawitz+ Translation

The following propositional quantification formalizes the condition for existential results over A to be witnessed by explicit solutions without restricting the logic to be intuitionistic:

$$(Exists\ A) \equiv [C:prop](A{\Rightarrow}C){\Rightarrow}C$$

where "$\equiv$" denotes definitional equality. The above construction encodes *explicit* values of type A since in order to prove it, we have to possess a closed term of type A. For example, if *nat* is the type of natural numbers in the system **F**,

$$nat \equiv [x:prop](x{\Rightarrow}x){\Rightarrow}x{\Rightarrow}x$$

then (*Exists nat*) encodes integers. To prove the proposition (*Exists nat*), we have to have a closed term of type *nat*. Such a term represents an integer n, i.e., it is a numeral:

$$\bar{n} \equiv \Lambda X{:}prop.\lambda z{:}X.\lambda s{:}X{\Rightarrow}X.s(s(s...(sz)...))$$

with n occurrences of s.

This translation can be extended to structured types (i.e., pairs, sums), namely

$$P\ \&\ Q \equiv [C:prop](P{\Rightarrow}Q{\Rightarrow}C){\Rightarrow}C$$

$$P\ or\ Q \equiv [C:prop](P{\Rightarrow}C){\Rightarrow}(Q{\Rightarrow}C){\Rightarrow}C$$

which encode *explicit* pairs and sums and at the same time formalize the condition for cut-free proofs of conjunction and disjunction, respectively. Similarly, the following construction is an example of an encoded inductive type schema, a list of objects of type A:

$$(List\ A) \equiv [C:prop](A{\Rightarrow}C{\Rightarrow}C){\Rightarrow}C{\Rightarrow}C.$$

We shall refer to such encodings as types of *implicit* values.

As we have already observed universal abstractions confuse data types and algorithms. This means that the algorithms are coded in terms of data type objects. Through the Curry-Howard isomorphism, this implies that the natural deduction proofs corresponding to the coded algorithms are cut-free. The ambiguity of the meaning of the second-order universal quantification is known to be algorithmically consistent [Gir70]. Its negative side is that the algorithms are either evaluated "by-values" only or are coded, i.e., they are different from original ones. The first is not economical, the second is bad for a programmer who wants to have an actual algorithm, not its code.

We shall refer to the encodings of propositional connectives, extended with the encodings of existential witness and with inductive type schemas as **Prawitz+ translation.**

3 Data-Types vs. Types

Data types will be distinguished from control structures in order to allow the introduction of programs which are *not* computed "by values" only but whose results are always computable to explicit values. The separation of control structures from data types is accomplished by "forgetting" the internal structure (or removing the uniformity problem) of proofs of universal quantifications representing data types in F^1. This way basic types (basic sets) of a classically founded program development system - Classical Type Theory (CTT) are obtained.

More precisely, let $\vdash M:T$ be provable in the system F and let M correspond to a data type constructor. *Forgetting* the internal structure of T and M is expressed by *naming* the constructions M and T. If ' ' is a map from terms in F to strings, then let strings "m" and "t" be results of application of that map to M and T written in mixfixed notation as 'M' and 'T'. Introducing constants *constr* and *Type* to CTT, such that

$$constr \equiv "c"$$
$$Type \equiv "t"$$

yields a fundamental form of a *judgement* in CTT (a mathematical assertion)

$$constr \in Type.$$

This judgement expresses that an object *constr* is an element of type *Type*. The epsilon notation represents the relation between an object and its type.

The symbol *prop* in F played the role of a type of propositions. Let constant *data* be the type of *data types* in CTT. Data-types correspond to universal quantifications derivable in F. Then, there is a judgement that *data is* a type:

$$data \text{ type.}$$

The judgement "$A \in data$" is rendered in words "A *is a data-type*". *data* is a set of all *data-types* of computer science[2]. By a *data-type* we mean here a type whose values (but not expressions) do not

---

1. Removing the uniformity problem of universal abstractions corresponds to taking their *dual*. More precisely, the introduction of propositional existential quantification (a dual of propositional universal quantification) in the system F

$$<U,v> \equiv \Lambda Y.\lambda x:[X:prop]\,V \Rightarrow Y.xUv,$$

consists of a data type object given *together* with its type. In other words, the type symbols and the terms are to be generated simultaneously. By dualizing universal abstractions we shall find a dual of their operational interpretation, i.e., of the second-order β-reduction. This dual is the true operational interpretation of "classical" programs.

contain any unevaluated continuations, i.e., its *values* are explicit. Expressions of such types can be always computed to explicit values. The objects of data-types can serve as the range of a variable. For example, if

$$Nat \equiv \text{'}[x{:}prop](x{\Rightarrow}x){\Rightarrow}x{\Rightarrow}x\text{'}$$

$$0 \equiv \text{'}\lambda x{:}prop.\lambda s{:}x{\Rightarrow}x.\lambda z{:}x.z\text{'}$$

then the following judgements are introduced into CTT:

$$Nat \in data$$

$$0 \in Nat.$$

The inductive types have functional constructors definable in **F**. For instance, the successor function is represented in **F** by the following construction:

$$succ \equiv \lambda n{:}[X{:}prop](X{\Rightarrow}X){\Rightarrow}X{\Rightarrow}X.\Lambda Y{:}prop.\lambda s{:}Y{\Rightarrow}Y.\lambda z{:}Y.s(nYsz)$$

of type $nat{\Rightarrow}nat$. The term *succ*, the natural iterator, when applied to a natural number, reduces to a universal abstraction. The successor in CTT is the function constant $S == \text{'}succ\text{'}$ of type $Nat{\rightarrow}Nat$. In contrast, Sn is in *canonical*, i.e., non-reducible form.

The constants k_n will be introduced in CTT to represent integers n:

$$k_n \equiv \text{'}\Lambda X{:}prop.\lambda z{:}X.\lambda s{:}X{\Rightarrow}X.s(s(s...(sz)...))\text{'}$$

with n occurrences of s.

We shall call simple and inductive data types (like booleans, integers, trees, lists, etc.) *ground* types in order to distinguished them from structured data types (product, sum, existential witness type). Once we have introduced terms of ground types, we shall do the same for structured types. The propositional schemas (*Exists A*), (*P or Q*), and (*P & Q*) define the proof methods provided by existential quantification, disjunction, and conjunction, respectively. Such proof methods are simple mechanical operations: "recover the same", "recover either of two", "recover both." The last operation, however, is not "primitive", i.e., it can be defined in terms of the other two as "recover the same and recover the other." These operations when combined with intuitionistic semantics of ground objects will produce new complex operations. We shall refer to such operations as *mechanical*.

The propositions (*Exists A*), (*P & Q*), and (*P or Q*) are definable only on the level of meta-notation of the second-order λ-calculus, i.e., they are propositional schemas. To internalize these definitions in CTT, we have to distinguish between *data-types*, i.e., members of *data*, and what we shall call "types". The judgement "*A* type" is rendered in words *A is a set*. Not just any set can be a "type" but only a "completely presented set", i.e., a set whose members carry the necessary "witnessing data" by means of a *mechanical operation* of their membership [Bee80]. In other words, a "type" is what is meant by it in constructive type theories (e.g., Martin-Löf's type theories) except that the meaning of the "mechanical

---

2. *Types* in Martin-Löf's type theories [ML82, ML84] are also viewed as "data types" of computer science but the associated programming languages are purely functionals (i.e., languages without assignments).

operation" is different from the one defined above. Our meaning is narrower.

Operational semantics of the underlying language of Classical Type Theory is defined by a set of *reduction* or *evaluation* rules. Each type has terms of two categories: *canonical* and *noncanonical*. A term is canonical when it is not a redex, otherwise it is noncanonical. The difference with data-types is that there are noncanonical terms and evaluation rules *only* for *explicit values* of data-types.

In CTT the canonical (non-reducible) objects of a "type" represent *implicit values* of "classical" programs, i.e., values that are computable to *explicit values* by some mechanical operation.

Extracting the computational content of Prawitz+ "translation" is a three-step process. First, the primitive operations are introduced as evaluation rules for types referred to as existential witness and disjunctive witness types. Next, in order to define operational content of conjunction, the primitive operations are internalized as terms of the theory. The types introduced in this step will be the types of *top-level* (i.e., empty context level) continuations of conjunction, disjunction, and existential quantification types. In the third step, all local control structures and programs can be introduced and some examples will be shown in this paper.

4 Dependent Function Type

To internalize mechanical operations as terms of CTT, a dependent function type is necessary. With such a type the *top-level* continuations of classical types can be represented. The following hypothetical judgement is used to introduce dependent function types:

$$B(x) \text{ type } [x \in A].$$

Let 'Π' be the dependent function type constructor. The following is a formation rule for dependent function types:

$$\frac{A \in data, \quad B(x) \text{ type } [x \in A]}{\Pi x \in A.B(x) \text{ type}} \qquad \text{(Π-form)}$$

If B doesn't depend on A, then $\Pi x \in A.B(x)$ reduces to $A \to B$ with "B type" or $B \in data$. An object of a dependent function type represents an *implicit* function, whose values are implicit, but when that function is supplied with an explicit value, it always returns an explicit value. The objects of a Π-type are introduced by the rule:

$$\frac{b(x) \in B(x) \quad [x \in A]}{\lambda x \in A.b(x) \in \Pi x \in A.B(x)} \qquad \text{(Π-intro)}$$

The expression $\lambda x \in A.b(x)$ is *fully evaluated* in CTT. More precisely, an expression is "fully" or "completely" evaluated in CTT when it is in canonical form and all its binding-free intermediate subterms are fully evaluated. Hence, $b(x)$ in $\lambda x \in A.b(x)$ is not to be evaluated since doing so would have been like trying to execute a program which expects an input but the input data is not provided.

The *noncanonical* constant for the Π-type is the application represented by the juxtaposition of an object of a $\Pi x \in A.B(x)$-type and an object of A:

$$\frac{f \in \Pi x \in A.B(x), \quad a \in A}{fa \in B(a)} \qquad \text{(}\Pi\text{-elim)}$$

The evaluation rule associated with the Π-type is defined by the following one-step ($\rightarrow_1$) reduction rule:

$$\frac{b(x) \in B(x) \ [x \in A], \quad a \in A}{(\lambda x \in A.b(x))a \rightarrow_1 b[a/x]} \qquad \text{(}\Pi\text{-red)}$$

5 Primitive Operations

5.1 Existential Witness Type

The quantifier "$\exists$" can be interpreted in the second-order predicate logic [Pra65] as follows

$$Sig(R,Q) \equiv \forall^2 X.(\forall x:R.Q(x) \supset X) \supset X.$$

But the operational interpretation of the witness of existential quantification can be formalized in the second-order propositional logic, namely by the quantification

$$(Exists \ P) \equiv [C:prop](P \Rightarrow C) \Rightarrow C.$$

The quantification $(Exists \ P)$ encodes the condition for the existential results to be witnessed by explicit solutions. The explicit solutions to existential formulas imply that their proofs are cut-free[3]. In other words, the quantification $(Exists \ P)$ encodes the constructive meaning of existential quantification without restricting the logic to be intuitionistic.

The following deduction in the system F represents the introduction rule for $(Exists \ P)$:

$$[P:prop][x:P] \ \vdash \ \Lambda C:prop.\lambda u:P \Rightarrow C.(u \ x) \qquad \text{(Wit)}$$

The context (i.e., the sequence of bindings on the left-hand side of deduction sign) of the deduction (Wit) *describes* what is needed to prove the proposition $(Exists \ P)$. Let 'Σ' be the symbol of existential witness type. The following rules internalize the schema (Wit) into the formalism of CTT[4]:

$$\frac{A \in data, \quad a \in A}{\Sigma(A,a) \ type} \qquad \text{(}\Sigma\text{-form)}$$

$$\frac{C \in data, \quad c \in C}{[c]_C \in \Sigma(Id,\iota)} \qquad \text{(}\Sigma\text{-intro)}$$

$$\frac{d \in \Sigma(A,a)}{split_C(d) \in C} \qquad \text{(}\Sigma\text{-elim)}$$

$$\frac{C \in data, \quad c \in C}{split_C([c]_C) \rightarrow_1 c} \qquad \text{(}\Sigma\text{-red)}$$

---

3. When we have a cut-free proof of " $\vdash A$ " then the last rule applied in deduction must be a logical rule (vs. structural rule in sequent calculus), and this has immediate consequences, e.g., if A is $\exists y \ B$, then $B(t)$ has been proved for some t, etc. [Gir89].

4. With $Id \equiv$ '$[X:prop]X \Rightarrow X$' and $\iota \equiv$ '$[X:prop][x:X]x$' represent non-emptiness of a type.

The Σ-type tells what is the form of constructive witness of existential quantification over A, namely a value of the data-type A. It does *not* tell, however, how this value is computed. The object $[c]_C$ represents an implicit value of an arbitrary data-type C. The evaluation of $split_C([c]_C)$ to c represents an *arbitrary computation* which simply recovers the same arbitrary value c.

We obtain the specialized existential witness type when the propositional universal quantification in (*Exists R*) is instanciated to R^5:

$$\frac{a \in A}{[a] \in \Sigma_A(A,a)} \qquad (\Sigma_A\text{-}intro)$$

$$\frac{d \in \Sigma_A(A,a)}{split(d) \in A} \qquad (\Sigma_A\text{-}elim)$$

$$\frac{a \in A}{split([a]) \to_1 a} \qquad (\Sigma_A\text{-}red)$$

The type $\Sigma_A(A,a)$ expresses the judgement that $a \in A$. The object $[a]$ represents an implicit value of data-type A. For instance, if A is *Nat*, $[0]$ is an implicit integer from which the explicit value 0 is simply "read off". The evaluation of $split([0])$ to 0 represents an *empty computation* which simply recovers the same value a program evaluated to. The Σ-type formalizes the primitive operation of identity, i.e., of "recovering the same".

5.2 Disjunctive Witness Type

The impredicative construction of the operational interpretation of disjunction is as follows:

$$P \ or \ Q \equiv [C:prop](P{\Rightarrow}C){\Rightarrow}(Q{\Rightarrow}C){\Rightarrow}C$$

This interpretation of disjunction is in agreement with its intuitionistic semantics. The proposition $P \vee Q$ is a *method* of proving any proposition C provided one has either a proof that C follows from P or a proof that C follows from Q. The following deductions of the system **F** introduce disjunction:

$$[P:prop][Q:prop][x:P] \ \vdash \ \Lambda C:prop.\lambda u:P{\Rightarrow}C.\lambda v:Q{\Rightarrow}C.(u \ x) : (P \ or \ Q) \qquad (Inl)$$

$$[P:prop][Q:prop][y:Q] \ \vdash \ \Lambda C:prop.\lambda u:P{\Rightarrow}C.\lambda v:Q{\Rightarrow}C.(v \ y) : (P \ or \ Q) \qquad (Inr)$$

The type schema $(P \ or \ Q)$ encodes the condition for cut-free proofs of disjunction[6].

The contexts of the introduction rules for disjunction describe what is needed to prove $(P \ or \ Q)$. To internalize the primitive operation corresponding to the proof method by disjunction, the propositional

---

5. We note that the type $(R{\Rightarrow}R){\Rightarrow}R$, where $R{:}prop$ represents a ground type in the system **F**, corresponds to double-negation translation of R, with the term $\lambda x{:}R.x$ as the top-level continuation of type R [Mur90].

6. When we have a cut-free proof of $\vdash A{\vee}B$ then the last rule applied in deduction must be a logical rule. This has immediate consequence, namely that A has been proved or that B has been proved, and that there is a tag telling us which disjunct we were getting evidence for.

universal quantification in (P *or* Q) is instanciated either to P or Q^7. Two types are necessary to specify whether the evidence of (P *or* Q) comes from the evidence of P or of Q. Let 'V_0' and 'V_1' be the symbols of "left" and right" disjunctive witness type respectively. The following rules define disjunctive witness types:

$$\frac{A \in data,\ B \in data,\ a \in A}{V_0^A (A,B,a)\ \text{type}} \qquad (V_0^A \text{-form})$$

$$\frac{A \in data,\ B \in data,\ b \in B}{V_1^B (A,B,b)\ \text{type}} \qquad (V_1^B \text{-form})$$

$$\frac{a \in A}{inl(a) \in V_0^A (A,B,a)} \qquad (V_0^A \text{-intro})$$

$$\frac{b \in B}{inr(b) \in V_1^B (A,B,b)} \qquad (V_1^B \text{-intro})$$

$$\frac{d \in V_0^A (A,B,a)}{outl(d) \in A} \qquad (V_0^A \text{-elim})$$

$$\frac{d \in V_1^B (A,B,b)}{outr(d) \in B} \qquad (V_1^B \text{-intro})$$

$$\frac{a \in A}{outl(inl(a)) \rightarrow_1 a} \qquad (V_0^A \text{-red})$$

$$\frac{b \in B}{outr(inr(b)) \rightarrow_1 b} \qquad (V_1^B \text{-red})$$

The types $V_0^A (A,B,a)$ and $V_1^B (A,B,b)$ define the primitive operation provided by disjunction: either a value of data-type A or a value of data-type B, a program evaluated to, is recovered.

5.3 Disjunctive-Conjunctive Witness Type

The following second-order propositional quantification expresses the operational interpretation of conjunction:

$$P \ \& \ Q \equiv [C{:}prop](P \Rightarrow Q \Rightarrow C) \Rightarrow C$$

---

7. We shall skip the uninteresting case of disjunctive witness, when its intended domain is arbitrary. This case would have corresponded to, what we may call, an *arbitrary sum*.

This *procedural* interpretation of conjunction is in agreement with its intuitionistic semantics. The proposition $P \wedge Q$ is a *method* of proving any proposition C provided one has a proof that C follows from P and Q. The type schema $(P \ \& \ Q)$ encodes the condition for proofs of conjunction to be cut-free[8].

The following deduction in the system **F** represents the introduction rule for conjunction:

$$[P{:}prop][Q{:}prop][x{:}P][y{:}Q] \ \vdash \ \wedge C{:}prop.\lambda h{:}P{\Rightarrow}Q{\Rightarrow}C.(h \ x \ y) \ : \ (P \ \& \ Q) \qquad \text{(Pair)}$$

The context of the deduction (Pair) *describes* what is needed to prove the proposition $(P \ \& \ Q)$, namely a proof of P and a proof Q. In order to formalize a proof method by conjunction, the propositional universal quantification in $(P \ \& \ Q)$ is instanciated to either P or Q. Let $\wedge_A$ and $\wedge_B$ be the symbols for the type "disjunctive-conjunctive" witness specifying whether the "intended domain" of conjunctive witness is A or B. The following rules define disjunctive-conjunctive witness types:

$$\frac{A \in data, \ B \in data, \ b \in B, \ a \in A}{\wedge_A(A,B,a,b)} \qquad (\wedge_A\text{-}form)$$

$$\frac{A \in data, \ B \in data, \ b \in B, \ a \in A}{\wedge_B(A,B,a,b)} \qquad (\wedge_B\text{-}form)$$

$$\frac{a \in A}{(a,\textit{tt}) \in \ \wedge_A(A,Id,a,\textit{tt})} \qquad (\wedge_A\text{-}intro)$$

$$\frac{b \in B}{(\textit{tt},b) \in \ \wedge_B(Id,B,\textit{tt},b)} \qquad (\wedge_B\text{-}intro)$$

$$\frac{d \in \ \wedge_A(A,B,a,b)}{fst(d) \in A} \qquad (\wedge_A\text{-}elim)$$

$$\frac{d \in \ \wedge_B(A,B,a,b)}{snd(d) \in B} \qquad (\wedge_B\text{-}elim)$$

$$\frac{a \in A}{fst((a,\textit{tt})) \ \rightarrow_1 \ a} \qquad (\wedge_A\text{-}red)$$

$$\frac{b \in B}{snd((\textit{tt},b)) \ \rightarrow_1 \ b} \qquad (\wedge_B\text{-}red)$$

As we pointed out before, the proof method by conjunction is not a primitive operation. It is definable in terms of the primitive operations provided by the notions of existential and disjunctive witness. The $\wedge_A$- and $\wedge_B$-types formalize only the "disjunctive" part of the definition (hence, *disjunctive-conjunctive* witness type). The type $\wedge_A(A,Id,a,\textit{tt})$ takes as its second component an object of an arbitrary data-

---

8. When we have a cut-free proof of $\vdash A{\wedge}B$ then the last rule applied in deduction must be a logical rule, and this has immediate consequence, namely that A has been proved and that B has been proved.

type. Similarly, the $\wedge_B$-type defines a pair which "is arbitrary" in its first component. The evaluation rule for $\wedge_A$-type represents how to recover ("read off") the first component of a pair, and the evaluation rule for $\wedge_B$-type represents how to recover ("read off") the second component. Thus, the operation of "reading off" both values of types A and B has not been yet completely formalized. In order to formalize the proof method by conjunction completely, the primitive operations have to be internalized as terms of CTT.

6 Top-Level Continuations of Classical Types

In the previous section we have formalized the primitive operations. We have to internalize these operations as terms of CTT to allow definition of complex operations, in particular pairing. We shall accomplish this by "fixing" the intended domain of primitive operations, i.e., instanciating the propositional universal quantifications defining the operational content of logical constants to a specific proposition. We shall see that the terms defining *simple* operations (i.e., primitive operations and pairing) represent *top-level continuations* of corresponding classical types. By a *top-level continuation* we mean a value with a store such that when a concrete data is "plugged" in the store, a concrete data is return at the top-level (empty context level) of computation. The returned data is a result of the *entire* computation. It is rather a general believe that the result of computation is a ground value. The notion of top-level continuation provides the operational semantics of complete evaluation for structured types, whose terms are evaluated in a *lazy* manner internally but *eagerly* at the top-level. That is, the notion of *explicit value*, i.e., a value which can be a result of computation, is extended to structured types. We shall see later that the top-continuation semantics is the true operational meaning of "classical" programs.

6.1 Top-Level Continuation of Disjunction Type

Let "+" be the symbol of the type whose terms represent top-level continuations for disjunction type. Its definition is as follows:

$$\frac{C \in data, \ A \in data, \ B \in data}{A +_C B \ \text{type}} \qquad (\text{+-}form)$$

$$\frac{f \in C{\to}V_0^A(A,B,a), \ g \in C{\to}V_1^B(A,B,b)}{<f,g> \ \in \ A +_C B} \qquad (\text{+-}intro)$$

$$\frac{d \in A +_C B}{d^0 \ \in \ C{\to}A, \ d^1 \in \ C{\to}B} \qquad (\text{+-}elim)$$

$$\frac{f \in C{\to}V_0^A(A,B,a), \ g \in C{\to}V_1^B(A,B,b)}{<f,g>^0 \ \to_1 \ \lambda c \in C.outl(fc), \ <f,g>^1 \ \to_1 \ \lambda c \in C.outr(gc)} \qquad (\text{+-}red)$$

The term $<f,g>$ represents *either* a top-level continuation of type A *or* a top-level continuation of type B, i.e., it is a top-level continuation of a sum of A and B.

The type $A +_C B$ *internalizes* in CTT the proof method by disjunction by associating a value of type A and a value of type B with (*not* necessarily the same) value of the same type C. This internalizes the choice of recovering either a value of type A or a value of type B. The computation of either value is lazy which is expressed by the fact that the terms $\lambda c \in C.outl(fc)$ and $\lambda c \in C.outr(gc)$ *are* fully evaluated. The definition of +-type assures that the subterms $outl(fc)$ and $outr(gc)$ are reducible to a and b respectively. In other words, the term $<f,g>$ is an explicit pair of the two top-level continuations.

When C is A in $A +_C B$ then specialized type $A +B$ is obtained and is defined as follows:

$$\frac{A \in data, \ B \in data}{A +B \ \text{type}} \qquad (+_A\text{-}form)$$

$$\frac{f \in \Pi a \in A.\mathsf{v}_0^A(A,B,a), \ g \in A {\rightarrow} \mathsf{v}_1^B(A,B,b)}{<f,g> \ \in \ A +B} \qquad (+_A\text{-}intro)$$

$$\frac{d \in A+B}{d^0 \in \ A{\rightarrow}A, \ d^1 \in \ A{\rightarrow}B} \qquad (+_A\text{-}elim)$$

$$\frac{f \in \Pi a \in A.\mathsf{v}_0^A(A,B,a), \ g \in A {\rightarrow} \mathsf{v}_1^B(A,B,b)}{<f,g>^0 \rightarrow_1 \lambda a \in A.a, \ <f,g>^1 \rightarrow_1 \lambda a \in A.outr(fa)} \qquad (+_A\text{-}red)$$

The type $A +B$ internalizes the operation of either recovering the original *ground* value a or of "reading off" a value b of data-type B a program evaluated to. The data-type A is a ground type because its values are being "read off" by an empty computation. In other words, the term $<f,g>$ of type $A +B$ represents *either* a top-level continuation of a *ground* type A *or* a top-level continuation of a data-type B.

6.2 Top-Level Continuation of Conjunction Type

The operation defined by conjunction is "reading off" *both* a value of data-type A and a value of data-type B. This corresponds to associating both a value of type A and a value of type B with the *same* value of data-type C. To define this procedure purely mechanically, i.e., in terms of primitive operations, we have not only to "fix" the intended domain of conjunctive witness to C but take C to be either A or B.

Let " $\times$ " be the symbol of a type whose terms represent top-level continuations for conjunction type. If we choose A as the intended domain for conjunctive witness, the definition of $\times$-type is as follows:

$$\frac{A \in data, \ B \in data}{A \times_A B \ \text{type}} \qquad (\times_{LR}\text{-}form)$$

$$\frac{h \in \Pi a \in A.\wedge_B(A,B,a,b)}{\pi(h) \in \ A \times_A B} \qquad (\times_{LR}\text{-}intro)$$

$$\frac{c \in A \times_A B}{c_0 \in \ A{\rightarrow}A, \ c_1 \in \ A{\rightarrow}B} \qquad (\times_{LR}\text{-}elim)$$

$$\frac{h \in \Pi a \in A.\wedge_B(A,B,a,b)}{(\pi(h))_0 \to_1 \lambda a \in A.a, \quad (\pi(h))_1 \to_1 \lambda a \in A.snd(ha)} \quad (\times_{LR}\text{-}red)$$

The type $A \times_A B$ internalizes the operation of recovering the *ground* value a and "reading off" a value b of data-type B a program evaluated to. The term $\pi(h)$ represents *both* empty continuation of type A and a top-level continuation of type B. In other words, the values of conjunction of two types can be returned as results of computation only when one of the types (here the *first*) is a ground type. When the store a is filled with an explicit ground value, first, the same value is recovered, and second, a value of data-type B a program evaluated to is "read off". Classical conjunctive type "fixes" the order of complete evaluation either from left-to-right or right-to-left for pairs of terms. The type $A \times_A B$ establishes the order of evaluation from left-to-right which is denoted by the subscript "LR" in the rule names above.

6.3 Implicit Top-Level Continuation of Existential Quantification Type

We shall first introduce a type whose terms represent "implicit" top-level continuations for existential quantification types, and then explain its definition. Let '$\{\ \}$' be the symbol of the type in question. Its definition is as follows:

$$\frac{A \in data, \quad B \in data}{\{A\}_B \ type} \quad (\{\ \}\text{-}form)$$

$$\frac{h \in B \to \Sigma_A(A,\alpha)}{\rho(h) \in \{A\}_B} \quad (\{\ \}\text{-}intro)$$

$$\frac{d \in \{A\}_B}{d_0 \in B \to A} \quad (\{\ \}\text{-}elim)$$

$$\frac{h \in B \to \Sigma_A(A,\alpha)}{(\rho(h))_0 \to_1 \lambda b \in B.split(hb)} \quad (\{\ \}\text{-}red)$$

The type $\{A\}_B$ formalizes the operation provided by the proof method by existential quantification over A which is "reading off" a value of data-type A as a result of *entire* computation. Since the expression $\lambda b \in B.split(hb)$ is *fully evaluated* in CTT, the evaluation of α is lazy internally. However, being the result of entire computation, α has to be an *explicit value*. The Greek, instead of Roman letter α is used to denote that the value is explicit. This is a completely *abstract* way of assuring that α is an explicit value. What we need is the *actual* algorithm (i.e., a "classical" program) for computing α. In other words, the term $\rho(h)$ is only an *implicit* top-level continuation for existential quantification type. If B is A, then h can be an identity $\lambda x \in A.x$ on A which is a top-level continuation for a *ground* type A and expresses *empty* computation.

7 Natural Function Iterator

By combining simple operations with intuitionistic semantics of ground types, new local control operators can be defined. In this section we shall derive a *natural function iterator* by combining the primitive operation defined by disjunction type with the intuitionistic semantics of natural numbers.

We take the type A in $A+B$ to be *Nat*. We will show below that the type $Nat+A$ yields an *abstract control operator* which completely characterizes the logic of numerical reasoning.

According to (+-*elim*), the rule of type $Nat \rightarrow A$ is defined from $d \in Nat+A$. We know that $d = <f,g>$ where

$$f \in \Pi n \in Nat.\vee_0^{Nat}(Nat,A,n)$$
$$g \in Nat \rightarrow \vee_1^A(Nat,A,a)$$

According to ($\rightarrow$-*intro*),

$$<f,g>^0 \rightarrow_1 \lambda n \in Nat.n,$$
$$<f,g>^1 \rightarrow_1 \lambda n \in Nat.snd(gn)$$

We shall show, under the assumption $[n \in Nat]$, the n-step reduction

$$snd(gn) \rightarrow_n a$$

by discharging the assumption by the structural induction on *Nat*. First, let $n = 0$:

$$\frac{a \in A}{inr(a) \in Nat+A} \qquad (+_{Nat}^0\text{-}intro)$$

The rule $(+_{Nat}^0-intro)$ is an introduction rule for disjunction $Nat+A$ with *Nat* being defined according to its intuitionistic interpretation. The defining constructor *constr* is denoted by a superscript "*constr*" in the name of the introduction rule.

Second, let $n = Sm$ and suppose $f \in A \rightarrow A$:

$$\frac{\vdots}{\dfrac{inr(b) \in Nat+A}{inr(fb) \in Nat+A}} \qquad (+_{Nat}^S\text{-}intro)$$

Thus, according to (+-*red*), the evaluation rule $k \in Nat \rightarrow A$ for the right inject of the type $Nat+A$ is defined as follows:

$$k0 \rightarrow_1 a, \qquad kS \rightarrow_1 fk.$$

We shall introduce an abstract local control operator I into CTT:

$$\frac{A \in data, \ a \in A, \ f \in A \rightarrow A}{IAaf \in Nat \rightarrow A} \qquad \textbf{(Iter)}$$

This is the elimination rule for disjunction $Nat+A$. The function iterator I is defined by the following rules of one-step reduction:

$$\frac{A \in data, \ a \in A, \ f \in A \rightarrow A, \ n \in Nat}{IAaf(Sn) \rightarrow_1 f(IAafn)} \qquad (+_{Nat}^S red)$$

$$\frac{A \in data, \ a \in A, \ f \in A \rightarrow A}{IAaf0 \rightarrow_1 a} \qquad (+_{Nat}^0 red)$$

($IAfan$) is definitionally equal to the following "for-loop", i.e., flowchart in a programming language:

$$z{:}{=}a; \text{ for } i{=}0 \text{ to } n{-}1 \text{ do } z{:}{=}fz.$$

8 Primitive Recursion Operator

In this section we shall derive a *primitive recursion operator* by combining pairing with the structural induction on natural numbers.

We take the type B in $B \times A$ to be *Nat*. According to ($\times$-*elim*), the rule of type $Nat \rightarrow A$ is defined from $d \in Nat \times A$. We know that $d = \pi(h)$ where

$$h \in \Pi n \in Nat. \wedge_A (Nat, A, n, a).$$

According to the $\rightarrow - intro$,

$$\pi(h)_0 \rightarrow_1 \lambda n \in Nat.n$$
$$\pi(h)_1 \rightarrow_1 \lambda n \in Nat.snd(hn).$$

We shall show that under the assumption $[n \in Nat]$,

$$snd(hn) \rightarrow_n a$$

by discharging the assumption according to the structural induction on *Nat*. First, let $n = 0$:

$$\frac{a \in A}{(0, a) \in Nat \times A} \qquad (\times_{Nat}^0 \text{-}intro)$$

Second, let $n = Sm$ and suppose $f \in Nat \rightarrow A \rightarrow A$:

$$\frac{\vdots}{(m, b) \in Nat \times A} \over {(Sm, fmb) \in Nat \times A} \qquad (\times_{Nat}^S \text{-}intro)$$

We shall introduce a new local control operator R into CTT:

$$\frac{A \in data, \quad a \in A, \quad f \in Nat \rightarrow A \rightarrow A}{RAaf \in Nat \rightarrow A} \qquad (Rec)$$

This is the elimination rule for conjunction $Nat \times A$. The operational semantics of primitive recursion operator R is defined by the following one-step reduction rules:

$$\frac{A \in data, \quad a \in A, \quad f \in Nat \rightarrow A \rightarrow A, \quad n \in Nat}{RAaf(Sn) \rightarrow_1 fn(RAafn)} \qquad (\times_{Nat}^S \text{-}red)$$

$$\frac{A \in data, \quad a \in A, \quad f \in Nat \rightarrow A \rightarrow A}{RAaf0 \rightarrow_1 a} \qquad (\times_{Nat}^0 \text{-}red)$$

($RA fan$) is definitionally equal to the following "for-loop" in a programming language:

$$z := a; \text{ for } i = 0 \text{ to } n\text{-}1 \text{ do } z := f(i, z);$$

which makes an *explicit use* of an *implicit* value of i.

9 Top-Level Continuation of Existential Quantification over *Nat*

So far we have managed to define the *empty* computation $\lambda x \in A.x$ for ground type programs. All non-empty computation rules are defined implicitly as specified by $\{\}$-type. A "classical" program which computes to a value of type A corresponds to a top-level continuation of existential quantification over A. A value of type A can be just "read off" when a top-level continuation of existential quantification over A is supplied with a concrete data.

We shall define a program schema by combining pairing with natural function iterator. We shall take $C = B \times_B A$ in $\{A\}_C$ and specialize B further to a concrete ground type, namely to *Nat*. According to ($\{\}$-*elim*), the rule of type $(Nat \times A) \rightarrow A$ is defined from $d \in \{A\}_{Nat \times A}$. We know that $d = \rho(h)$ where

$$h \in \Pi p \in Nat \times A.\Sigma_A(A,\alpha).$$

According to the ($\{\}$-*red*),

$$\rho(h)_0 \rightarrow_1 \lambda p \in Nat \times A.split(hp).$$

We shall show below that under the assumption $[p \in Nat \times A]$, the $fst(p)$-step reduction

$$split(hp) \rightarrow_{fst(p)} a$$

will take place by discharging the assumption by the structural induction on $Nat \times A$. We define one-step $k \in (Nat \times A) \rightarrow (Nat \times A)$ as follows:

$$k(n,a) \rightarrow_1 (Sn,fna)$$

where $f \in Nat \rightarrow A \rightarrow A$. In order to define a computation schema $g \in Nat \rightarrow A$, we apply the iterator to $Nat \times A$:

$$I(Nat \times A)(0,a)k.$$

We can define g as "by-value" primitive recursion, i.e., $gn \equiv snd(hn)$, "by-value" primitive recursion. $I(Nat \times A)k(0,a)n$ is definitionally equal to the following "while-loop" which implements the flowchart defined by R:

$$z:=a;\ i:=0;\ \textbf{while}\ i<n\ \textbf{do}\ i:=i+1;\ z:=f(i,z);$$

The definition of primitive recursion: $\quad g(0) = a, \quad g(n+1) = f(n,(gn))$

in terms of iteration and pairing implies that the second equation for g is satisfied by values only, i.e., for each n separately. However, by introducing a special notation t^D for integer terms, i.e., terms computable to numerals, the second equation could be computed *lazily*. The following prefix notation can be introduced for the above expression:

$$natrec(n^D;a;i^D,y.f(i^D,y)).$$

The operational interpretation of *natrec* is given in terms of following reduction rules:

$$natrec(0;a;x^D,y.f(x^D,y)) \rightarrow_1 a$$
$$natrec(k_n;a;x^D y.f(x^D,y)) \rightarrow_1 f(k_n,natrec(k_{n-1};a;x^D,y.f(x^D,y))).$$

In summary, the explicit top-level continuation of existential quantification type on natural numbers is simply the primitive recursion computation schema. Such a schema is represented in CTT by *natrec*, a term with a *classical* (i.e., free) variable n^D which plays a role of a store. Thus classically, a term t of type T, which depends on a variable x of type U, is no longer seen as the result of *substituting* for x the terms u of type U but as a typed **plugging** instruction[9]. Classical variable symbolizes a place in which one can plug *input data* of appropriate type. The term t itself can be plugged in any variable of type T appearing in another term. Hence, program execution can be seen as a symmetrical input/output process. This is the true operational semantics of "classical" programs.

In this way we have identified a class of sentences of Peano Arithmetic (PA) whose classical proofs provide evidence in a constructive sense. These are Π_2^0 sentences, i.e., sentences of the form

---

9. We almost repeat here the words of Girard, who, in his search for *universal dynamics*, is the foremost propagator of true *sense* of mathematical sentences [Gir89, GLT89].

$\forall x \exists y.R(x,y)$, where R is a decidable relation. Such a relation can always be written as $f(x,y)=0$, where f is a primitive recursive function.

Examples. As an example, let us define an algorithm for computing *predecessor* function:
$$pred(0) = 0, \; pred(n+1) = n.$$
The program for predecessor is as follows:
$$natrec(n^D;0;x^D,y.x^D-1) \equiv n^D-1.$$

As another example consider *factorial* defined by the following equations:
$$fact(0) = 1, \; fact(n+1) = (n+1)*fact(n).$$
Factorial is computed lazily by
$$natrec(n^D;1;x^D,y.x^D*y).$$

10 Exceptions

The impredicative quantification $[X:Prop]X$ represents the notion of the *emptiness* of a type. We shall introduce the corresponding *absurdity* type into CTT:
$$\nabla \in data$$

Absurdity is characterized by a unique function $\{\}_A \in \nabla \rightarrow A$, for each $A \in data$:
$$\frac{A \in data}{\{\}_A \in \nabla \rightarrow A}$$

The uniqueness of $\{\}_A$ is defined by the following reduction rule:
$$\frac{A \in data, \; b \in \nabla \rightarrow A}{b \rightarrow_1 \{\}_A} \tag{Empty}$$

The meaning of **(Empty)** is the **absence** of data of type A. From the algorithmic point of view ∇ is the type of exceptions. Each ground type A can be extended with an *exceptional value* as follows:
$$A^\perp \equiv \nabla + A$$
By assuming $\perp: [X:prop]X$ one can define $\perp_A \equiv \,'(\perp P)\,'$, where $A \equiv \,'P\,'$, such that
$$inl(\perp_A) \in \nabla + A$$
This corresponds to raising an exception in a program of type A. Each exceptional value requires an extra storage allocation in order that an algorithm with exceptions be executable.

11 The Smallest Element of a Non-Empty List

It can be easily shown that terms of other inductive data-types, e.g., lists, trees, etc., have no control side-effects, i.e., their top-level continuation is an empty computation. Such types can serve as domains of quantification of classical sentences. The types defined through inductive type schemas can be the types of results of computation when the terms of the base types are explicit values.

In other words, we can easily extend the result for integers as a domain of quantification for classical sentences to encompass other simple and inductive types like booleans, lists, trees, etc.

As an example, we present the development of an algorithm for the search of the smallest element in a non-empty list of integers.

Lists. We shall first introduce *lists*. The impredicative definition of the type of list of elements of data-type P is the following propositional schema:

$$(list\ P) \equiv [X:Prop]X \Rightarrow (P \Rightarrow X \Rightarrow X) \Rightarrow X$$

Its constructors *nil* (empty list constructor) and *cons* (non-empty list constructor) are defined below

$$nil \equiv \Lambda X:Prop.\lambda n:X.\lambda c:P \Rightarrow X \Rightarrow X.n$$

$$(cons\ e\ l) \equiv \Lambda X:Prop.\lambda n:X.\lambda c:P \Rightarrow X \Rightarrow X.(c\ e\ (l\ X\ n\ c))$$

The impredicative definition of lists confuses the data type for lists with iteration over lists. To disambiguate this identification, the uniformity of the above constructions will be "forgot", leaving only their intuitionistic interpretation. As a consequence, the judgements

$$List_A \in data, \qquad Nil \in List_A, \qquad Cons \in A \rightarrow List_A \rightarrow List_A$$

are introduced to CTT.

The following is the computation schema associated with lists:

$$listrec(l^L;a;x^L,y^A,z.f(x^L,y^A,z))$$

where superscripts L and A annotate terms that represent explicit values.

The operational interpretation of *listrec* is defined by the following one-step reductions:

$$listrec(Nil;a;x^L,y^A,z.f(x^L,y^A,z)) \rightarrow_1 a$$

$$listrec(b.rest;a;x^L,y^A,z.f(x^L,y^A,z)) \rightarrow_1 f(b,rest,listrec(rest;a;x^L,y^A,z.f(x^L,y^A,z)))$$

where the "dot" notation $a.l$ is for *explicit* lists.

The task is to define an algorithm that computes for any non-empty list of integers its smallest element. The specification for the algorithm is a paradigm for plugging in an integer in places of its requests. The following **plugging instruction** based on the structure of the input is the program specification:

Nil: If the list is empty, an exception $\perp_{Nat}$ is raised.

n.Nil: If the list contains one element n, the value returned as a result of the entire program is that element.

n.m.rest: If the list contains more than one element, the following primitive operation is iterated: either m or n is chosen depending on whether either $m \leq n$ or $n < m$.

This plugging instruction is the program for our task. Several operations have to be introduced in order to define the program.

There are two operations defined on lists

$$hd(a.rest) \rightarrow_1 a \qquad tl(a.rest) \rightarrow_1 rest$$

which recover the first element of a list and a list without its first element respectively. To define the program we have to introduce the "boolean" data type (a type with two objects *true* and *false*), and the local control operator *case* associated with it:

$$Bool \equiv `[X:prop]X \Rightarrow X \Rightarrow X'$$

$$case(true;a;b) \rightarrow_1 a \qquad case(false;a;b) \rightarrow_1 b.$$

Let us assume that we have an operation *less* which returns smaller of two integers,

$$less(m,n) \rightarrow_1 m\ \text{if}\ m < n \qquad less(m,n) \rightarrow_1 n\ \text{if}\ m \geq n.$$

The program *is_Nil* checks if a list is empty:

$$is\_Nil(l^L) \equiv listrec(l^L;true;x^L,y^L,z.false).$$

The computation scheme $listrec^\perp$ of type $\{\nabla + Nat\}_{List_{Nat}}$ for searching for a minimum element in a list is defined as follows:

$$MinL(l^L) \equiv listrec^\perp(l^L;case(is\_Nil(l^L);\perp_{Nat};hd(l^L));x^D,y^L,z.less(x^D,z)))$$

Its operational content is defined by the following reduction rules:

$$listrec^\perp(Nil;\perp_{Nat};x^D,y^L,z.f(x^D,y^L,z))) \to_1 inl(\perp_{Nat})$$
$$listrec^\perp(a.Nil;a;x^D,y^L,z.less(x^D,z))) \to_1 inr(a)$$
$$listrec^\perp(c.b.l;c;x^D,y^L,z.less(x^D,z))) \to_1 inr(less(c,listrec^\perp(b.l;b;x^D,y^L,z.less(x^D,z))))))$$

There are two requests for a list, one in the first subterm of the computation schema $MinL(l^L)$ and another in the first subterm of $is\_Nil(l^L)$ which itself is a subterm of $MinL(l^L)$. The second request is necessary to process the exceptional element. In other words, the implementation of the algorithm $MinL(l)$ requires an extra storage allocation for data needed to process the exception.

The above example shows a program specification of type A seen as a paradigm for satisfying a **request** for data of type A. Hence, specifications in CTT can be seen as **plugging instructions** [Gir89]. The termination of programs corresponds to memory **"crash-free"** program execution.

12 Conclusion

A top-level continuation style translation on classical types which is the computational content of Prawitz+ translation, defines programs as *lazy value transformers*. This can be contrasted with the view of programs as *eager continuation* transformers, obtained through double-negation translations [Mur90]. In other words, Prawitz+ translation corresponds to *lazy value passing style* translation (LVPS-translation) on programs which seems to be the true meaning of top-level continuation.

Through Prawitz+ translation, we could define a top-level continuation for existential type *only* implicitly. This is because Prawitz+ translation works *exactly* for Π_2^0 sentences. Let us recapitulate why. As it was shown in this paper, a "program" of existential type expects an input and it cannot be executed until the input is supplied. This is to say that existential quantification type is a type of program *continuations*, not of programs. This implies immediate extension to Π_2^0 sentences as types of actual programs.

When a top-level continuation is defined as a function applied to a store and a value which is a parallel of Kolmogorov translation on types, the extracted programs are not "purely" classical. This is because in order to make continuation transformers eager at the top level, non-classical operations are used. In particular, *constructive* pairing operation is different from *classical* pairing. As we have seen in the paper, the classical pairing requires that one of the two values is of ground type. That is, classical conjunction type "fixes" the order of complete evaluation either from left-to-right or right-to-left for pairs. The constructive pairing, on the other hand, is nondeterministic, i.e., two choices of top-level continuation are always possible [Mur90]. This implies that with Kolmogorov translation, the evaluation strategy has to be determined on the meta-level in order to make the entire program execution deterministic, i.e., classical.

In summary, we have seen how to extract *directly* computational evidence from classical proofs. The soundness of extraction is a direct consequence of LVPS-translation on types. The extraction procedure is not an extension of standard extraction procedure for constructive proofs [Mur90]. In other words, the language of classical proofs is not "guessed" and then verified through translation, but *derived* from Prawitz+ translation. Extracted programs may return values of structured types, i.e., pair, sums, and functions from values to values.

From a program specification perspective, CTT is a classically founded program development system - a total-correctness reasoning system for non-functional programs. In comparison with other work, e.g., [Mur90], we have extended the result for PA to theories of finite sequences of integers, booleans, trees, etc. (with endless possibilities).

First-order iteration does not yield all computable functions that might be constructed on natural numbers. This applies also to computation schemas associated with other first-order theories. Yet such constructions must count as a part of reasoning about numbers, lists, trees, etc. CTT can be extended with constructive predicates (and relations), i.e., terms of type $A \rightarrow data$, with $A \in data$ and $\rightarrow$ being second-order constructive implication. This would allow to introduce second-order control structures (i.e., data-type valued operations) associated with data-types. In the case of natural numbers, for example, we could extend the above results to Σ_2^0 sentences with this restricted reasoning.

References

[Bee85]   Beeson, M.J., *Foundations of Constructive Mathematics, Metamathematical Studies*, Springer-Verlag, 1985.

[Con85]   Constable, R. et al., The semantics of evidence, Tech. Rep. TR 85-684, Cornell University, 1985.

[Fri78]   Friedman, H., Classically and intuitionistically provable recursive functions, in *Higher Set Theory*, Scott, D. S. and Muller, G. H. (ed.), *Lecture Notes in Mathematics*, vol. 699, Springer-Verlag, 21-28.

[GLT89]   Girard, J.Y., Lafont, Y., Taylor, P., *Proofs and Types*, Cambridge University Press, 1989.

[Gir70]   Girard, J.Y., Une extension de l'interprétation de Gödel á l'analyse, et son application á l'élimination des coupures dans l'analyse et la théorie des types, *Proceedings of the Second Scandinavian Logic Symposium*, Ed. J.E. Fenstad, North Holland, 1970.

[Gir89]   Girard, J.Y., Towards Geometry of Interaction, *Contemporary Mathematics* **92**, 1989.

[ML82]   Martin-Löf, P., Constructive Mathematics and Computer Programming, *Proc. Sixth International Congress for Logic, Methodology and Philosophy of Science*, North-Holland, Amsterdam, 1982, 153-175.

[ML84]   Martin-Löf, P., *Intuitionistic type theories*, Bibliopolis, (Napoli, 1984).

[Mur90]   Murthy, C., Extracting Constructive Content from Classical Proofs, Ph.D. thesis, Cornell University, 1990.

[Pra65]   Prawitz, D., *Natural Deduction*, Almqvist and Wiksell, Stockholm, 1965.

[Rey74]   Reynolds, J. C., Towards theory of type structure, *Paris Colloquium on Programming*, LNCS **19**, Springer-Verlag (1974), 408-425.

Axiomatization of Calculus of Constructions

Yong SUN[1][2]
Department of Computer Science, University of York,
Heslington, York YO1 5DD, UK

Abstract

This paper is a companion of [S89, S90b] and it demonstrates that FBO is powerful enough to axiomatize CC-like calculi [HHP87, CH88] (where FBO is short for a Framework for Binding Operators [S89, S90b], and CC stands for the Calculus of Constructions).

More specifically, we introduce an extra universe **kind** *above the original universe* **type** *in CC. But we allow neither quantification over* **kind** *nor introduction of y :* **kind**. *Another operator $\Rightarrow$ is added for typing Π types in FBO, since we are only considering a single-sorted FBO. Also, every CC term is typed in FBO. For example, $\Pi x : M.N$ in CC is translated as $\Pi y.u : t \Rightarrow v$ in FBO, where x, M and N correspond to y, t and u respectively under the translation.*

Hence, as a result of our axiomatization of CC in FBO, we know that a countably infinite hierarchy of type universes for CC may be not necessary.

1   Introduction

The Calculus of Constructions (CC) [CH88] is a higher-typed formalism for constructive proofs in natural deduction style [P65], which extends Intuitionistic Higher Order Logic and Girard-Reynold's second-order Lambda Calculus [G70, R74]. For characterizations of the expressiveness of the limited CC, i.e. second-order Lambda Calculus, we can consult [FLO83]. Nevertheless, CC is a result from many years' research in constructive mathematics by many mathematicians [B80, C86, CW87, H80, ML84, GL88].

A Framework for Binding Operators (FBO) [S89, S90b] is a higher-ordered algebraic framework, which extends the framework of classical first order universal algebra(s) [BS81] with arbitrary finite binding primitives. It is known that FBO provides a unified framework for First Order Logic, Lambda Calculus and Combinatory Logic, Calculus of Communicating Systems, i.e. they can be axiomatized in this framework [S90b].

In this paper, we show that CC can also be axiomatized in FBO. This further demonstrates that equational logics can provide a unified basis for computer science. On the other hand, equational logics

---

[1]E-mail: yong@uk.ac.york.minster; fax: +44-904-432-767; telephone: +44-904-432-762.

[2]The author would like to thank M. O'Donnell and the other attendants of the Symposium on Constructivity in Computer Science (Trinity University, San Antonio, Texas, USA, June 19-22, 1991) for their comments; to thank M. Atkins and A. Dix for proof-readings and helpful comments on the presentation of this paper; to thank the Department of Computer Science, University of York (UK), for providing a part of travel expense to attend the Symposium.

can also be used for implementation, say as programming languages [O85].

For the reason of limited space, we are going to present the main idea of the axiomatization and leave more technical treatment to the Appendix.

In what follows, Section 2 gives a brief introduction to CC. An introduction to FBO is provided in Section 3. A set Ax^{cc} of axioms to axiomatize CC in FBO is described in Section 4. The proof that Ax^{cc} is a right and faithful set of axioms which axiomatize CC in FBO, and it is briefly drafted in Section 5. More details of the proof are given in the Appendix. Final comments are given in Section 6.

2   Calculus of Constructions (CC)

This calculus contains context Γ, term M, and conversion $\cong$ along with their judgements. For readability, we slightly modify the original notation in [CH88], e.g. instead of using natural numbers as variables, we use $x, y, z, ...,$ etc, and replace $*$ by $type$.

There are three kinds of judgements in CC:

(2.a) $\Gamma \vdash^{cc} M$ between a context and a term,

(2.b) $\Gamma \vdash^{cc} M : N$ between a context and two terms, and

(2.c) $\Gamma \vdash^{cc} M \cong N$ of conversions.

Suppose V is a set of variables and there is a linear order among them. Then, we let $T^{cc}(V)$ be the set of CC-terms defined in Definition 8.1.1 of the Appendix, and let M, N range over it.

For judgements of CC, let $x \in FV(M)$ mean x occurs free in M, $x \in FV(\Gamma)$ be the evident extension of $x \in FV(M)$ to a list of terms, and $[N/x]M$ be the result of substituting N for free x in M.

Definition 2.1 ($\Gamma \vdash^{cc} M$ and $\Gamma \vdash^{cc} M : N$): *For judgements in the forms of $\Gamma \vdash^{cc} M$ and $\Gamma \vdash^{cc} M : N$, we have the following:*

(1) (cc-type)

$$\vdash^{cc} type,$$

(2) (cc-type-intr)

$$\frac{\Gamma \vdash^{cc} M}{\Gamma, x : M \vdash^{cc} type},$$

where x is the least variable such that $x \notin FV(\Gamma)$ and $x \notin FV(M)$;

(3) (cc-context)

$$\frac{\Gamma, x : M, \Gamma' \vdash^{cc} type}{\Gamma, x : M, \Gamma' \vdash^{cc} x : M},$$

(4) (cc-opn-intr)

$$\frac{\Gamma \vdash^{cc} M : type;}{\Gamma, x : M \vdash^{cc} type},$$

where x is the least variable such that $x \notin FV(\Gamma)$ and $x \notin FV(M)$;

(5) (cc-Π-kind)
$$\frac{\Gamma, x : M \vdash^{cc} N}{\Gamma \vdash^{cc} \Pi x : M.N} \text{ ,}$$

(6) (cc-Π-type)
$$\frac{\Gamma, x : M \vdash^{cc} N : type}{\Gamma \vdash^{cc} \Pi x : M.N : type} \text{ ,}$$

(7) (cc-λ-Π)
$$\frac{\Gamma, x : M \vdash^{cc} N' : N}{\Gamma \vdash^{cc} \lambda x : M.N' : \Pi x : M.N} \text{ ,}$$

(8) (cc-app-β)
$$\frac{\Gamma \vdash^{cc} M : \Pi x : M'.N'; \Gamma \vdash^{cc} N : M'}{\Gamma \vdash^{cc} (MN) : [N/x]N'} \text{ ,}$$

(9) (cc-lift-opn)
$$\frac{\Gamma \vdash^{cc} M : N; \Gamma \vdash^{cc} N : L}{\Gamma \vdash^{cc} M : L} \text{ .}$$

Furthermore, there is judgement for conversions in CC of the form of $\Gamma \vdash^{cc} M \cong N$, which is given in Definition 8.1.2 of the Appendix.

So far, in this section we have presented three kinds of judgement in CC; i.e. those of the forms (2.a), (2.b) and (2.c). Before going on to present FBO, we should point out that rule (9) (which does not appear in [CH88] and the readers are advised to see the Added Comments in the Appendix) is used to replace rules (9a) and (9b) below in [CH88]:

(9a)
$$\frac{\Gamma \vdash^{cc} N : L}{\Gamma \vdash^{cc} N \cong L} \text{ ,}$$

(9b) (cc-type-conv)
$$\frac{\Gamma \vdash^{cc} M : N; \Gamma \vdash^{cc} N \cong L}{\Gamma \vdash^{cc} M : L} \text{ .}$$

In other words, (9) is derivable by (9a) and (9b). The advantage of this replacement is that it enables us to separate the judgement systems for $\Gamma \vdash^{cc} M : N$ and $\Gamma \vdash^{cc} M \cong N$.

This separation is crucial. And later after obtaining the axiomatization of this simplified CC in FBO, as a consequence of this axiomatization, we have that rules (10) to (16) in Definition 8.1.2 are redundant and also that (9b) is redundant. For (9a), we can make it an extra axiom to be put into the axiomatization, i.e. into Ax^{cc} in Definition 4.1.

Another missing rule in Definition 8.1.2 is (cc-β) below, which can also be made into another extra axiom to be put into Ax^{cc}.

(17) (cc-β)
$$\frac{\Gamma, x : M \vdash^{cc} N : L; \Gamma \vdash^{cc} N' : M}{\Gamma \vdash^{cc} (\lambda x : M.NN') \cong [N'/x]N}$$

Then, after these adjustment we have the full derivation power of the original CC in [CH88] axiomatized in FBO (see Section 6).

So, from now on, we concentrate on this simplified CC given in Definition 2.1.

3   A Framework for Binding Operators (FBO)

This framework has signature Σ, binding term p, and equality $\simeq$. Since arbitrary finite binding primitives are available, this FBO is of second order. For the purpose of this paper, a single-sorted FBO is sufficient, in which an m-ary binding primitive is written as $\langle \underbrace{\_, \_, ..., \_}_{m \ times} : \_ \rangle$, and it indicates that the ordinary variables (the first m "holes") are bound in the term (the last "hole").

For a single-sorted FBO, the indices of signature Σ are elements in $Nat^* \times Nat$ where Nat is the set of natural numbers and Nat^* is the set of all arbitrary finite length sequences of natural numbers. Suppose X is a set of ordinary variables with a linear order among them, and $\vec{F}$ is a family of sets F_m of function variables with arity $m \in Nat$. We have binding terms $BT(X, \vec{F})$ as a pair of $< T(X, \vec{F}), FT(X, \vec{F}) >$ where $T(X, \vec{F})$ is the set of ordinary terms, and $FT(X, \vec{F})$ is the family of sets $FT_m(X, \vec{F})$ of function terms with arity $m \in Nat$. They are formally defined in Definition 8.2.1 of the Appendix.

For example, the binding operators in signature Σ^{cc} for CC are as follows: **type** $\in \Sigma^{cc}_{<\epsilon,0>}$; $\Pi, \lambda \in \Sigma^{cc}_{<1,0>}$; and **app**, $\boxed{:} \in \Sigma^{cc}_{<\epsilon,2>}$; where ϵ is a special symbol for the empty list. The usual lambda term $\lambda x.t$ and application term tu are convenient abbreviations of binding terms $\lambda(\langle x : t \rangle)$ and **app**(t, u), respectively.

Basic axioms in FBO are *binding equations* of the form $p \simeq q$, where either $p, q \in T(X, \vec{F})$ or $p, q \in FT_m(X, \vec{F})$ for some $m \in Nat$. This can be further extended to include

(3.a) *dependent binding equations* of the form $\gamma \mapsto p \simeq q$, where γ is a set of binding equations,

(3.b) *quasi-dependent binding equations* of the form $\gamma \hookrightarrow p \simeq q$, and

(3.c) *universal binding equations* of the form $\{\gamma^{(i)} \hookrightarrow p^{(i)} \simeq q^{(i)} | i \in I\} \mapsto (\gamma \hookrightarrow p \simeq q)$, where I is a set indexing quasi-dependent binding equations. For convenience, we will use Ω to refer an universal binding equation.

We should mention that classical first order algebras [BS81], $\mathcal{P}\omega$-model of Lambda Calculus [P72, S76], and qualitative domains [G86] are special examples of (extensional) Binding Algebras (eBAs) (see comments after Definition 8.3.3 in the Appendix).

Let $p, q \in T(X, \vec{F})$ or $p, q \in FT_m(X, \vec{F})$ for some $m \in Nat$; $p \simeq q$ be a binding equation; **B** be an (extensional) binding algebra (see [S89, S90b] or Definition 8.3.3 in the Appendix); ρ be an ordinary variable assignment; $\vec{\phi}$ be a family of function variable assignments ϕ_m with arity $m \in Nat$; and γ be a set of binding equations.

Definition (satisfaction $\models$) 3.1: *The satisfaction relation $\models$ for universal binding equations as well*

as other forms is defined by:

(assignment) $(\mathbf{B}, < \rho, \vec{\phi} >) \models p \simeq q$ iff $B[\![p]\!](\rho, \vec{\phi}) = B[\![q]\!](\rho, \vec{\phi})$;

(binding equation) $\mathbf{B} \models p \simeq q$ iff $(\mathbf{B}, < \rho, \phi >) \models p \simeq q$ for every $< \rho, \vec{\phi} >$;

(dependent binding equation) $\mathbf{B} \models \gamma \mapsto p \simeq q$ iff $\mathbf{B} \models p' \simeq q'$ for each $p' \simeq q' \in \gamma$ implies $\mathbf{B} \models p \simeq q$;

(quasi-dependent binding equation) $\mathbf{B} \models \gamma \hookrightarrow p \simeq q$ iff $(\mathbf{B}, < \rho, \vec{\phi} >) \models p' \simeq q'$ for each $p' \simeq q' \in \gamma$
   implies $(\mathbf{B}, < \rho, \vec{\phi} >) \models p \simeq q$ for every $< \rho, \vec{\phi} >$;

(universal binding equations) $\mathbf{B} \models \Omega$ iff $\mathbf{B} \models \gamma^{(i)} \hookrightarrow p^{(i)} \simeq q^{(i)}$ for each $i \in I$ implies $\mathbf{B} \models \gamma \hookrightarrow p \simeq q$.

From the above definition of satisfaction, we should notice that when $I = \emptyset$ (empty set), Ω is a (pure) quasi-dependent binding equation; furthermore when $\gamma = \emptyset$, it is a (pure) binding equation; when all γ's in Ω are empty, Ω becomes a (pure) dependent equation; and when $I = \emptyset$, it is a (pure) binding equation.

Also, careful readers may discover that dependent binding equations and quasi-dependent binding equations correspond to "equational implications" and "conditional equations" (or "quasi-equations" or "universal Horn Clauses") in the literature respectively. The reason for choosing such terminology is to unify different terminology, and to indicate their relationships at a conceptual level [S90a, S91]. We will not follow this line further.

There are four calculi corresponding to binding equations, dependent binding equations, quasi-dependent binding equations and universal binding equations. Since a universal binding equation is a unified form for the other three forms, we introduce the calculus corresponding to this form in Definition 8.2.2 of the Appendix.

So, we can use universal binding equations to axiomatize CC in FBO. However, for simplicity we sometimes use

(3.i) $p \simeq q$ for $\emptyset \mapsto (\emptyset \hookrightarrow p \simeq q)$,

(3.ii) $\gamma \mapsto p \simeq q$ for $\{\emptyset \hookrightarrow p' \simeq q' | p' \simeq q' \in \gamma\} \mapsto (\emptyset \hookrightarrow p \simeq q)$, and

(3.iii) $\gamma \hookrightarrow p \simeq q$ for $\emptyset \mapsto (\gamma \hookrightarrow p \simeq q)$, respectively.

We can further define a consequence relation as well as models of given axioms as follows.

Definition 3.2 (models and consequences): Given a set Δ of axioms (universal binding equations), we have the following:

(model) a binding algebra $\mathbf{B}$ is a model of Δ iff $\mathbf{B} \models \Omega$ for each $\Omega \in \Delta$;

(consequence) Ω is a consequence of Δ, written as $\Delta \models \Omega$, iff $\mathbf{B} \models \Omega'$ for each $\Omega' \in \Delta$ implies $\mathbf{B} \models \Omega$ for every binding algebra $\mathbf{B}$.

Note that we adopt the common practice of "abusing" consequence relations with satisfaction relations for convenience.

Now, we can have a calculus $\vdash^{fbo}$ for universal binding equations given in Definition 8.2.2 of the Appendix, which is sound and complete. Details of the lengthy proofs for soundness and completeness are omitted from this paper, and they can be found in [S90b].

Let Δ be a set of universal binding equations; Ω range over universal binding equations; γ range over sets of binding equations; $p, q, r \in T(X, \vec{F})$ or $p, q, r \in FT_m(X, \vec{F})$ for some $m \in Nat$; $t, u \in T(X, \vec{F})$ and $ft, fu \in FT_k(X, \vec{F})$ for some $k \in Nat$; I be a set for indexing. Thus, we have the following.

Theorem 3.3 (soundness and completeness of $\vdash^{cc}$): *Calculus $\vdash^{fbo}$ defined in Definition 8.2.2 of the Appendix is sound and complete with respect to universal binding equations, i.e $\Delta \models \Omega$ iff $\Delta \vdash^{fbo} \Omega$.*

Actually, there are two substitution rules: one is restricted to the substitutions of function variables for semi-closed function terms (i.e. the function terms without free ordinary variables), and the other one is not. If we name these two calculi as $\vdash^{closed-fbo}$ and $\vdash^{adm-fbo}$ respectively, then we have two completeness results. The latter one is named as *admissible completeness* in [S90b]. For the sake of the present paper, these two calculi coincide since function variables are not under our consideration [S90b]. So, we use $\vdash^{fbo}$ instead.

Also, axiom (fbo-ξ^{-1}) is not needed in the present paper since there is no axiom associated with function terms appearing in Ax^{cc} of Definition 4.1.

4 Axiomatization of CC in FBO

Firstly, let us recall that the binding operators in signature Σ^{cc} contains $\textbf{type} \in \Sigma^{cc}_{<\epsilon,0>}$; $\boxed{:}$, $\textbf{app} \in \Sigma^{cc}_{<\epsilon,2>}$; and $\lambda, \Pi \in \Sigma^{cc}_{<1,0>}$. To present CC in FBO, we need to introduce three more operations, i.e. $\textbf{true}, \textbf{kind} \in \Sigma^{cc}_{<\epsilon,0>}$ and $\boxed{\Rightarrow} \in \Sigma^{cc}_{<\epsilon,2>}$.

Secondly, for convenience, we use infix notation instead of prefix notation, e.g. using $t : u$ instead of $\boxed{:}(t, u)$, using $t \Rightarrow u$ instead of $\boxed{\Rightarrow}(t, u)$, and using t instead of $t \simeq \textbf{true}$.

Definition 4.1 (Ax^{cc}): *Let t, u, v range over $T(X, \vec{F})$, $FV(t)$ mean the set of free ordinary variables (and free function variables) occurring in t, and γ be a set of binding equations. Thus, the set Ax^{cc} contains the following instances:*

(1) (fbo-kind)

$$\textbf{type} : \textbf{kind}$$

(2a) (fbo-type-intr)

$$\{\gamma \hookrightarrow t : \textbf{kind}\} \mapsto (\gamma \cup \{x : t\} \hookrightarrow \textbf{type} : \textbf{kind})$$

(2b) (fbo-imp)

$$\{\gamma \hookrightarrow v : t \Rightarrow u\} \mapsto (\gamma \cup \{v : t \Rightarrow u, y : v\} \hookrightarrow \textbf{type} : \textbf{kind}),$$

where $y \notin FV(\gamma) \cup FV(v) \cup FV(t) \cup FV(u)$;

(3) (fbo-type-opn)

$$\{\gamma \cup \{y : t\} \cup \gamma' \hookrightarrow \textbf{type} : \textbf{kind}\} \mapsto (\gamma \cup \{y : t\} \cup \gamma' \hookrightarrow y : t)$$

(4) (fbo-opn-intr)

$$\{\gamma \hookrightarrow t : \textbf{type}\} \mapsto (\gamma \cup \{y : t\} \hookrightarrow \textbf{type} : \textbf{kind}),$$

where $y \notin FV(\gamma) \cup FV(t)$;

(5a) (fbo-Π-kind)

$$\{\gamma \cup \{y : t\} \hookrightarrow u : \textbf{kind}\} \mapsto (\gamma \hookrightarrow \Pi y.u : t \Rightarrow u)$$

(5b) (fbo-Π-imp-kind)

$$\{\gamma \cup \{y : t\} \hookrightarrow u : v \Rightarrow v'\} \mapsto (\gamma \hookrightarrow \Pi y.u : t \Rightarrow (v \Rightarrow v'))$$

(6) (fbo-Π-type)

$$\{\gamma \cup \{y : t\} \hookrightarrow u : \textbf{type}\} \mapsto (\gamma \hookrightarrow \Pi y.u : t \Rightarrow \textbf{type})$$

(7) (fbo-λ-Π)

$$\{\gamma \cup \{y : t\} \hookrightarrow u' : u \mid u \neq \textbf{kind}\} \mapsto (\gamma \cup \{\Pi y.u : t \Rightarrow \textbf{type}\} \hookrightarrow \lambda y.u' : \Pi y.u)$$

(8) (fbo-app-β)

$$\{\gamma \hookrightarrow t : \Pi y.u', \gamma \hookrightarrow \Pi y.u' : t' \Rightarrow v, \gamma \hookrightarrow u : t'\} \mapsto (\gamma \cup \{u'[y := u] : v[y := u]\} \hookrightarrow \textbf{app}(t, u) : u'[y := u])$$

(9) (fbo-lift-opn)

$$\{\gamma \hookrightarrow t : u, \gamma \hookrightarrow u : v \mid v \neq \textbf{kind}\} \mapsto (\gamma \hookrightarrow t : v)$$

From Theorem 3.3, we know that calculus $\vdash^{fbo}$ is very powerful, e.g. (2a), (2b) and (4) in Ax^{cc} of Definition 4.1 are redundant. So, in order to capture the "true" derivability of $\vdash^{cc}$ in FBO. Let us limit the calculus to $\vdash^{fbo}_{cc}$, which only contains one rule (fbo-d-ctr) in Definition 8.2.2 of the Appendix with Ax^{cc}. We are going to show that this very limited calculus $\vdash^{fbo}_{cc}$ is equivalent to $\vdash^{cc}$.

5 Proof of the Axiomatization

Let $R^{cc} = \{1, 2, 3, 4, 5, 6, 7, 8, 9\}$ in which each number corresponds to rules in Definition 2.2.

A *proof* $\vec{P}$ in CC is a list of pairs $P_i = \; <r_i, P_i'>$, the ith element of list $\vec{P}$, such that $r_i \in R^{cc}$, and P_i' is a form of either $\Gamma \vdash^{cc} M : N$ or $\Gamma \vdash^{cc} M$ and there exists $P_j = \;<r_j, P_j'>$ and $j < i$ (and $P_{j'} = \;<r_{j'}, P_{j'}'>$ and $j' < i$) and P_i' is a result of applying rule (r_i) in Definition 2.2 to P_j' (and $P_{j'}'$).

Suppose Prf^{cc} is the set of proofs in CC, we say that $\Gamma \vdash^{cc} M : N$ ($\Gamma \vdash^{cc} M$) is *derivable in CC* iff there is a proof $\vec{P} \in Prf^{cc}$ such that the last element $P_{|\vec{P}|}$ of proof $\vec{P}$ is a pair $< r, \Gamma \vdash^{cc} M : N >$ ($< r, \Gamma \vdash^{cc} M >$) and $r \in R^{cc}$ (where $|\vec{s}|$ is the length of list $\vec{s}$).

From abstract syntax, we know that derivable judgments in CC have the following forms:

(5.1) for $\Gamma \vdash^{cc} M : N$, there are at least five patterns, i.e.

1. $\Gamma \vdash^{cc} x : M$,

2. $\Gamma \vdash^{cc} M : type$,

3. $\Gamma \vdash^{cc} \Pi x : M.N : type$,

4. $\Gamma \vdash^{cc} \lambda x : M.N : \Pi x : M.L$, and

5. $\Gamma \vdash^{cc} (MN) : L$.

(5.2) for $\Gamma \vdash^{cc} M$, there are only two possible patterns, i.e.

1. $\Gamma, x : M \vdash^{cc} type$, and

2. $\Gamma \vdash^{cc} \Pi x : M.N$.

Let $A^{fbo} = \{1, 2a, 2b, 3, 4, 5a, 5b, 6, 7, 8, 9\}$ in which each element corresponds to an axiom in Definition 4.1. Also, the numerical numbers between R^{cc} and A^{fbo} show their correspondences under translations.

A *proof* $\vec{Q}$ in the very restricted FBO $\vdash_{cc}^{fbo}$ is a list of pairs $Q_i = \;<a_i, Q_i'>$ such that $a_i \in A^{fbo}$, and Q_i' is a form of $Ax^{cc} \vdash_{cc}^{fbo} \gamma \hookrightarrow t : u$ and there exists $Q_j' = \;<a_j, Q_j'>$ and $j < i$ (and $Q_{j'} = \;<a_{j'}, Q_{j'}'>$ and $j' < i$) and Q_i' is a result of applying (fbo-d-ctr) to Q_j' (and $Q_{j'}'$) with the axiom a_j in Ax^{cc} of Definition 4.1.

Suppose $Prf^{fbo}(Ax^{cc})$ is the set of proofs in the very restricted FBO, we say that $Ax^{cc} \vdash_{cc}^{fbo} \gamma \hookrightarrow p \simeq q$ is *derivable in FBO with Ax^{cc}* iff there is a proof $\vec{Q}$ such that the last element $Q_{|\vec{Q}|}$ of proof $\vec{Q}$ is a pair $< a, Ax^{cc} \vdash \gamma \hookrightarrow p \simeq q >$ and $a \in A^{fbo}$, where $Ax^{cc} \vdash \gamma \hookrightarrow p \simeq q$ is short for $Ax^{cc} \vdash \emptyset \mapsto (\gamma \hookrightarrow p \simeq q)$.

Then, from abstract syntax, we know that there are at least seven possible patterns for derivable quasi-dependent binding equations:

(5.3)

    1. $\gamma \hookrightarrow y : t$, where $t \neq \mathbf{kind}$;

    2. $\gamma \hookrightarrow t : \mathbf{type}$;

    3. $\gamma \hookrightarrow \Pi y.u : t \Rightarrow \mathbf{type}$;

    4. $\gamma \cup \{\Pi y.v : t \Rightarrow \mathbf{type}\} \hookrightarrow \lambda y.u : \Pi y.v$, where $v \neq \mathbf{kind}$;

    5. $\gamma \hookrightarrow \mathbf{app}(t, u) : v$, where $v \neq \mathbf{kind}$;

(5.4)

    1. $\gamma \cup \{y : t\} \hookrightarrow \mathbf{type} : \mathbf{kind}$;

    2. $\gamma \hookrightarrow \Pi y.u : t \Rightarrow v$ and $u \neq \mathbf{kind}$.

Note that when $v = \mathbf{type}$, the third one in (5.3) is the same as the second one in (5.4).

Furthermore, for derivable $\gamma \hookrightarrow t : u$, we have the following property.

Property 5.1 ($\vdash_{cc}^{fbo}$):

(a) If $\gamma \hookrightarrow t : u$ is derivable, then $t \neq \mathbf{kind}$ and either $u \neq \mathbf{kind}$ or $t : u = \mathbf{type} : \mathbf{kind}$;

(b) suppose $\gamma \hookrightarrow t : u$ is derivable, and further

    (b.1) if $u = u' \Rightarrow v$ then $v \neq \mathbf{kind}$,

    (b.2) if $t = \Pi y.t'$ then $t' \neq \mathbf{kind}$ (and $u \neq \mathbf{kind}$),

    (b.3) if $u = \Pi y.u'$ then $\Pi y.u' : v \Rightarrow v' \in \gamma$ for some v and v', and

    (b.4) further if $t' : t'' \in \gamma$ then neither $t' = \mathbf{kind}$ nor $t'' = \mathbf{kind}$;

This property can be verified by induction on the lengths of proofs, and we leave the verification to the interested readers.

In order to achieve the equivalence between $\vdash^{cc}$ and $\vdash_{cc}^{fbo}$ (Theorem 5.3), we need to have two translations $[\bullet]$ and $[\bullet]^{-1}$ (Definition 8.4.1 and Definition 8.4.3). One is the reversed version of the other such that the following property holds.

Property 5.2 ($\vdash^{cc}$ and $\vdash_{cc}^{fbo}$):

1. For translation $[\bullet]$, let γ be $[\Gamma]$, t be $[M]$, u be $[N]$ and v be $[L]$ respectively, we have:

    a). if $\Gamma \vdash M$ is derivable then is $\gamma \hookrightarrow t : u$

        where $t : u = \mathbf{type} : \mathbf{kind}$ when $M = type$,

        and $M = \Pi x : N.L$ and $u = v \Rightarrow v'$ for some N, L, v and v' when $M \neq type$;

b). if $\Gamma \vdash M : type$ is derivable then is $\gamma \hookrightarrow t : \textbf{type}$;

c). if $\Gamma \vdash M : N$ is derivable then is $\gamma \hookrightarrow t : u$ and $u \neq \textbf{kind}$;

d). if $\Gamma \vdash M : \Pi x : N.L$ is derivable then is $\gamma \hookrightarrow t : \Pi y.v$ and $\Pi y.v : u \Rightarrow v' \in \gamma$;

e). if $\Gamma \vdash \Pi x : M.N : L$ is derivable then is $\gamma \hookrightarrow \Pi y.u : t \Rightarrow v$.

2. For translation $[\bullet]^{-1}$, let Γ be $[\gamma]^{-1}$, M be $[t]^{-1}$, N be $[u]^{-1}$ and L be $[v]^{-1}$ respectively, we have the reversed versions of a) to e) above.

Two translations with Property 5.2 are give in the Appendix (Definition 8.4.1 and Definition 8.4.3). So, from this we have the following:

Theorem 5.3 (equivalence between $\vdash^{cc}$ and $\vdash^{fbo}_{cc}$):

1. For $\vec{P} \in Prf^{cc}$, there is a $\vec{Q} \in Prf^{fbo}(Ax^{cc})$ such that $[\vec{P}] = \vec{Q}$;

2. Conversely, for $\vec{Q} \in Prf^{fbo}(Ax^{cc})$, there is a $\vec{P} \in Prf^{cc}$ and $\vec{P} = [\vec{Q}]^{-1}$.

3. Furthermore, if $V = X$ then we have $[\,[\vec{P}]\,]^{-1} = \vec{P}$ and $[\,[\vec{Q}]^{-1}\,] = \vec{Q}$.

6 Remarks

By Theorem 5.3, we know that $\vdash^{cc}$ in Definition 2.1 is equivalent to $\vdash^{fbo}_{cc}$. Now, we consider extending the equivalence to include the judgement of conversions in Definition 8.1.2. For this purpose, if we interprete $\cong$ as $\simeq$, then we know that

1. (10), (11), (12) are special cases of (iii), (iv) and (v) respectively;

2. (9b), (15) and (16) are special cases of (xii);

3. the translated version of (13) is derivable by either (fbo-Π-kind), or (fbo-Π-imp-kind) or (fbo-Π-type) with (xii) and (v);

4. the translated version of (14) is derivable by (fbo-λ-Π) with (xii) and (v).

Further along this line, we can introduce two possible axioms into Ax^{cc} to capture (9a) and (cc-β) in FBO, viz

(fbo-9a)

$$\{\gamma \hookrightarrow u : v \,|\, v \neq \textbf{kind}\} \mapsto (\gamma \hookrightarrow u : v),$$

(fbo-β)

$$\{\gamma \cup \{y : t\} \hookrightarrow u : v, \gamma \hookrightarrow u' : t\} \mapsto (\gamma \cup \{v[y := u'] : \textbf{type}\} \hookrightarrow \text{app}(\lambda y.u, u') \simeq u[y := u']).$$

Finally, the axiomatization presented in this paper is done by introducing an extra universe **kind** above the original universe *type* in CC (or **type** in FBO), over which we can not quantify. This is explicitly expressed by (5a) in Definition 4.1, in contrast with (6) for **type**. Thus, we eliminate the need for a countably infinite hierarchy of type universes.

Furthermore, we observe that our proof of the axiomatization can be generalized to axiomatize other logics as long as the following translation scheme from inference rules to universal Binding Equations is sound,

(inf-rule-schema)

$$\frac{\{\Gamma_i \vdash N_i | i \in I\}}{\Gamma \vdash M}$$

(univ-b-eq)

$$\{\gamma_i \hookrightarrow u_i \simeq \mathbf{true} | i \in I\} \mapsto \{\gamma \hookrightarrow t \simeq \mathbf{true}\}$$

where γ, γ_i, u_i, t are the results of the translation from Γ, Γ_i, N_i and M ($i \in I$) respectively.

7 References

[B80] N. G. De Bruijn, "A survey of the project Automath", in *To H.B. Curry: Essays on Combinatory Logic, Lambda Calculus and Formalism*, eds. J. P. Seldin and J.R. Hindley, Academic Press, 1980.

[BS81] S. Burris and H.P. Sankappanavar, "A course in universal algebra", GTM vol.78, Springer-Verlag, 1981.

[CW87] L. Cardelli and P. Wagner, "On understanding types, data abstraction and polymorphism", the ACM *computing survey*, 1987.

[C86] R.L.Constable et al, "Implementing Mathematics in the NuPrl System", Prentice-Hall, Englewood Cliff, N.J. 1986.

[CH88] Th. Coquand and G. Huet, "The calculus of constructions", Information and Computation, vol.76, No.2/3, pp.95-120, 1988.

[FLO83] S. Fortune, D. Leivant and M. O'Donnell, "The expressiveness of simple and second-order type structures", JACM vol.30, No.1, pp151-185, 1983.

[GL88] A. Avron, R. Harper, F. Honsell, I. Mason and G. Plotkin (eds.), "Workshop on General Logic – Edinburgh February 1987", LFCS report ECS-LFCS-88-52, University of Edinburgh, 1988.

[G70] J.Y.Girard, "Une extension de l'interpretation de Gödel à l'analyse et son application à l'elimination des coupures dans l'analyse et la théorie des types", Proceedings of the 2nd Scaninavian Logic Symposium, J. E. Fenstad (ed), pp.63-92, North-Holland, 1970.

[G86] J.Y. Girard, "System F of variable types, fifteen years later", TCS vol.45, pp.159-192, 1986.

[HHP87] R. Harper, F. Honsell and G. Plotkin, "A framework for defining logics", proceedings of IEEE 2nd Symposium on Logic in Computer Science, Ithaca, New York, USA, June 1987.

[H80] W.A. Howard, "The formulae-as-types notion of construction", in *To H.B. Curry: Essays on Combinatory Logic, Lambda Calculus and Formalism*, eds. J. P. Seldin and J.R. Hindley, Academic Press, 1980.

[ML84] "P. Martin-Löf, "Intuitionistic type theory", studies in proof theory, lecture notes, Bibliopolis, Naples, 1984.

[O85] M. O'Donnell, "Equational logic as a programming language", MIT press, 1985.

[P65] Dag Prawitz, "Natural deduction: a proof-theoretical study", Almquist & Wiksell, Stockholm, 1965.

[P72] G. D. Plotkin, "Application structures", technical report, Dept. of Artificial Intelligence, University of Edinburgh, 1972.

[R74] J. C. Reynolds, "Towards a theory of type structure", in "Programming Symposium, Paris", pp.408-425, LNCS vol.19, Springer-Verlag, 1974.

[S76] D. Scott, "Data types as lattices", SIAM Journal of Computing, vol.5, pp.522-587, 1976.

[S89] Y. Sun, "Equational characterization of binding", talk presented in European Typed Lambda Calculus workshop (Jumelage Meeting), Edinburgh, Sepetmber 1989 (LFCS report series, ECS-LFCS-89-94, 1989).

[S90a] Y. Sun, "Equational logics (Birkhoff method revisited)", proceedings of 2nd international workshop on Conditional and Typed Rewriting Systems (CTRS '90), Concordia University, Montreal Canada, June 1990.

[S90b] Y. Sun, "A framework for binding operators", Ph.D. thesis, Dept. of Computer Science, University of Edinburgh, forthcoming.

[S91] Y. Sun, "Equational Logics", to appear in the proceedings of the workshop on Static Analysis of Equational, Functional, and Logic Programs, Bordeaux, France, October 9-11, 1991.

8 Appendix

Added Comments

The author is very grateful to Jonathan P. Seldin, who points out the following to the author after completion of this paper, i.e. rule (9a) in Section 2 should be rule (9a') due to a misprint in the original paper (see the rule labelled by (∗) in [p.101, CH88]).

(9a')

$$\frac{\Gamma \vdash N : L}{\Gamma \vdash N \cong N}$$

Subsequently, rule (9) in Section 2 should be dropped out, and axiom (9) in Section 4 should also be dropped out. So, the simplified CC should only contain rules from (1) to (8) in Section 2, and the set A^{cc} of axioms in Section 4 should also only contain axioms from (1) to (8). Similarily, translations between CC and FBO defined in this appendix should be modified accordingly. After these modifications, Theorem 5.3 should be valid (note that the correspondences between rules in Section 2 and axioms in Section 4 are indicated by sharing a same numerical number). The extension to full CC, i.e. rule (9a') corresponds to axiom (9') and the others should be the same as previously commented in Section 6, should work as well.

(9')

$$\{\gamma \hookrightarrow u : v | v \neq \mathbf{kind}\} \mapsto (\gamma \hookrightarrow u \simeq u)$$

In fact, axiom (9') would be redundant because of (iii) and (vii) in Definition 8.2.2. Also interestingly, if rule (9) is added to the simplified CC, it may not destroy the Normalization of CC but CC may lose the uniqueness of typing by such an adding.

8.1 Details of Definitions for CC

Suppose V is a set of variables and there is a linear order among them.

Definition 8.1.1 (terms of CC): *The set $T^{cc}(V)$ of terms are defined as the least closure of the following inductive definitions:*

(type term)

$$type \in T^{cc}(V),$$

(variable term)

$$\frac{x \in V}{x \in T^{cc}(V)},$$

(Π-term)

$$\frac{M, N \in T^{cc}(V); \ x \in V}{\Pi x : M.N \in T^{cc}(V)},$$

(λ-term)

$$\frac{M, N \in T^{cc}(V); \ x \in V}{\lambda x : M.N \in T^{cc}(V)},$$

(application term)

$$\frac{M, N \in T^{cc}(V)}{(MN) \in T^{cc}(V)}.$$

For judgements of conversions in CC, we have the following.

Definition 8.1.2 ($\Gamma \vdash^{cc} M \cong N$): *For judgement in the form of $\Gamma \vdash^{cc} M \cong N$, we have the following:*

(10) (cc-id)

$$\frac{\Gamma \vdash^{cc} M}{\Gamma \vdash^{cc} M \cong M} \text{ , }$$

(11) (cc-sym)

$$\frac{\Gamma \vdash^{cc} M \cong N}{\Gamma \vdash^{cc} N \cong M} \text{ , }$$

(12) (cc-trs)

$$\frac{\Gamma \vdash^{cc} M \cong M'; \Gamma \vdash^{cc} M' \cong M''}{\Gamma \vdash^{cc} M \cong M'} \text{ , }$$

(13) (cc-Π-conv)

$$\frac{\Gamma \vdash^{cc} M \cong M'; \Gamma, x : M \vdash^{cc} N \cong N'}{\Gamma \vdash^{cc} \Pi x : M.N \cong \Pi x : M'.N'} \text{ , }$$

(14) (cc-λ-conv)

$$\frac{\Gamma \vdash^{cc} M \cong M'; \Gamma, x : M \vdash^{cc} N \cong N'; \Gamma, x : M \vdash^{cc} N : L}{\Gamma \vdash^{cc} \lambda x : M.N \cong \lambda x : M'.N'} \text{ , }$$

(15) (cc-app-fun-conv)

$$\frac{\Gamma \vdash^{cc} (M \ N) : L; \Gamma \vdash^{cc} M \cong M'}{\Gamma \vdash^{cc} (M \ N) \cong (M' N)} \text{ , }$$

(16) (cc-app-obj-conv)

$$\frac{\Gamma \vdash^{cc} (M N) : L; \Gamma \vdash^{cc} N \cong N'}{\Gamma \vdash^{cc} (M N) \cong (M N')} .$$

8.2 Details of Definitions for FBO

We have bindig terms $BT(X, \vec{F})$ as a pair of $< T(X, \vec{F}), FT(X, \vec{F}) >$ where $T(X, \vec{F})$ is the set of ordinary terms, and $FT(X, \vec{F})$ is the family of sets $FT_m(X, \vec{F})$ of function terms with arity $m \in Nat$. They are formally defined below.

Definition 8.2.1 (binding terms): $T(X, \vec{F})$ and $FT(X, \vec{F}) = \{FT_m(X, \vec{F}) | m \in Nat\}$ *are inductively defined as follows:*

(ordinary variable)

$$\frac{x \in X}{x \in T(X, \vec{F})} \text{ , }$$

(function variables)

$$\frac{f \in F_m; \ t_j \in T(X, \vec{F})}{f(\vec{t}) \in T(X, \vec{F})} \text{ , }$$

where $\bullet_j$ is the jth element in list $\vec{\bullet}$;

(binding primitive)

$$\frac{x_j \in X; \ x_i \neq x_j; \ t \in T(X, \vec{F})}{(\vec{x} : t) \in FT_m(X, \vec{F})}$$

(binding operator)

$$\frac{ft_i \in FT_{m_i}(X, \vec{F});\ u_j \in T(X, \vec{F})}{\sigma(\vec{ft}, \vec{u}) \in T(X, \vec{F})}$$

where $\sigma \in \Sigma_{<\vec{m}, n>}$, $|\vec{ft}| = |\vec{m}|$ and $|\vec{u}| = n$.

Let Δ be a set of universal binding equations; Ω range over universal binding equations; γ range over sets of binding equations; $p, q, r \in T(X, \vec{F})$ or $p, q, r \in FT_m(X, \vec{F})$ for some $m \in Nat$; $t, u \in T(X, \vec{X})$ and $ft, fu \in FT_k(X, \vec{F})$ for some $k \in Nat$; I be a set for indexing. We can define the calculus $\vdash^{fbo}$ below.

Definition 8.2.2 (calculus $\vdash^{fbo}$): *Calculus $\vdash^{fbo}$ contains the following axioms and rules:*

(i) (fbo-idn)

$$\{\Omega\} \vdash^{fbo} \Omega$$

(ii) (fbo-wkn)

$$\frac{\Delta \vdash^{fbo} \Omega}{\Delta \cup \Delta' \vdash^{fbo} \Omega}$$

(iii) (fbo-refl)

$$\vdash^{fbo} p \simeq p$$

(iv) (fbo-sym)

$$\vdash^{fbo} \{p \simeq q\} \hookrightarrow q \simeq p$$

(v) (fbo-trs)

$$\vdash^{fbo} \{p \simeq q, q \simeq r\} \hookrightarrow p \simeq r$$

(vi) (fbo-d-ctr)

$$\frac{\Delta \vdash^{fbo} \Omega';\ \{\Delta \vdash^{fbo} \Omega^{(j)} | j \in J\}}{\Delta \vdash^{fbo} \Omega}$$

where $I \cap J = \emptyset$,

$\Omega = \{\gamma^{(i)} \hookrightarrow p^{(i)} \simeq q^{(i)} | i \in I\} \mapsto (\gamma \hookrightarrow p \simeq q)$,

$\Omega^{(j)} = \{\gamma^{(i)} \hookrightarrow p^{(i)} \simeq q^{(i)} | i \in I\} \mapsto (\gamma^{(j)} \hookrightarrow p^{(j)} \simeq q^{(j)})$ for $j \in J$, and

$\Omega' = \{\gamma^{(j)} \hookrightarrow p^{(j)} \simeq q^{(j)} | j \in J\} \mapsto (\gamma \hookrightarrow p \simeq q)$;

(vii) (fbo-q-ctr)

$$\frac{\Delta \vdash^{fbo} \Omega';\ \{\Delta \vdash^{fbo} \Omega^{(p' \simeq q')} | p' \simeq q' \in \gamma'\}}{\Delta \vdash^{fbo} \Omega}$$

where

$\Omega = \{\gamma^{(i)} \hookrightarrow p^{(i)} \simeq q^{(i)} | i \in I\} \mapsto (\gamma \hookrightarrow p \simeq q)$,

$\Omega^{(p' \simeq q')} = \{\gamma^{(i)} \hookrightarrow p^{(i)} \simeq q^{(i)} | i \in I\} \mapsto (\gamma \hookrightarrow p' \simeq q')$ for $p' \simeq q' \in \gamma'$, and

$\Omega' = \{\gamma^{(i)} \hookrightarrow p^{(i)} \simeq q^{(i)} | i \in I\} \mapsto (\gamma' \hookrightarrow p \simeq q)$;

(viii) (fbo-α)

$$\vdash^{fbo} \langle \vec{x} : t \rangle \simeq \langle \vec{y} : t[\vec{x} := \vec{y}] \rangle$$

(ix) (fbo-ξ⁻¹)

$$\vdash^{fbo} \{ \langle \vec{z} : t \rangle \simeq \langle \vec{y} : u \rangle \} \hookrightarrow t[\vec{z} := \vec{x}] \simeq u[\vec{y} := \vec{x}]$$

where $\{\vec{x}\} \cap FV(\langle \vec{z} : t \rangle) \cup FV(\langle \vec{y} : u \rangle) = \emptyset$

(x) (fbo-ξ)

$$\vdash^{fbo} \{ t \simeq u \} \mapsto \langle \vec{x} : t \rangle \simeq \langle \vec{x} : u \rangle$$

(xi) (fbo-cmp-1)

$$\vdash^{fbo} \{ t_j \simeq u_j | 1 \le j \le m \} \hookrightarrow f(\vec{t}) \simeq f(\vec{u})$$

where $f \in F_m$

(xii) (fbo-cmp-2)

$$\vdash^{fbo} \{ ft_i \simeq fu_i, t_j \simeq u_j | 1 \le i \le |\vec{m}|, 1 \le j \le n \} \hookrightarrow \sigma(\vec{f}t, \vec{t}) \simeq \sigma(\vec{f}u, \vec{u})$$

where $\sigma \in \Sigma_{<\vec{m},n>}$

(xiii) (fbo-eq-d)

$$\frac{\Delta \cup \gamma \vdash^{fbo} p \simeq q}{\Delta \vdash^{fbo} \gamma \mapsto p \simeq q}$$

(xiv) (fbo-q-d)

$$\frac{\Delta \vdash^{fbo} \gamma \hookrightarrow p \simeq q}{\Delta \vdash^{fbo} \gamma \mapsto p \simeq q}$$

(xv) (fbo-skm)

$$\frac{\Delta \vdash^{fbo} \{ p'[\vec{x} := \vec{c}] \simeq q'[\vec{x} := \vec{c}] | p' \simeq q' \in \gamma \} \mapsto p[\vec{x} := \vec{c}] \simeq q[\vec{x} := \vec{c}]}{\Delta \vdash^{fbo} \gamma \hookrightarrow p \simeq q}$$

where $\{\vec{x}\} \supseteq FV(p) \cup FV(q) \cup (\bigcup\{FV(p') \cup FV(q') | p' \simeq q' \in \gamma\})$ and $\vec{c}$ is a list of fresh constants
(not available in signature Σ);

(xvi) (fbo-q-u)

$$\frac{\Delta \cup \{ \gamma^{(i)} \hookrightarrow p^{(i)} \simeq q^{(i)} | i \in I \} \vdash^{fbo} \gamma \hookrightarrow p \simeq q}{\Delta \vdash^{fbo} \{ \gamma^{(i)} \hookrightarrow p^{(i)} \simeq q^{(i)} | i \in I \} \mapsto (\gamma \hookrightarrow p \simeq q)}$$

(xvii) (fbo-sub)

$$\frac{\Delta \vdash^{fbo} \{ \gamma^{(i)} \hookrightarrow p^{(i)} \simeq q^{(i)} | i \in I \} \mapsto (\gamma \hookrightarrow p \simeq q)}{\Delta \vdash^{fbo} \{ \gamma^{(i)} \hookrightarrow p^{(i)} \simeq q^{(i)} | i \in I \} \mapsto (\gamma\psi \hookrightarrow p\psi \simeq q\psi)}$$

where ψ is a substitution map;

(xviii) (fbo-cut)

$$\frac{\Delta' \vdash^{fbo} \Omega; \{ \Delta \vdash^{fbo} \Omega' | \Omega' \in \Delta' \}}{\Delta \vdash^{fbo} \Omega}$$

Note that rule (xv) can be viewed as a kind of "skolemization" technique in logic.

8.3 (Extensional) Binding Algebras (eBAs)

To introduce (extensional) binding alegebras (see Definition 8.3.3), we proceed as follows. Let B be a set, $\mathcal{F}_m$ be a subset of function space $B^m \rightarrow B$ for $m \in Nat$, and $\mathcal{F}$ be a pair $< B, \{\mathcal{F}_m | m \in Nat\} >$.

Definition 8.3.1 (explict closedness): $\mathcal{F}$ *is said to be* explictly closed *iff*

(constant functions) for each $m \in Nat$ and every $a \in B$, there is a unique function $C_{m,a} \in \mathcal{F}_m$ such that
$$C_{m,a}(\vec{b}) = a \text{ for } \vec{b} \in B^m;$$

(projection functions) for each $m \in Nat$ and $m \neq 0$, there is a unique function $\pi_{m,i} \in \mathcal{F}_m$ for every
$1 \leq i \leq m$ *such that $\pi(\vec{a}) = a_i$ for $\vec{a} \in B^m$;*

(compositions) for each $m \in Nat$ and every $k \in Nat$, given $g \in \mathcal{F}_m$ and $h_i \in \mathcal{F}_k$, there is a unique
function $h' \in \mathcal{F}_k$ such that $h'(\vec{a}) = g(h_1(\vec{a}), h_2(\vec{a}), ..., h_m(\vec{a}))$ for $\vec{a} \in B^k$ or $h' = g o < \vec{h} >$; sometimes,
it can be further abbreviated as $g(\vec{h})$ leaving k being decided by context;

For any binding operator $\sigma \in \Sigma_{<\vec{m},n>}$ $(\ell = |\vec{m}|)$, an interpretation B_σ of σ in $\mathcal{F} =< B, \{\mathcal{F}_m | m \in Nat\} >$ is a functional $B_\sigma : \mathcal{F}_{m_1} \times \mathcal{F}_{m_2} \times ... \times \mathcal{F}_{m_\ell} \times B^n \rightarrow B$ (or $\mathcal{F}_{\vec{m}} \times B^n \rightarrow B$).

Definition 8.3.2 (uniformity): *We say that functional B_σ is* uniform over $\mathcal{F}$ *iff for each $k \in Nat$,*
given $g_i \in \mathcal{F}_{k+m_i}$ and $h_j \in \mathcal{F}_k$, there is a unique function $h' \in \mathcal{F}_k$ such that for $\vec{a} \in B^k$, $h'(\vec{a}) = B_\sigma(\vec{g}', \vec{b})$
where $b_j = h_j(\vec{a})$ and $g_i'(\vec{c^i}) = g_i(\vec{a}, \vec{c^i})$ for $\vec{c^i} \in B^{m_i}$.

Now, we are able to define an extensional *binding algebra* $\mathbf{B}$ as follows.

Definition 8.3.3 (eBA): *A* binding algebra $\mathbf{B}$ *is a pair of $< \mathcal{F}, \mathcal{B} >$ where*

(explicit closedness) $\mathcal{F} =< B, \{\mathcal{F}_m | m \in Nat\} >$ is explicitly closed; and

(uniformity) $\mathcal{B} = \{B_\sigma | \sigma \in \Sigma\}$ and for each $\sigma \in \Sigma_{<\vec{m},n>}$, functional B_σ is uniform over $\mathcal{F}$.

For example, classical first order algebras are special cases of binding algebras. More specifically, let $\mathbf{A}$ $= < A, \mathcal{A} >$ be a classical first order algebra with signature $\Sigma^1 = \bigcup_{n \in Nat} \Sigma_n^1$, where $\sigma \in \Sigma_n^1$ means that σ is an operation with arity n; $\mathbf{B} =< B, \mathcal{F} >$ along with $\mathcal{B}$ be a binding algebra with signature $\Sigma^2 = \bigcup \Sigma_{<\vec{m},n>}^2$ for $< \vec{m}, n >\in Nat^* \times Nat$, where

(8.3.a) the signature $\Sigma_{<\epsilon,n>}^2 = \Sigma_n^1$ and $\Sigma_{<\vec{m},n>}^2 = \emptyset$ if $\vec{m} \neq \epsilon$,

(8.3.b) carrier $B = A$ and function carrier $\mathcal{F}_m$ is the full function space from carrier A^m to carrier A with $m \in Nat$,

(8.3.c) $B_\sigma = A_\sigma$ for $\sigma \in \Sigma^1$.

Obviously, B_σ is uniform over full function space $\mathcal{F}$.

Another example of (extensional) binding algebras is $\mathcal{P}\omega$-model of Lambda Calculus in [P72, S76]. To see this more clearly, let $\mathcal{F}_m$ be the continuous function space from m product of $\mathcal{P}\omega$ to $\mathcal{P}\omega$. Then, given

$k \in Nat$ we have that $[\![\lambda]\!] \circ < curry_{k,m}(g), h >$ and $[\![app]\!] \circ < h_1, h_2 >$ are continuous for each $g \in \mathcal{F}_{k+1}$ and every $h, h_1, h_2 \in \mathcal{F}_k$. This implies that $[\![\lambda]\!]$ and $[\![app]\!]$ are uniform over $\{\mathcal{F}_m | m \in Nat\}$.

A third example of (extensional) binding algebras is Girard's *qualitative domains* in F system [G86]. More specifically, let A be a qualitative domain, and $\mathcal{F}_m$ be the stable function space from A^m to A. Theorem 1.11 (i) and (ii) in [G86] guarantees that $[\![\lambda]\!]$ (i.e. (i)) and $[\![app]\!]$ (i.e. (ii)) are uniform over stable function spaces.

Let ρ be an ordinary variable assignment, $\vec{\phi}$ be a family of function variable assignment ϕ_m with arity $m \in Nat$. Then, for any $a \in A$, $\rho[a/x]$ is defined by

$$\rho[a/x](y) = \begin{cases} a & if\ y = x \\ \rho(y) & otherwise \end{cases}$$

For a list of distinct ordinary variables $\vec{x}$, we can define $\rho[\vec{a}/\vec{x}]$ as $\rho[a_1/x_1][a_2/x_2]...[a_k/x_k]$ where $\vec{a} \in B^k$ and $k = |\vec{x}| = |\vec{a}|$.

Definition 8.3.4 (interpretations): *Let* **B** *be a binding algebra, and* $< \rho, \vec{\phi} >$ *is a pair of an ordinary variable assignment and a family of function variable assignment. An interpretation* $\mathcal{B}$ *of binding terms* $BT(X, \vec{F})$ *over* **B** *is defined inductively by:*

(ordinary variable) $\mathcal{B}[\![x]\!](\rho, \vec{\phi}) = \rho(x)$ *for* $x \in X$;

(function variable) $\mathcal{B}[\![f(\vec{t})]\!](\rho, \vec{\phi}) = \phi_m(f)(\vec{a})$, *where* $f \in F_m$ *and* $a_j = \mathcal{B}[\![t_j]\!](\rho, \vec{\phi})$;

(function term) $\mathcal{B}[\![(\vec{x} : t)]\!](\rho, \vec{\phi}) = g$, *where* $g(\vec{a}) = \mathcal{B}[\![t]\!](\rho[\vec{a}/\vec{x}], \vec{\phi})$ *for* $\vec{a} \in B^k$ *and* $|\vec{x}| = k$;

(binding operator) $\mathcal{B}[\![\sigma(\vec{f}t, \vec{u})]\!](\rho, \vec{\phi}) = B_\sigma(\vec{g}, \vec{a})$, *where* $\sigma \in \Sigma_{<\vec{m}, n>}$, $g_i = \mathcal{B}[\![ft_i]\!](\rho, \vec{\phi})$ *with arity* m_i, $a_j = \mathcal{B}[\![u_j]\!](\rho, \vec{\phi})$, $|\vec{g}| = |\vec{m}|$ *and* $|\vec{a}| = n$.

More detailed treatment with regards to FBO and its applications can be found in [S89, S90b].

8.4 Translations between proofs in CC and proofs in FBO

Now, we are to define a (partial) translation $[\![\bullet]\!]$ from judgment forms in CC, to quasi-dependent binding equations in FBO. This translation is inductively defined on the lengths of judgements in CC. We will write $\vdash$ for both $\vdash^{cc}$ and $\vdash^{fbo}_{cc}$ for convenience. Possible confusions are resolved by context.

Definition 8.4.1 (translation $[\![\bullet]\!]$): *Suppose there is a linear order among variables in* X.

1.

$$[\![< 1; \vdash type >]\!] = < 1;\ Ax^{cc} \vdash type : kind >$$

2.

$$[\![< 1; \vdash type >< 2;\ x : type \vdash type >]\!] = [\![< 1; \vdash type >]\!] < 2a;\ Ax^{cc} \vdash \{y : type\} \hookrightarrow type : kind >$$

where y is the least $y \in X$;

3.

$$[\![\vec{P} < 2; \Gamma, x : M \vdash type >]\!] = \begin{cases} [\![\vec{P}]\!] < 2a; \ Ax^{cc} \vdash \gamma \cup \{y : \mathbf{type}\} \hookrightarrow \mathbf{type} : \mathbf{kind} > & (*) \\ [\![\vec{P}]\!] < 2b; \ Ax^{cc} \vdash \gamma \cup \{t : v \Rightarrow v', y : \mathbf{type}\} \hookrightarrow \mathbf{type} : \mathbf{kind} > & (**) \end{cases}$$

(where $(*)$ is $t : u = \mathbf{type} : \mathbf{kind}$ and $(**)$ is $u = v \Rightarrow v'$),

when γ, y, t, u, v and v' satisfy the following condition: for the least i such that $P_i = <r_i; \Gamma \vdash M >$,
we have

$$[\![\vec{P} \lceil_{i-1} < r_i; \Gamma \vdash M >]\!] = [\![\vec{P} \lceil_{i-1}]\!] < a_i; \ Ax^{cc} \vdash \gamma \hookrightarrow t : u >$$

(where $\vec{\bullet} \lceil_{i-1}$ is the prefix sub-list of list $\vec{\bullet}$ with length $i - 1$) and let y be the least variable in X such that $y \notin FV(\gamma)$ and $y \notin FV(t)$;

4.

$$[\![\vec{P} < 3; \Gamma, x : M, \Gamma' \vdash x : M >]\!] = [\![\vec{P}]\!] < 3; \ Ax^{cc} \vdash \gamma \cup \{y : t\} \cup \gamma' \hookrightarrow y : t >$$

when $u : v = \mathbf{type} : \mathbf{kind}$ where γ, y, t, γ' and v satisfy the following:

for the least i such that $P_i = <r_i; \Gamma, x : M, \Gamma' \vdash type >$, we have

$$[\![\vec{P} \lceil_{i-1} < r_i; \Gamma, x : M, \Gamma' \vdash type >]\!] = [\![\vec{P} \lceil_{i-1}]\!] < a_i; \ Ax^{cc} \vdash \gamma \cup \{y : t\} \cup \gamma' \hookrightarrow u : v >;$$

5.

$$[\![\vec{P} < 4; \Gamma, x : M \vdash type >]\!] = [\![\vec{P}]\!] < 4; \ Ax^{cc} \vdash \gamma \cup \{y : t\} \hookrightarrow \mathbf{type} : \mathbf{kind} >$$

when γ, y and t satisfy the following:

for the least i such that $P_i = <r_i; \Gamma \vdash M : type >$, we have

$$[\![\vec{P} \lceil_{i-1} < r_i; \Gamma \vdash M : type >]\!] = [\![\vec{P} \lceil_{i-1}]\!] < a_i; \ Ax^{cc} \vdash \gamma \hookrightarrow t : \mathbf{type} >$$

and let y be the least variable in X such that $y \notin FV(\gamma)$ and $y \notin FV(t)$;

6.

$$[\![\vec{P} < 5; \Gamma \vdash \Pi x : M.N >]\!] = \begin{cases} [\![\vec{P}]\!] < 5a; \ Ax^{cc} \vdash \gamma \hookrightarrow \Pi y.u : t \Rightarrow u > & (*) \\ [\![\vec{P}]\!] < 5b; \ Ax^{cc} \vdash \gamma \hookrightarrow \Pi y.u : t \Rightarrow (u' \Rightarrow v') > & (**) \end{cases}$$

(where $(*)$ is $u : v = \mathbf{type} : \mathbf{kind}$ and $v = u' \Rightarrow v'$),

when γ, y, u, t, v, u' and v' satisfy the following:

for the least i such that $P_i = <r_i; \Gamma, x : M \vdash N >$, we have

$$[\![\vec{P} \lceil_{i-1} < r_i; \Gamma, x : M \vdash N >]\!] = [\![\vec{P} \lceil_{i-1}]\!] < a_i; \ Ax^{cc} \vdash \gamma \cup \{y : t\} \hookrightarrow u : v >$$

7.

$$[\vec{P} < 6;\ \Gamma \vdash \Pi x : M.N : type >] = [\vec{P}] < 6;\ Ax^{cc} \vdash \gamma \hookrightarrow \Pi y.u : t \Rightarrow \textbf{type} >$$

when γ, y, u and t satisfy the following:

for the least i such that $P_i = < r_i;\ \Gamma, x : M \vdash N : type >$, we have

$$[\vec{P}\lceil_{i-1} < r_i;\ \Gamma, x : M \vdash N : type >] = [\vec{P}\lceil_{i-1}] < a_i;\ Ax^{cc} \vdash \gamma \cup \{y : t\} \hookrightarrow u : \textbf{type} >;$$

8.

$$[\vec{P} < 7;\ \Gamma \vdash \lambda x : M.N' : \Pi x : M.N >] = [\vec{P}] < 7;\ Ax^{cc} \vdash \gamma \cup \{\Pi y.u : t \Rightarrow \textbf{type}\} \hookrightarrow \lambda y.u' : \Pi y.u >$$

when γ, y, u, t and u' satisfy the following:

for the least i such that $P_i = < r_i;\ \Gamma, x : M \vdash N' : N >$, we have

$$[\vec{P}\lceil_{i-1} < r_i;\ \Gamma, x : M \vdash N' : N >] = [\vec{P}\lceil_{i-1}] < a_i;\ Ax^{cc} \vdash \gamma \cup \{y : t\} \hookrightarrow u' : u >$$

and $u \neq \textbf{kind}$;

9.

$$[\vec{P} < 8;\ \Gamma \vdash (MN) : N' >] = [\vec{P}] < 8;\ Ax^{cc} \vdash \gamma \cup \{u'[y := u] : v[y := u]\} \hookrightarrow \textbf{app}(tu) : u'[y := u] >$$

when γ, y, u, t, v and u' satisfy the following:

for the least i, j such that $P_i = < r_i;\ \Gamma \vdash M : \Pi x : M'.N' >$ and $P_j = < r_j;\ \Gamma \vdash N : M' >$, we have

$$[\vec{P}\lceil_{i-1} < r_i;\ \Gamma \vdash M : \Pi x : M'.N' >] = [\vec{P}\lceil_{i-1}] < a_i;\ Ax^{cc} \vdash \gamma \hookrightarrow t : \Pi y.u' >,$$

and

$$[\vec{P}\lceil_{j-1} < r_j;\ \Gamma \vdash N : M' >] = [\vec{P}\lceil_{j-1}] < a_j;\ Ax^{cc} \vdash \gamma \hookrightarrow u : t' >,$$

and $\Pi y.u' : t' \Rightarrow v \in \gamma$;

10.

$$[\vec{P} < 9;\ \Gamma \vdash M : L >] = [\vec{P}] < 9;\ Ax^{cc} \vdash \gamma \hookrightarrow t : v >$$

when γ, v, t satisfy the following:

for the least i, j such that $P_i = < r_i;\ \Gamma \vdash M : N >$ and $P_j = < r_j;\ \Gamma \vdash N : L >$, we have

$$[\vec{P}\lceil_{i-1} < r_i;\ \Gamma \vdash M : N >] = [\vec{P}\lceil_{i-1}] < a_i;\ Ax^{cc} \vdash \gamma \hookrightarrow t : u >,$$

and

$$[\vec{P}\lceil_{j-1} < r_j;\ \Gamma \vdash N : L >] = [\vec{P}\lceil_{j-1}] < a_j;\ Ax^{cc} \vdash \gamma \hookrightarrow u : v >,$$

and either $v \neq \textbf{kind}$ or $u : v = \textbf{type} : \textbf{kind}$.

We can further verify that this translation is a total function from proofs in CC to proofs in FBO. ormally,

Lemma 8.4.2: $\llbracket \bullet \rrbracket$ *is a total function from* Prf^{cc} *to* $Prf^{fbo}(Ax^{cc})$.

Now, we turn to give a definition for a (partial) function from quasi-dependent binding equations to adgements in CC, i.e. a reversed translation of $\llbracket \bullet \rrbracket$.

Definition 8.4.3 (translation $\llbracket \bullet \rrbracket^{-1}$): *Translation* $\llbracket \bullet \rrbracket^{-1}$ *from quasi-dependent binding equations to* adgements in CC is inductively defined by the following:

1.

$$\llbracket < 1;\ Ax^{cc} \vdash \textbf{type} : \textbf{kind} > \rrbracket^{-1} =< 1;\ \vdash type >$$

2.

$$\llbracket < 1;\ Ax^{cc} \vdash \textbf{type} : \textbf{kind} >< 2a;\ Ax^{cc} \vdash \{y : \textbf{type}\} \hookrightarrow \textbf{type} : \textbf{kind} > \rrbracket^{-1}$$

$$= \llbracket < 1;\ Ax^{cc} \vdash \textbf{type} : \textbf{kind} > \rrbracket^{-1} < 2;\ x : type \vdash type >$$

where x is the least $x \in V$;

3.

$$\llbracket \vec{Q} < 2a;\ Ax^{cc} \vdash \gamma \cup \{y : \textbf{type}\} \hookrightarrow \textbf{type} : \textbf{kind} > \rrbracket^{-1} = \llbracket \vec{Q} \rrbracket^{-1} < 2;\ \Gamma, x : M \vdash type >$$

when Γ, x and M satisfying the following:

for the least i such that $Q_i =< a_i;\ Ax^{cc} \vdash \gamma \hookrightarrow t : \textbf{kind} >$. we would have

$$\llbracket \vec{Q}\lceil_{i_1} < a_i;\ Ax^{cc} \vdash \gamma \hookrightarrow t : \textbf{kind} > \rrbracket^{-1} = \llbracket \vec{Q}\lceil_{i_1} \rrbracket^{-1} < r_i;\ \Gamma \vdash M >$$

and let x be the least $x \in V$ such that $x \notin FV(\Gamma) \cup FV(M)$;

4.

$$\llbracket \vec{Q} < 2b;\ Ax^{cc} \vdash \gamma \cup \{t : v \Rightarrow v', y : \textbf{type}\} \hookrightarrow \textbf{type} : \textbf{kind} > \rrbracket^{-1} = \llbracket \vec{Q} \rrbracket^{-1} < 2;\ \Gamma, x : M \vdash type >$$

when Γ, x, M satisfy the following condition: for the least i such that

$$Q_i =< a_i;\ Ax^{cc} \vdash \gamma \cup \{v : t \Rightarrow u, y : v\} \hookrightarrow \textbf{type} : \textbf{kind} >,$$

we have

$$\llbracket \vec{Q}\lceil_{i-1} < a_i;\ Ax^{cc} \vdash \gamma \cup \{v : t \Rightarrow u, y : v\} \hookrightarrow \textbf{type} : \textbf{kind} > \rrbracket^{-1} = \llbracket \vec{Q}\lceil_{i_1} \rrbracket^{-1} < r_i;\ \Gamma \vdash M >$$

and let x be the least variable in V such that $x \notin FV(\Gamma)$ and $x \notin FV(M)$;

5.

$$\llbracket \vec{Q} < 3;\ Ax^{cc} \vdash \gamma \cup \{y : t\} \cup \gamma' \hookrightarrow y : t > \rrbracket^{-1} = \llbracket \vec{Q} \rrbracket^{-1} < 3;\ \Gamma, x : M, \Gamma' \vdash x : M >$$

when $u : v = \textbf{type} : \textbf{kind}$ where Γ, x, M, and Γ' satisfy the following:

for the least i such that $Q_i =< a_i;\ Ax^{cc} \vdash \gamma \cup \{y : t\} \cup \gamma' \hookrightarrow u : v >$ we have

$$\llbracket \vec{Q}\lceil_{i-1} < a_i;\ Ax^{cc} \vdash \gamma \cup \{y : t\} \cup \gamma' \hookrightarrow u : v > \rrbracket^{-1} = \llbracket \vec{Q}\lceil_{i-1} \rrbracket^{-1} < r_i, \Gamma, x : M, \Gamma' \vdash type >$$

6.

$$[\vec{Q} < 4;\ Ax^{cc} \vdash \gamma \cup \{y : t\} \hookrightarrow \text{type} : \text{kind} >]^{-1} = [\vec{Q}]^{-1} < 4;\ \Gamma, x : M \vdash type >$$

when Γ, x and M satisfy the following: for the least i such that

$Q_i =< a_i;\ \Gamma \vdash M : type >$, we have

$$[\vec{Q}\lceil_{i-1} < a_i;\ Ax^{cc} \vdash \gamma \hookrightarrow t : \text{type} >]^{-1} = [\vec{Q}\lceil_{i-1}]^{-1} < r_i;\ \Gamma \vdash M : type >$$

and let x be the least variable in V such that $x \notin FV(\Gamma)$ and $x \notin FV(M)$;

7.

$$[\vec{Q} < 5a;\ Ax^{cc} \vdash \gamma \hookrightarrow \Pi y.u : t \Rightarrow u >]^{-1} = [\vec{Q}]^{-1} < 5;\ \Gamma \vdash \Pi x : M.N >$$

when Γ, x, N and M satisfy the following:

for the least i such that $Q_i =< a_i;\ Ax^{cc} \vdash \gamma \cup \{y : t\} \hookrightarrow u : v >$, we have

$$[\vec{Q}\lceil_{i-1} < a_i;\ Ax^{cc} \vdash \gamma \cup \{y : t\} \hookrightarrow u : v >]^{-1} = [\vec{P}\lceil_{i-1}]^{-1} < r_i;\ \Gamma, x : M \vdash N >$$

8.

$$[\vec{Q} < 5b;\ Ax^{cc} \vdash \gamma \hookrightarrow \Pi y.u : t \Rightarrow (u' \Rightarrow v') >] = [\vec{Q}]^{-1} < 5;\ \Gamma \vdash \Pi x : M.N >$$

when Γ, x, N, and M satisfy the following:

for the least i such that $Q_i =< a_i;\ Ax^{cc} \vdash \gamma \cup \{y : t\} \hookrightarrow u : v >$, we have

$$[\vec{Q}\lceil_{i-1} < a_i;\ Ax^{cc} \vdash \gamma \cup \{y : t\} \hookrightarrow u : v >]^{-1} = [\vec{P}\lceil_{i-1}]^{-1} < r_i;\ \Gamma, x : M \vdash N >$$

9.

$$[\vec{Q} < 6;\ Ax^{cc} \vdash \gamma \hookrightarrow \Pi y.u : t \Rightarrow \text{type} >]^{-1} = [\vec{Q}]^{-1} < 6;\ \Gamma \vdash \Pi x : M.N : type >]$$

when Γ, x, M and N satisfy the following:

for the least i such that $Q_i =< a_i;\ Ax^{cc} \vdash \gamma \cup \{y : t\} \hookrightarrow u : \text{type} >$, we have

$$[\vec{Q}\lceil_{i-1} < a_i;\ Ax^{cc} \vdash \gamma \cup \{y : t\} \hookrightarrow u : \text{type} >]^{-1} = [\vec{Q}\lceil_{i-1}]^{-1} < r_i;\ \Gamma, x : M \vdash N : type >]$$

10.

$$[\vec{Q} < 7;\ Ax^{cc} \vdash \gamma \cup \{\Pi y.u : t \Rightarrow \text{type}\} \hookrightarrow \lambda y.u' : \Pi y.u >]^{-1}$$

$$= [\vec{Q}]^{-1} < 7;\ \Gamma \vdash \lambda x : M.N' : \Pi x : M.N >$$

when Γ, x, N, M and N' satisfy the following:

for the least i such that $Q_i =< a_i;\ Ax^{cc} \vdash \gamma \cup \{y : t\} \hookrightarrow u : u >$, we have

$$[\vec{Q}\lceil_{i-1} < a_i;\ Ax^{cc} \vdash \gamma \cup \{y : t\} \hookrightarrow u' : u >]^{-1} = [\vec{Q}\lceil_{i-1}]^{-1} < r_i;\ \Gamma, x : M \vdash N' : N >]$$

and $u \neq \text{kind}$;

11.

$$[\vec{Q} < 8;\; Ax^{cc} \vdash \gamma \cup \{u'[y := u] : v[y := u]\} \hookrightarrow \mathbf{app}(tu) : u'[y := u] >]^{-1}$$

$$= [\vec{Q}]^{-1} < 8;\; \Gamma \vdash (MN) : N' >$$

when Γ, x, N, M, and N' satisfy the following:

for the least i, j such that $Q_i =< a_i;\; Ax^{cc} \vdash \gamma \hookrightarrow t : \Pi y.u' >$ and $Q_j =< a_j;\; Ax^{cc} \vdash \gamma \hookrightarrow u : t' >$, we have

$$[\vec{Q}\lceil_{i-1} < a_i;\; Ax^{cc} \vdash \gamma \hookrightarrow t : \Pi y.u' >]^{-1} = [\vec{Q}\lceil_{i-1}]^{-1} < r_i;\; \Gamma \vdash M : \Pi x : M'.N' >$$

and

$$[\vec{Q}\lceil_{j-1} < a_j;\; Ax^{cc} \vdash \gamma \hookrightarrow u : t' >]^{-1} = [\vec{Q}\lceil_{j-1}]^{-1} < r_j;\; \Gamma \vdash N : M' >$$

and $\Pi y.u' : t' \Rightarrow v \in \gamma$;

12.

$$[\vec{Q} < 9;\; Ax^{cc} \vdash \gamma \hookrightarrow t : v >]^{-1} = [[\vec{Q}]^{-1} < 9;\; \Gamma \vdash M : L >$$

when Γ, L, and M satisfy the following: for the least i, j such that $Q_i =< a_i;\; Ax^{cc} \vdash \gamma \hookrightarrow t : u >$ and

$Q_j =< a_j;\; Ax^{cc} \vdash \gamma \hookrightarrow u : v >$, we have

$$[\vec{Q}\lceil_{i-1} < a_i;\; Ax^{cc} \vdash \gamma \hookrightarrow t : u >]^{-1} = [\vec{Q}\lceil_{i-1}]^{-1} < r_i;\; \Gamma \vdash M : N >]$$

and

$$[\vec{Q}\lceil_{j-1} < a_j;\; Ax^{cc} \vdash \gamma \hookrightarrow u : v >]^{-1} = [\vec{P}\lceil_{j-1}] < r_j;\; \Gamma \vdash N : L >$$

and either $v \neq \mathbf{kind}$ or $u : v = \mathbf{type} : \mathbf{kind}$.

We can further verify that this translation is a total function from proofs in CC to proofs in FBO. Formally,

Lemma 8.4.4: $[\bullet]^{-1}$ is a total function from $Prf^{fbo}(Ax^{cc})$ to Prf^{cc}.

A Logical View of Assignments

Vipin Swarup[*]

The MITRE Corporation
Burlington Road
Bedford, MA 01730.
E-mail: swarup@mitre.org

Uday S. Reddy[†]

Dept. of Computer Science
University of Illinois
at Urbana-Champaign
Urbana, IL 61801.
E-mail: reddy@cs.uiuc.edu

Abstract

Imperative lambda calculus (ILC) is an abstract formal language obtained by extending the typed lambda calculus with imperative programming features, namely references and assignments. The language shares with typed lambda calculus important properties such as the Church-Rosser property and strong normalization. In this paper, we describe the logical symmetries that underlie ILC by exhibiting a constructive logic for which ILC forms the language of constructions. Central to this formulation is the view that references play a role similar to that of variables. References can be used to range over values and instantiated to specific values. Thus, we obtain a new form of universal quantification that uses references instead of variables. The essential term forms of ILC are then obtained as the constructions for the introduction and elimination of this quantifier. While references duplicate the role of variables, they also have important differences. References are *semantic* values whereas variables are syntactic entities and, secondly, references are *reusable*. These differences allow references to be used in a more flexible fashion leading to efficiency in constructions and algorithms.

1 Introduction

Reasoning about imperative programs has proven hard to formalize. While Hoare Logic [Hoa69] made an impressive beginning, it proved unsuitable for treating higher-order functions, pointer structures and a variety of other popular features. Reynolds extended Hoare Logic in a fundamental way in his formulation of Specification Logic [Rey82, Ten89], but the complexity of this system is deterring. Further, Specification Logic is unable to handle pointer structures. Other approaches, such as dynamic logic [Pra76] and algorithmic logics [Eng75, MS87], are close to Hoare logic and suffer from the same limitations.

In our recent work [SRI91], we attempted a new approach based on the insights obtained from applicative programming over the last two decades. Our formulation, called *Imperative Lambda Calculus* (ILC), extends typed lambda calculus

---

[*]Supported by NASA grant NAG-1-613 (while at the University of Illinois at Urbana-Champaign).

[†]Supported by a grant from Motorola Corporation.

with references and assignments, and uses a layered type system to prohibit all forms of side effects. We were able to show that, in spite of the radical extensions, the language retains the salient properties of typed applicative languages such as the Church-Rosser property and strong normalization.

In this paper, we report on the fundamental logical symmetries brought forth by this formulation of assignments. References play a role similar to that of variables in logic. They can be used to range over sets of values and instantiated to specific values. Thus, we obtain a new form of universal quantification that uses references instead of variables. The essential operators of imperative lambda calculus can be treated as the proof terms for the introduction and elimination rules of this quantifier. Thus, imperative lambda calculus can be regarded as the language of constructions for a suitably formulated constructive logic.

After a brief overview of imperative lambda calculus, we discuss the basic concepts involved in such a formulation in Section 3. In Section 4, we present a formal system called *observation type theory* which incorporates these concepts.

2 Overview of Imperative Lambda Calculus

Imperative lambda calculus (ILC) [SRI91] extends the simply typed lambda calculus with first-class references and assignments. The fundamental principle underlying ILC is that the only manipulation needed for states is *observation* (i.e. inspection). Consider the fact that in typed lambda calculus, environments are implicitly extended and observed (via the use of variables), but are never explicitly manipulated. Similarly, in ILC, states are implicitly extended and observed (via the use of references), but are never explicitly manipulated. Thus, in a sense, *the world exists only to be observed.*

The types of ILC are stratified into three layers as follows:

$$
\begin{array}{llll}
\text{(Applicative types)} & \tau & ::= & \beta \mid \tau_1 \times \tau_2 \mid \tau_1 \to \tau_2 \\
\text{(Storage types)} & \theta & ::= & \tau \mid \mathbf{Ref}\ \theta \mid \theta_1 \times \theta_2 \mid \theta_1 \to \theta_2 \\
\text{(Imperative types)} & \omega & ::= & \theta \mid \mathbf{Obs}\ \tau \mid \omega_1 \times \omega_2 \mid \omega_1 \to \omega_2
\end{array}
$$

All three layers are closed under product and function space constructions. *Applicative* types are those of typed lambda calculus. *Storage* types include reference types of the form $\mathbf{Ref}\ \theta$; such types contain references that range over values of type θ. *Imperative* types include observation types of the form $\mathbf{Obs}\ \tau$; such types contain terms (called *observers*) that observe the state and return values of applicative type τ.

The terms belonging to these types have the following syntax:

$$
\begin{array}{llll}
\text{Terms} & e & ::= & k \mid x \mid v^* \mid \lambda x.e \mid e_1(e_2) \mid \langle e_1, e_2 \rangle \mid e.1 \mid e.2 \\
& & & \mathtt{letref}\ v^* := e\ \mathtt{in}\ t \mid \mathtt{get}\ x \Leftarrow l\ \mathtt{in}\ t \mid l := e\ ;\ t
\end{array}
$$

where l, t also range over these terms, l ranging over reference-valued terms and t ranging over observer terms. Terms use two countable sets of variables: *conventional variables* and *reference variables*. Conventional variables x are the usual variables of the typed lambda calculus. Reference variables v^* are a new set of variables that have all the properties of conventional variables and, in addition, the property that distinct reference variables always denote distinct references in any term. This property permits us to reason about the equality of references without recourse to reference constants (which are absent from the language).

In addition to the conventional term forms of typed lambda calculus, there are three new forms which are used to construct observer terms. An observer of type Obs τ observes a state to yield a value of type τ. Trivially, any term of type τ is a term of type Obs τ. In addition:

- an observer (letref $v^* := e$ in t) allocates a new reference that is not used in the current environment, binds it to the variable v^*, initializes it to the value of e and evaluates the observer t in the extended environment and state.

- an observer (get $x \Leftarrow l$ in t) extends the environment by binding the value referred to by l in the state to the variable x, and evaluates the observer t in the extended environment.

- An observer ($l := e$; t) updates the state so that reference l refers to the value of e, and evaluates the observer t in the resultant state.

Note that "$l := e$" is not a term by itself (in contrast to Algol-like imperative languages). The modification of l is observable only within t, and there are no side effects produced by an assignment observer. The type system of ILC ensures that l and e are not state-dependent terms, thus ensuring that the state is used in a single-threaded fashion. The state can thus be implemented efficiently as a store.

The typing rules, denotational semantics, and reduction semantics of the language are presented in [SRI91, Swa91]. It is also proved that

- Well-typed terms are strongly normalizing, and

- Reduction has the Church-Rosser property.

The latter means that an outermost evaluation order can be used and so this form of assignment can be used with lazy functional languages like Haskell [HW90].

3 Logical Concepts

In constructive logic, propositions are not simply deemed to be true or false, but their provability issue is considered. For a proposition to be considered true, it must be provable. Conversely, a false proposition is one which has no proof. This view-point stands in opposition to classical logic and gives rise to a separate branch of mathematics, *viz.*, constructive mathematics. From a programming point of view, constructive propositions may be thought of as *specifications* of computations and their proofs as *programs* for the specifications [Con86]. A programming language that forms the language of constructions (proofs) for a constructive logic exhibits logical symmetries and supports reasoning about programs.

The foremost constructive logic is intuitionistic logic, and functional programming forms the language of constructions for it [Abr90, ML82]. That is, the salient constructs of functional programming are obtained as the proof terms for the introduction and elimination rules of intuitionistic logical operators. In a similar fashion, it is appropriate to ask whether imperative programming forms the language of constructions for some constructive logic. We answer this in the affirmative and exhibit a constructive logic called *Observation Type Theory* for which ILC forms the language of constructions. This theory is presented in Section 4. First, we examine the logical concepts underlying the theory.

Central to the theory is a certain correspondence between variables and references. Ever since von Neumann [GvN47], researchers have noted a degree of similarity between variables and references. In fact, in common imperative programming parlance, references are simply called variables. Though the notion of "variance" suggested here is different from that of variables in logic, it is nevertheless true that techniques applicable to variables are often applied to references. For instance, in Hoare logic, references are treated just as variables of logic.

In our formulation, references share many of the important properties of variables. These are that

- variables range over the values of a type,
- they allow quantification of propositions, *e.g.*, $(\forall x : \tau)P$,
- they allow quantified propositions to be specialized,
- they allow abstraction of terms to form "methods", *e.g.*, $\lambda x.t$, and
- they allow methods to be applied to values.

While references share all these properties, there are two important differences between variables and references. First, while variables are *lexical*, references are *semantic* values. This fact allows references to participate in constructions such as pairing (data structure building), functional abstraction (parameter passing) and observer abstraction (storing). Variables cannot participate in such semantic constructions. [1] Secondly, references are *reusable*. This permits data structures to be updated without copying.

3.1 Quantification of references

A proof of a universally quantified proposition $(\forall x : \tau)P$ is a *method* (function or rule) which, given a value x of type τ, yields a proof of P. The quantified proposition can be obtained by *generalization* from a judgement of the form $\Gamma, x : \tau \vdash P$. If ϕ is a proof of P, then $\lambda x.\, \phi$ is a proof method for $(\forall x : \tau)P$. Given any term $e : \tau$, the quantified proposition can be *specialized* to $P[e/x]$. The proof of this is $f(e)$ where f is any proof of $(\forall x : \tau)P$. Thus, function abstraction and application are based on the variables of logic.

In contrast, imperative programming is based on the idea of *references*. A reference is a kind of token that ranges over the values of a certain type. In this respect, references are similar to variables. But, whereas variables are lexical,

---

[1]Girard's Linear Logic [Gir87] exploits the duality between inputs (variables) and outputs (values). Thus, some effects of references can be obtained in this setting. Logic programming can also obtain some of these effects.

references are semantic values. They can be embedded in data structures, obtained as results of functions, or stored in other references. There are only two properties known about a reference: its identity and the type of values it ranges over. The latter is determined by the type of the reference itself: a reference of type $\mathtt{Ref}\ \theta$ ranges over values of type θ. But, the identity of a reference is a tricky issue, particularly because we don't wish to have reference constants. We assume that we have a separate class of variables called *reference variables* which have the property that any two distinct reference variables are always bound to distinct references. The reference variables are distinguished from ordinary variables in the formal treatment by an asterisk superscript, as in v^*.

References give us a new form of universal quantification written as

$$\mathtt{Get}\ x{:}\,\theta \Leftarrow l\ \mathtt{in}\ P$$

Here, quantification is over type θ and l must be a reference-valued term of type $\mathtt{Ref}\ \theta$. The quantified proposition is true iff, for all values v that l may refer to, $P[v/x]$ is true. Since l may refer to any value of type θ, this means the same as $(\forall x{:}\,\theta)P$ in classical (truth value) terms. The quantified proposition $\mathtt{Get}\ x{:}\,\theta \Leftarrow l\ \mathtt{in}\ P$ can be obtained by *generalization* from a judgement of the form $\Gamma, x{:}\,\theta \vdash P$. If ϕ is a proof of P, then the *observer method* $(\mathtt{get}\ x \Leftarrow l\ \mathtt{in}\ \phi)$ is a proof of $(\mathtt{Get}\ x{:}\,\theta \Leftarrow l\ \mathtt{in}\ P)$.

The role of the reference l in $(\mathtt{Get}\ x{:}\,\theta \Leftarrow l\ \mathtt{in}\ P)$ is roughly the same as that of the bound variable x in $(\forall x{:}\,\theta)P$. Both are used to range over the values of type θ. (The presence of the variable x in $\mathtt{Get}\ x{:}\,\theta \Leftarrow l\ \mathtt{in}\ P$ may be confusing in seeing this correspondence. Note that x plays a purely *local* role in this expression and the quantification is indeed over l. In contrast, the quantification in $(\forall x{:}\,\tau)P$ is over the variable x). The important difference between the two quantifications is that l may be a term whereas x must be a variable. As a special case, l may also be a reference variable of the form v^*.

A $\mathtt{Get}$-quantified proposition can be *specialized* in two ways. First, the reference l can be *assigned* the value of some term e. (This is similar to instantiating a variable, but since l is not a variable we use alternative terminology). If ϕ is a proof of $\mathtt{Get}\ x{:}\,\theta \Leftarrow l\ \mathtt{in}\ P$ then we can obtain $P[e/x]$ with proof $(l := e\ ;\ \phi)$. Note that the reference l is still retained. It is available for further generalizations and specializations. The crucial benefits of references are obtained from the fact that the *same* reference can be reused for many generalizations and specializations. In contrast, the variable x involved in $(\forall x{:}\,\tau)P$ is lost after a specialization.

The second method of $\mathtt{Get}$-specialization applies to the special case where l is a reference variable v^*. Here, we not only assign a value to v^*, but also discharge it so that it is not available for further generalizations. Given

$$\Gamma, v^*{:}\,\mathtt{Ref}\ \theta \vdash \mathtt{Get}\ x{:}\,\theta \Leftarrow v^*\ \mathtt{in}\ P$$

with evidence ϕ, we can derive

$$\Gamma \vdash P[e/x]$$

with proof $(\mathtt{letref}\ v^* := e\ \mathtt{in}\ \phi)$.

3.2 Linearity

The three constructions mentioned in the previous section, (get $x \Leftarrow l$ in ϕ), ($l := e$; ϕ), and (letref $v^* := e$ in ϕ) are called *observer terms*. They have the property that each has a single observer subterm. Thus, observers can only be composed linearly. This is an important property. If multiple observer subterms were allowed, separate copies of the state would be required for each observer (or, side effects must be tolerated).

To ensure that observer terms are linearly composed, we allow such terms to be proofs of only selected forms of propositions. Conventional propositions (of, say, intuitionistic logic) cannot have observer terms as proofs. Get-quantified propositions and, in addition, propositions of the form Obs P can have observer proofs. A proposition of the form Obs P means that a proof of P is observable in every state. Under the types-as-propositions correspondence, this is the same as the Obs type constructor of ILC. Further, Obs P is equivalent to every quantification of the form Get $x: \theta \Leftarrow l$ in P where x does not occur free in P. Thus, Get-quantification generalizes the Obs type constructor in much the same fashion as $\forall$ quantification generalizes the $\rightarrow$ type constructor of functional programs. [2]

3.3 Benefits of references

As noted above, references play essentially the same role as variables in conventional logic and Get-types play the same role as universal quantification. Why, then, do we need these new forms?

There are several reasons. First, the implications for storage reuse are already apparent. Whenever we use a variable x for generalization and instantiation via the $\forall$ quantifier, we are obliged to use a location for x that is different from the locations for all other variables. On the other hand, the same reference v^* can be used for multiple generalizations and instantiations, indicating to the language processor that the same storage location can be used in each case.[3]

Second, we are able to construct data structures with references. Given a data structure, we can reuse some of its references for new bindings and retain the old bindings of other references. This saves not only storage locations, but also the computation involved in binding. For example, let l be a pair of type Ref nat $\times$ Ref nat. Given a construction t of (Get $x:$ nat $\Leftarrow l.1$ in Get $y:$ nat $\Leftarrow l.2$ in $Q[x, y]$), we can form a construction for (Get $x:$ nat $\Leftarrow l.1$ in Get $y:$ nat $\Leftarrow l.2$ in $Q[x + 1, y]$) as (get $x \Leftarrow l.1$ in $l.1 := x + 1$; t). In contrast, if we consider the corresponding logical proposition $(\forall p:$ nat $\times$ nat$)Q[p.1, p.2]$ with a construction f, the construction for $(\forall p:$ nat $\times$ nat$)Q[p.1 + 1, p.2]$ would be the function $\lambda p. f(\langle p.1 + 1, p.2 \rangle)$. This involves consuming the input pair p and constructing a new pair $\langle p.1 + 1, p.2 \rangle$, whereas the construction for the Get-proposition involves only one binding and no construction of new data structures. Such saved effort can grow arbitrarily in large data structures. For instance, consider a specification that quantifies over all linked lists. The evidence for the specification would be

---

[2]The analogy with $\forall$ and $\rightarrow$ is not perfect. Obs cannot be *defined* to be a special case of Get because it is also needed in contexts where no references are available, but Get necessarily involves a reference.

[3]Compilers for functional languages use a variety of techniques to reuse storage locations allocated for function parameters, but it is doubtful if they can achieve all the storage reuse expressible in a language like ILC.

indifferent to the length of the linked list and its elements. In instantiating the evidence with a specific list, we can form the list from an existing linked list by extending it. In doing so, we save not only the recycling of storage, but also the *computation* of reconstructing the old portion of the linked list.

Third, references allow *sharing*. Consider a pair of the form $\langle v^*, v^* \rangle$. By instantiating v^*, we achieve the instantiation of *both* the elements of the pair. If such shared references are embedded deep in data structures, their shared instantiation can save arbitrary amounts of computation. For a concrete example, consider the unification problem. The specification is that for every pair of terms t and u, either there exists a most general common instance of t and u or there is no common instance. There would, in general, be multiple occurrences of a variable z in t. During unification, we have to ensure that all the occurrences z correspond to identical subterms of u. If term variables are modeled by references in the term representations, then mapping an occurrence of z to a subterm of u merely involves assigning the subterm to the reference representing z. This has the effect of constructing a new term t' with all the occurrences of z replaced by this subterm. Such effects cannot be achieved by variables of conventional logic.

In summary, variables of logic have certain properties that are essential to them. These are that variables range over values, propositions can be formed by quantifying over such variables, and quantified propositions can be instantiated with specific values. In the domain of constructions, one can abstract over variables to construct functions, and such functions can be applied to values to obtain specific constructions. All these properties are retained by references. They range over values. Propositions can be formed by quantifying over references (using Get), and such quantified propositions can be instantiated. In the domain of constructions, one can construct "observer methods" using get, in effect abstracting over references. Such methods can be applied to specific values using assignment or letref.

The essential difference between variables and references is that variables are syntactic entities whereas references are semantic values. Therefore, references can participate in semantic constructions like pairs and functions while variables cannot. This allows the various flexibilities outlined above.

3.4 State-assertions

Consider the difference between the following judgements:

$$\Gamma, x: \text{nat} \vdash P[x]$$
$$\Gamma \vdash (\forall x: \text{nat})P[x]$$

Both of them say that there is evidence for $P[x]$ for all natural numbers x (in the context of Γ). However, in the first judgement, the quantification over numbers is contained in the judgement whereas in the latter it is contained in the proposition itself. The former kind of judgements are unavoidable in logic because they form the raw material from which quantified propositions can be obtained by generalization. Formally, a judgement $\Gamma, x: \text{nat} \vdash P$ is valid iff there is a proof term ϕ such that, for all environments η satisfying $(\Gamma, x: \text{nat})$, $[\![\phi]\!]\eta$ is a proof of $[\![P]\!]\eta$. A judgement $\Gamma \vdash (\forall x: \text{nat})P$ is valid iff there is a proof method f such that, for all environments η satisfying Γ and all natural numbers i, $[\![f]\!]\eta i$ is a proof of $[\![P]\!]\eta[i/x]$.

When we consider observers and Get-propositions, judgements involve not only environments but also states. The latter map references to values. As usual, a judgement of the form $\Gamma \vdash Q$ is valid iff there is an observer proof method ϕ such that $[\![\phi]\!]\eta$ is a proof of $[\![Q]\!]\eta$, for all environments η satisfying Γ. However, $[\![\phi]\!]\eta$ as well as $[\![Q]\!]\eta$ are both functions of state. This means that the judgement is valid iff, for all environments η satisfying Γ and for all states σ, $[\![\phi]\!]\eta\sigma$ is a member of $[\![Q]\!]\eta\sigma$. Note that the states are not constrained by Γ. So, as for the proposition $(\forall x\colon \mathtt{nat})P$ above, the quantification over states is contained in the proposition rather than the judgement.

This suggests the need for judgements which involve quantification over states. Propositional expressions appearing in such judgements refer to propositions in specific states. We call such expressions *state-assertions* (*assertions* for short) and write them enclosed in braces as $\{Q\}$.[4] A judgement of the form $\Gamma \vdash \{Q\}$ is valid iff there is an observer proof method ϕ such that, for all environments η and states σ satisfying Γ, $[\![\phi]\!]\eta\sigma$ is a proof of $[\![Q]\!]\eta\sigma$. As an example of such a judgement, consider

$$\ldots, v^*\colon \mathtt{Ref}\ \mathtt{nat}, \{\mathtt{Get}\ y\colon \mathtt{nat} \Leftarrow v^*\ \mathtt{in}\ y = t\} \vdash \{\mathtt{Obs}\ (\exists z\colon \mathtt{nat})z = t\}$$

It states that in every state in which v^* refers to the value of t, it is possible to observe a natural number equal to t. If ϕ is the evidence for the hypothesis Get-assertion, then the evidence for the right hand side is $(\mathbf{get}\ y \Leftarrow v^*\ \mathbf{in}\ \langle y, \phi \rangle)$. Note that the corresponding judgement with state-universal propositions

$$\ldots, v^*\colon \mathtt{Ref}\ \mathtt{nat}, (\mathtt{Get}\ y\colon \mathtt{nat} \Leftarrow v^*\ \mathtt{in}\ y = t) \vdash \mathtt{Obs}\ (\exists z\colon \mathtt{nat})z = t$$

would be trivial because the hypothesis $(\mathtt{Get}\ y\colon \mathtt{nat} \Leftarrow v^*\ \mathtt{in}\ y = t)$ can have no evidence. (It means the same as $(\forall y\colon \mathtt{nat})y = t$).

State-assertions are common in informal reasoning about time-varying phenomena. For instance, consider the statement "a free-falling body accelerates at the rate of g". There are two possible interpretations of it:

1. For every body x, if x is falling freely in all states, then x accelerates at the rate of g in all states.

2. For every body x and every state σ of x, if x is falling freely in state σ, then x accelerates at the rate of g in σ.

Obviously, the second interpretation is the appropriate one because we want the statement to cover bodies which are not perpetually free-falling. The predicates "free-falling" and "accelerates at the rate of g" are state-dependent properties or *state-assertions*. However, the entire statement is not a state-assertion but is state-universal: it incorporates an implicit quantification over all states of x. In reasoning about imperative programs, we require similar quantifications of state-assertions and state-universal propositions.

---

[4]The notion of state-assertions closely corresponds to the notion of "assertions" in Hoare logic [Hoa69]. So, we use notation reminiscent of the latter.

The following inference rules capture the required forms of reasoning:

Assertion-introduction
$$\frac{\Gamma \vdash Q}{\Gamma \vdash \{Q\}}$$

Assertion-elimination
$$\frac{\Gamma \vdash \{Q\}}{\Gamma \vdash Q} \qquad \text{if } \Gamma \text{ has no state-assertions}$$

For the second rule, note that if Γ has no state-assertions and ϕ forms evidence for $\{Q\}$, ϕ is, in fact, evidence for $\{Q\}$ in all states. Thus, ϕ is also evidence for the proposition Q.

Now, state-assertions are rather like propositions except that they are state-dependent. For every logical operator on propositions, there is a corresponding operator for state-assertions. The rules for the state-assertional operators would be the same as those for propositional operators except that they deal with state-assertions. The copy of the standard logic used for state-assertions is called *assertion logic*.

3.5 Noninterference

We must now address a special concern that arises with the use of references. Consider a proposition of the form

$$\text{Get } x{:}\,\theta \Leftarrow l \text{ in Get } y{:}\,\theta \Leftarrow l \text{ in } Q[x, y]$$

This is a well-formed proposition. However, it uses the same reference twice for quantification. Even though it lexically appears to have two quantifiers, it is equivalent to the following proposition with one quantification:

$$\text{Get } x{:}\,\theta \Leftarrow l \text{ in } Q[x, x]$$

But, this kind of reduction of quantifications is not always possible. A formula of the form

$$\text{Get } x{:}\,\theta \Leftarrow l \text{ in Get } y{:}\,\theta \Leftarrow l' \text{ in } Q[x, y]$$

may use two different expressions l and l' which denote the same reference. Or, they may denote the same reference in some states and different references in others. Thus, the degree of quantification involved in a formula is not syntactically discernible.

In a sense, this is the cost we must pay for generalizing syntactic variables to semantic references. But, note that it is precisely this feature — the ability to write distinct expressions which may denote the same reference — that allows the flexibilities mentioned in Section 3.3.

The ambiguity involved in Get-quantification surfaces only when such quantifications are specialized by Get-elimination. The inference

$$\frac{\Gamma \vdash t : (\text{Get } x{:}\,\theta \Leftarrow l \text{ in } Q)}{\Gamma \vdash (l := e \,;\, t) : Q[e/x]}$$

is sound only if Q does not have further quantifications over the reference l. In that case, we say that l does not *interfere* with Q and write it as $l \sim Q$. (This notion of noninterference is similar to that used in Reynolds's Specification Logic [Rey82, Ten89]). The proposition $l \sim Q$ holds iff l is distinct from each reference used for Get-quantification in Q.

To reason about distinctness of references, we stipulate that all the reference variables in an environment denote distinct references. That is, each letref operator binds its reference variable to a reference that is not bound to any other reference variable in the environment. This corresponds to the notion of *block structure* in Algol-like languages [Rey81, Ten89].

Assertion logic, the copy of standard logic used for state-assertions, also contains introduction and elimination rules for Get. Therefore, we also need noninterference state-assertions of the form $\{l \sim Q\}$. Such an assertion means that l is distinct from all the references used for Get-quantification in the assertion $\{Q\}$, i.e., Q interpreted in the current state. To see the need for this, consider an assertion of the form:

$$\{\text{Get } y \colon \text{Ref } \theta \Leftarrow w^* \text{ in Get } z \colon \theta \Leftarrow y \text{ in } Q\}$$

We can say that a reference $v^* : \text{Ref } \theta$ does not interfere with this assertion only if w^* does not refer to v^* in the current state. Obviously, such noninterference does not hold for all states (since there are states in which w^* refers to v^*).[5]

4 Observation Type Theory

We now present a formal system based on the ideas presented in the previous section. Following Martin-Löf [ML84], we use the Curry-Howard isomorphism [CF58, How80] and identify propositions and types to achieve a uniform treatment. The constructive logic for which ILC forms the language of constructions is defined as a type theory called *observation type theory* (OTT). The type system of OTT extends the basic type system of ILC without altering the language of terms in a significant way.

4.1 Terms

The abstract syntax of *observation type theory* (OTT) is as follows:

---

[5]Specification logic does not allow state-specific noninterference assertions. As this example shows, full use of references (arrays, graphs, reference functions) cannot be made without them.

Terms	e	$::=$	$k \mid x \mid v^* \mid \lambda x.e \mid e_1(e_2) \mid \langle e_1, e_2 \rangle \mid e.1 \mid e.2 \mid \texttt{abort}(e) \mid$
			$\texttt{fact} \mid \texttt{letref } v^* := e \texttt{ in } t \mid \texttt{get } x \Leftarrow l \texttt{ in } t \mid l := e \; ; \; t$

Applicative types	τ	$::=$	$\beta \mid \texttt{void} \mid (\Pi x{:}\,\tau_1)\tau_2 \mid (\Sigma x{:}\,\tau_1)\tau_2 \mid e_1 = e_2 \texttt{ in } \tau$

Storage types	θ	$::=$	$\tau \mid \texttt{Ref } \theta \mid (\Pi x{:}\,\theta_1)\theta_2 \mid (\Sigma x{:}\,\theta_1)\theta_2 \mid e_1 = e_2 \texttt{ in } \theta \mid$
			$(e_1 =_{Ref} e_2)$

Assertion types	α	$::=$	$\tau \mid \texttt{Obs } \alpha \mid \texttt{Get } x{:}\,\theta \Leftarrow l \texttt{ in } \alpha \mid (\texttt{All } x{:}\,\alpha_1)\alpha_2 \mid$
			$(\texttt{Some } x{:}\,\alpha_1)\alpha_2 \mid e_1 = e_2 \texttt{ in } \alpha \mid (e_1 =_{Ref} e_2) \mid e \sim \alpha$

Imperative types	ω	$::=$	$\theta \mid \alpha \mid (\Pi x{:}\,\omega_1)\omega_2 \mid (\Sigma x{:}\,\omega_1)\omega_2 \mid e_1 = e_2 \texttt{ in } \omega$

The type system of ILC is extended with empty (void) and equality ($e_1 = e_2 \texttt{ in } T$) types. The product and function space constructions are generalized to dependent product and dependent function space constructions by the following correspondence:

$$T_1 \times T_2 \stackrel{\text{def}}{=} (\Sigma x{:}\,T_1)T_2 \quad \text{where } x \notin V(T_2)$$
$$T_1 \to T_2 \stackrel{\text{def}}{=} (\Pi x{:}\,T_1)T_2 \quad \text{where } x \notin V(T_2)$$

where $V(T)$ denotes the set of variables free in T. The Σ and Π operators represent existential and universal quantifications as in Martin-Löf type theory.

The Obs τ construct of ILC is enlarged into a new intermediate layer of *assertion types*. It includes, in addition to the Obs constructor, the Get-quantification discussed in section 3.1, the state-assertional quantifiers All and Some (which are the state-assertional analogs of the Π and Σ operators), a new reference equality type, and a noninterference type. The following abbreviations are defined for the quantifiers:

$$\alpha_1 \texttt{ and } \alpha_2 \stackrel{\text{def}}{=} (\texttt{Some } x{:}\,\alpha_1)\alpha_2 \quad \text{where } x \notin V(\alpha_2)$$
$$\alpha_1 \texttt{ implies } \alpha_2 \stackrel{\text{def}}{=} (\texttt{All } x{:}\,\alpha_1)\alpha_2 \quad \text{where } x \notin V(\alpha_2)$$

An α type-term can be used as a type as well as a state-assertion. In the latter role, we enclose it in braces $\{\alpha\}$.

The reference equality type $(e_1 =_{Ref} e_2)$ is well-formed only if e_1 and e_2 are of some reference types $\texttt{Ref } \theta$ and $\texttt{Ref } \theta'$. It denotes the proposition that e_1 and e_2 denote the same reference. Its main use is to reason about inequality of references in establishing noninterference.

The type formation rules needed to complete the description of the syntax of type terms may be found in [Swa91].

The type-terms of the α layer have a considerable degree of redundancy because they capture implicit dependence on the state. Figure 1 lists the equivalences imposed on these terms to eliminate this redundancy. The equivalences 1–4 state the relationship between Get and Obs operators. The equivalences 4–8 state that a Get operator in an argument position of All or Some can be pulled out and the equivalences 9 and 10 state that All and Some acting on applicative types are equivalent to Π and Σ.

$$(1) \qquad\qquad \text{Obs } \alpha \;=\; \text{Get } x{:}\theta \Leftarrow l \text{ in } \alpha \quad (\text{if } x \notin V(\alpha))$$

$$(2) \qquad\qquad \text{Obs } (\text{Obs } \alpha) \;=\; \text{Obs } \alpha$$

$$(3) \qquad \begin{array}{l} \text{Get } x_1{:}\theta_1 \Leftarrow l_1 \text{ in} \\ \quad(\text{Get } x_2{:}\theta_2 \Leftarrow l_2 \text{ in } \alpha) \end{array} \;=\; \begin{array}{l} \text{Get } x_2{:}\theta_2 \Leftarrow l_2 \text{ in} \\ \quad(\text{Get } x_1{:}\theta_1 \Leftarrow l_1 \text{ in } \alpha) \end{array}$$

$$(4) \quad \text{Get } x{:}\theta \Leftarrow l \text{ in } (\text{Get } y{:}\theta \Leftarrow l \text{ in } \alpha) \;=\; \text{Get } x{:}\theta \Leftarrow l \text{ in } \alpha[x/y]$$

$$(5) \quad (\text{All } x : (\text{Get } y{:}\theta \Leftarrow l \text{ in } \alpha_1))\alpha_2 \;=\; \text{Get } y{:}\theta \Leftarrow l \text{ in } (\text{All } x : \alpha_1)\alpha_2$$

$$(6) \quad (\text{All } x : \alpha_1)(\text{Get } y{:}\theta \Leftarrow l \text{ in } \alpha_2) \;=\; \text{Get } y{:}\theta \Leftarrow l \text{ in } (\text{All } x : \alpha_1)\alpha_2$$

$$(7) \quad (\text{Some } x : (\text{Get } y{:}\theta \Leftarrow l \text{ in } \alpha_1))\alpha_2 \;=\; \text{Get } y{:}\theta \Leftarrow l \text{ in } (\text{Some } x : \alpha_1)\alpha_2$$

$$(8) \quad (\text{Some } x : \alpha_1)(\text{Get } y{:}\theta \Leftarrow l \text{ in } \alpha_2) \;=\; \text{Get } y{:}\theta \Leftarrow l \text{ in } (\text{Some } x : \alpha_1)\alpha_2$$

$$(9) \qquad\qquad (\text{All } x : \tau_1)\tau_2 \;=\; (\Pi x : \tau_1)\tau_2$$

$$(10) \qquad\qquad (\text{Some } x : \tau_1)\tau_2 \;=\; (\Sigma x : \tau_1)\tau_2$$

Figure 1: Equivalence of α type-terms

4.2 Type inhabitance

The inference rules of the constructive logic are presented as rules for type inhabitance. These take the form of introduction and elimination rules for various operators. We only present the rules that have direct relevance to this paper. The full set of rules may be found in [Swa91].

The rules for void, Π, Σ and $=$ types (in all layers) are conventional. The rules for $=$ types are also conventional except that, for assertion types, equality is defined to be extensional over states. Ref θ types have no rules because their members are only available in the language by allocation of reference variables. The reference equality type has the single introduction rule

$$\frac{e_1 = e_2 \text{ in Ref } \theta}{e_1 =_{Ref} e_2}$$

Figure 2 presents the introduction and elimination rules for the α layer of OTT. As such, these represent the core of the logic.

Rules Obs-*intro* and Obs-*elim* are coercion rules between applicative terms and observers. Obs-*intro* prescribes that all applicative and state-dependent terms may be coerced to observers, while Obs-*elim* prescribes that state-independent observers may be coerced back to applicative types. These rules are directly carried over from ILC [SRI91].

Obs-intro

$$\frac{\Gamma \vdash t : \alpha}{\Gamma \vdash t : \mathsf{Obs}\ \alpha}$$

Obs-elim

$$\frac{\Gamma \vdash t : \mathsf{Obs}\ \tau}{\Gamma \vdash t : \tau} \quad \text{(if } \Gamma \text{ has only } \tau \text{ types)}$$

Dereference

$$\frac{\begin{array}{c}\Gamma \vdash l : \mathsf{Ref}\ \theta \\ \Gamma, x : \theta \vdash t : \mathsf{Obs}\ \alpha\end{array}}{\Gamma \vdash (\mathsf{get}\ x \Leftarrow l\ \mathsf{in}\ t) : (\mathsf{Get}\ x : \theta \Leftarrow l\ \mathsf{in}\ \alpha)}$$

Assignment

$$\frac{\Gamma \vdash l : \mathsf{Ref}\ \theta \quad \Gamma, x : \theta \vdash \mathsf{fact} : l \sim \alpha \quad \Gamma \vdash e : \theta \quad \Gamma \vdash t : (\mathsf{Get}\ x : \theta \Leftarrow l\ \mathsf{in}\ \alpha)}{\Gamma \vdash (l := e\ ;\ t) : \mathsf{Obs}\ \alpha[e/x]}$$

Creation   (if $v^* \notin V(\alpha)$)

$$\frac{\Gamma \vdash e : \theta \qquad \Gamma \vdash l_1 : \mathsf{Ref}\ \theta_1 \quad \ldots \quad \Gamma \vdash l_n : \mathsf{Ref}\ \theta_n \qquad \Gamma, v^* : \mathsf{Ref}\ \theta, \neg(v^* = l_1), \ldots, \neg(v^* = l_n) \vdash t : (\mathsf{Get}\ x : \theta \Leftarrow v^*\ \mathsf{in}\ \alpha)}{\Gamma \vdash (\mathtt{letref}\ v^* := e\ \mathsf{in}\ t) : \mathsf{Obs}\ \alpha[e/x]}$$

Figure 2: Inference rules for observation type theory

The *Dereference, Assignment,* and *Creation* rules provide for the introduction and elimination of the Get operator. These rules generalize the corresponding rules of ILC. The *Dereference* rule is similar to the introduction rule for Π. The difference is that rather than abstract over a variable, we abstract over the content of a reference.

The *Assignment* rule eliminates Get operators that represent state-dependence on unaliased references. The premise $t : (\mathsf{Get}\ x : \theta \Leftarrow l\ \mathsf{in}\ \alpha)$ means that t is an observer method that proves $\mathsf{Get}\ x : \theta \Leftarrow l\ \mathsf{in}\ \alpha$. That is, if t is executed in a state where l refers to, say, a, then it yields a value that belongs to $\alpha[a/x]$. Thus, assigning e to l before t provides a proof of $\alpha[e/x]$. The additional Obs operator in the conclusion $\mathsf{Obs}\ \alpha[e/x]$ takes care of the special case where α may be an applicative or storage type (which is not permitted to have an observer term as a member). The premise $l \sim \alpha$ ensures that α has no further Get-quantifications over the reference l. So, assigning to l does not alter the meaning of α.

The *Creation* rule is fairly similar to the *Assignment* rule described above. The primary difference is that the reference used for quantification in the premise is a reference variable v^* which is discharged by the inference. The proof term for the conclusion, *viz.*, ($\mathtt{letref}\ v^* := e\ \mathsf{in}\ t$), is responsible for creating a new reference for the discharged variable. The semantics of letref guarantees that the reference allocated for v^* is distinct from all other references. Hence, if $l_1, \ldots, l_n$ are any references available from the context Γ, the proof of (Get $x{:}\,\theta \Leftarrow v^*$ in α) can assume that v^* is distinct from them. Such assumptions are quite important because they are the only information available for proving noninterference of v^* in a use of the *Assignment* rule. A second difference from the *Assignment* rule is that the noninterference premise $l \sim \alpha$ is not present in the *Creation* rule. Since v^* is distinct from all references available from the context Γ (and α must be well-formed under Γ), indeed, v^* cannot interfere with α.

The noninterference type ($l \sim \alpha$) asserts that the meaning of the type α does not depend on the content of reference l. In our language, a Get is the only way for a type to access a reference's content. $l \sim \alpha$ requires that there be no meaningful Get's to the reference l in α. This condition is axiomatized by a collection of rules. We only describe the most interesting rule, namely $\sim$-intro-Get. If α' is of the form Get $x{:}\,\theta_2 \Leftarrow l'$ in α, the meaning of α' may depend on the content of reference l'. Thus, for l to not interfere with α', l should be distinct from the reference l':

$$\frac{\Gamma \vdash l : \mathrm{Ref}\ \theta_1 \quad \Gamma \vdash l' : \mathrm{Ref}\ \theta_2 \quad \Gamma \vdash \mathrm{fact} : \neg(l = l') \quad \Gamma, x : \theta_2 \vdash \mathrm{fact} : (l \sim \alpha)}{\Gamma \vdash \mathrm{fact} : (l \sim (\mathrm{Get}\ x{:}\,\theta_2 \Leftarrow l'\ \mathrm{in}\ \alpha))}$$

It is nontrivial to show that two references l and l' are distinct since l and l' are both expressions and may be aliases of the same reference. The only way to show that they are distinct is to show that they were created by separate instances of the letref construct. The inequalities introduced by the *Creation* rule can be used for this purpose.

The Some and All operators of the α layer are state-assertional operators, *i.e.*, both the arguments to such operators are interpreted in the same state. These operators are handled by a separate set of inference rules collectively referred to as *assertion logic*.

Any assertion type α can be enclosed in braces to form a state-assertion $\{\alpha\}$. A judgement that has a state-assertion on either side of "$\vdash$" is called a *state-judgement*. State-judgements quantify over all states and every state-assertion appearing in such a judgement is interpreted with respect to the state pervasive in the judgement. Thus, a judgement of the form $\{\alpha_1\} \vdash \{\alpha_2\}$ means that, in all states satisfying α_1, there is evidence for α_2. Types appearing in state-judgements are, however, interpreted in their usual fashion. Assertion logic deals with inferences for state-judgements.

Figure 3 presents the important rules of assertion logic. *Assertion-intro* asserts that if α holds in all states, then α holds in states satisfying Γ. *Assertion-elim* asserts that if α holds in all states that satisfy Γ, and if Γ has no state-assertions (and hence is satisfied by all states), then α holds in all states. These rules coerce between the interpretations of α terms as state-assertions and types, and are essentially specialization and generalization rules for the implicit quantification over the state.

The remaining rules are the counterparts of the type inferences rules of Fig. 2 for assertion logic. Some of these rules are stronger than the corresponding type inference rules. For instance, the major premise of the *Dereference* rule assumes

Assertion-intro
$$\frac{\Gamma \vdash t : \alpha}{\Gamma \vdash t : \{\alpha\}}$$

Assertion-elim
$$\frac{\Gamma \vdash t : \{\alpha\}}{\Gamma \vdash t : \alpha} \quad \text{(if } \Gamma \text{ has no state-assertions)}$$

Obs-intro
$$\frac{\Gamma \vdash t : \{\alpha\}}{\Gamma \vdash t : \{\text{Obs } \alpha\}}$$

Obs-elim
$$\frac{\Gamma \vdash t : \{\text{Obs } \tau\}}{\Gamma \vdash t : \{\tau\}} \quad \text{(if } \Gamma \text{ has only } \tau \text{ types)}$$

Dereference
$$\frac{\Gamma \vdash l : \text{Ref } \theta \qquad \Gamma, x : \theta, \textbf{fact} : \{\text{Get } z : \theta \Leftarrow l \text{ in } x = z\} \vdash t : \{\text{Obs } \alpha\}}{\Gamma \vdash (\text{get } x \Leftarrow l \text{ in } t) : \{\text{Get } x : \theta \Leftarrow l \text{ in } \alpha\}}$$

Assignment
$$\frac{\Gamma \vdash l : \text{Ref } \theta \quad \Gamma, x : \theta \vdash \textit{fact} : \{l \sim \alpha\} \qquad \Gamma \vdash e : \theta \quad \Gamma \vdash t : (\text{Get } x : \theta \Leftarrow l \text{ in } \alpha)}{\Gamma \vdash (l := e \, ; \, t) : \{\text{Obs } \alpha[e/x]\}}$$

Creation   if $v^* \notin V(\alpha)$
$$\frac{\Gamma \vdash e : \theta \qquad \Gamma \vdash l_1 : \text{Ref } \theta_1 \quad \ldots \quad \Gamma \vdash l_n : \text{Ref } \theta_n \qquad \Gamma, v^* : \text{Ref } \theta, \neg(v^* = l_1), \ldots, \neg(v^* = l_n) \vdash t : (\text{Get } x : \theta \Leftarrow v^* \text{ in } \alpha)}{\Gamma \vdash (\text{letref } v^* := e \text{ in } t) : \{\text{Obs } \alpha[e/x]\}}$$

Figure 3: Inference rules for state-assertions in OTT

that x is equal to the content of l in the pervasive state. The *Dereference* rule of Fig. 2, in contrast, requires the premise to be established for all states. Similarly, the *Assignment* rule of assertion logic requires the noninterference premise to be only a state-assertion $\{l \sim \alpha\}$, i.e., it needs to be proved only for the pervasive state, not for all states. This is a powerful rule which is required for reasoning about state-dependent references. Suppose $A : \text{nat} \to \text{Ref nat}$ is an "array". If α is of the form

$$\text{Get } x : \text{nat} \Leftarrow l \text{ in Get } y : \text{nat} \Leftarrow A(x) \text{ in } T$$

then $A(0) \sim \alpha$ is not provable because $\neg(A(0) =_{Ref} A(x))$ does not hold for all states. (There is a state in which l refers to 0). However, the corresponding state-assertion $\{A(0) \sim \alpha\}$ is provable for states in which l does not refer to 0. Thus, reasoning about array-subscripted references (and pointers etc.) requires state-dependent noninterference assertions.

The rules for Some and All are similar to those for the Σ and Π operators. However, note that in a type of the form (Some $x : \alpha_1)\alpha_2$, both α_1 and α_2 should be interpreted in the *same* state. Correspondingly, the construction for the type, which would be a pair of the form $\langle t_1, t_2 \rangle$, should also interpret both its components in the same state. Since each component may involve assignments to references, this means that the state must be copied and passed to each component. This is not computationally practical. However, in practice, only one of the components is of computational significance. Typically, t_1 has computational content and t_2 is merely a verification of the fact that t_1 satisfies α_2. Thus, the constructions produced for assertion types can be tested for their computational feasibility. Alternatively, different type constructors, such as the subset type constructor of Nuprl [Con86], can be used to suppress the noncomputational constructions.

The following rules express the inferences required for Some and All in assertion logic.

Some *-introduction*

$$\frac{\Gamma \vdash s : \{S\} \quad \Gamma \vdash t : \{T[s/x]\}}{\Gamma \vdash \langle s , t \rangle : \{(\text{Some } x : S)T\}}$$

Some *-elimination*

$$\frac{\Gamma \vdash e : \{(\text{Some } x : S)T\}}{\Gamma \vdash e.1 : \{S\}}$$

$$\frac{\Gamma \vdash e : \{(\text{Some } x : S)T\}}{\Gamma \vdash e.2 : \{T[e.1/x]\}}$$

All *-introduction*

$$\frac{\Gamma, x : \{S\} \vdash t : \{T\}}{\Gamma \vdash \lambda x.t : \{(\text{All } x : S)T\}}$$

All *-elimination*

$$\frac{\Gamma \vdash f : \{(\text{All } x : S)T\} \quad \Gamma \vdash s : \{S\}}{\Gamma \vdash f(s) : \{T[s/x]\}}$$

The underscored constructors: "$\underline{,}$", "$\underline{.}$", "$\underline{\lambda}$" and "$\underline{()}$", signify the fact that they are "state-indexical", *i.e.*, all their components are interpreted in the same state.

5 Modelling Hoare Logic

Hoare logic and its extension, Specification logic, can be modelled in observation type theory. Commands, which are conventionally treated as transformers of state, are treated in OTT as observer transformers. A command s is a function that maps an observer c to enhanced observer $s(c)$ which first carries out the actions of s and then continues with the observer c. With this reinterpretation, the Hoare triple $\{P\}s\{Q\}$ corresponds precisely to a judgement of the form

$$\Gamma \vdash \overline{s} : (Q \text{ implies } C) \to (P \text{ implies } C)$$

where $\overline{s}$ is a function from observers to observers. The argument of $\overline{s}$, when evaluated in a state satisfying Q, yields a value of some (polymorphic) type C. The observer returned by $\overline{s}$, when evaluated in a state satisfying P, yields a value of the same type C. The relationship between Hoare-triples and OTT judgements reflects the duality between traditional state-dependent values and ILC's observers: the value s transforms states satisfying P to states satisfying Q, while the observer transformer $\overline{s}$ transforms observers of states that satisfy Q to observers of states that satisfy P.

The Hoare assignment axiom

$$\{P[e]\}\, l := e\, \{P[l]\}$$

is modelled by the following inference using the *Assignment* rule:

$$\frac{\Gamma \vdash t : \text{Get } x{:}\theta \Leftarrow l \text{ in } P[x] \text{ implies } C}{\Gamma \vdash (l := e \, ; \, t) : \text{Obs } (P[e] \text{ implies } C)}$$

So, forward deduction using the rule **Assignment** is remarkably similar to the Floyd-Hoare technique of pushing assertions backwards through assignments.

Since judgements such as the above are no different from other judgements in OTT, it is possible to derive rules such as antecedent-strengthening. The main advantage of OTT over a Hoare-style logic is that its constructions do not have side-effects; state information is localized and we can reason about the state independent of the context. Moreover, since all state-dependencies need to be explicitly stated by means of dereferences, reasoning about aliasing of references is simplified.

6 Conclusion

The fundamental logical symmetries that underlie our formulation of references and assignments are remarkably similarly to the symmetries that underlie variables and functions. These symmetries are embodied in *observation type theory* (OTT), which is a strict extension of a Martin-Löf-style constructive type theory. OTT's language of constructions (ILC) satisfies all the important properties of type theory's language of constructions, such as confluence and strong normalization of the reduction relation.

Compared to conventional type theories, OTT appears somewhat complex. Much of the complexity is in the stratification of the type system into four layers and the need for a separate assertion logic. However, a comparison with other formulations of imperative programming logics, such as Specification logic, suggests that some of this complexity may be inevitable. We continue to investigate further refinement of the present theory.

Much work remains to be done regarding the practical application of the theory for carrying out proofs and program derivations. Some examples may be found in [Swa91]. But, much experience needs to be obtained and it is still too early to pass this as the final word on constructive formulation of imperative programs. The most rewarding aspect of the present formulation is the insight it brings to the nature of references, observer methods, and assignment as a form of method application. It would be worthwhile to study programming paradigms weaker than references to obtain a better understanding of their nature, *e.g.*, "logic variables" of logic programming [Lin85, WPP77] and the negative types of linear logic [Abr90, Gir87].

The issue of aliasing of references needs to be examined further. Since OTT requires all state dependencies to be specified explicitly by means of dereference operators, it may be possible to reason about noninterference even in the presence of aliasing. This is likely to be aided by modularizing data structures based on their interference characteristics — a systematic study of this would be beneficial. The state-indexical constructions of state-assertions also bear further investigation since they require the state to be copied. For efficiency considerations, their use should be restricted to parts of the proof that do not have meaningful computational content.

References

[Abr90] S. Abramsky. Computational interpretations of linear logic. Research Report DOC 90/20, Imperial College, London, Oct 1990.

[CF58] H. B. Curry and R. Feys. *Combinatory Logic*. North-Holland, Amsterdam, 1958.

[Con86] R. L. Constable, *et. al. Implementing Mathematics with the Nuprl Proof Development System*. Prentice-Hall, New York, 1986.

[Eng75] E. Engeler. Algorithmic logic. In J. W. de Bakker, editor, *Foundations of Computer Science*, pages 57–85. Mathematisch Centrum, Amsterdam, 1975. (Mathematical Centre Tracts 63).

[Gir87] J.-Y. Girard. Linear logic. *Theoretical Comp. Science*, 50:1–102, 1987.

[GvN47] H. H. Goldstine and J. von Neumann. Planning and coding problems for an electronic computing instrument, Part II. In *John von Neumann, Collected Works, Vol. V*, pages 80–151. Pergamon Press, 1963, Oxford, 1947.

[Hoa69] C. A. R. Hoare. An axiomatic basis for computer programming. *Communications of the ACM*, 12:576–583, 1969.

[How80] W. A. Howard. The formulae-as-types notion of construction. In J. R. Hindley and J. P. Seldin, editors, *To H. B. Curry: Essays on Combinatory Logic, Lambda Calculus and Formalism*, pages 479–490. Academic Press, New York, 1980.

[HW90] P. Hudak and P. Wadler (eds). Report on programming language Haskell, A non-strict purely functional language (Version 1.0). Technical Report YALEU/DCS/RR777, Yale University, Apr 1990.

[Lin85] G. Lindstrom. Functional programming and the logical variable. In *ACM Symp. on Princ. of Program. Languages*, 1985.

[ML82] P. Martin-Löf. Constructive mathematics and computer programming. In L. J. Cohen, J. Los, H. Pfeiffer, and K.-P. Podewski, editors, *Proc. Sixth Intern. Congress for Logic, Methodology and Philosophy of Science*, pages 153–175, Amsterdam, 1982. North-Holland.

[ML84] Per Martin-Löf. *Intuitionistic Type Theory*. Studies in Proof Theory. Bibliopolis, Napoli, 1984.

[MS87] G. Mirkowska and A. Salwicki. *Algorithmic Logic*. PWN - Polish Scientific Publishers, Warzwawa, Poland, 1987.

[Pra76] V. Pratt. Semantic considerations on Floyd-Hoare logic. In *Symp. on Foundations of Computer Science*, pages 109–121. IEEE, 1976.

[Rey81] J. C. Reynolds. The essence of Algol. In J. W. de Bakker and J. C. van Vliet, editors, *Algorithmic Languages*, pages 345–372. North-Holland, 1981.

[Rey82] J. C. Reynolds. Idealized Algol and its specification logic. In D. Neel, editor, *Tools and Notions for Program Construction*, pages 121–161. Cambridge Univ. Press, 1982.

[SRI91] V. Swarup, U. S. Reddy, and E. Ireland. Assignments for applicative languages. In R. J. M. Hughes, editor, *Conf. on Functional Program. Lang. and Comput. Arch.* Springer-Verlag, Berlin, 1991. (Earlier versions: University of Illinois at Urbana-Champaign, July 1990, Revised Nov 1990, Feb 1991).

[Swa91] V. Swarup. *Type Theoretic Properties of Assignments*. PhD thesis, Univ. Illinois at Urbana-Champaign, 1991. (to appear).

[Ten89] R. D. Tennent. Denotational semantics of Algol-like languages. In S. Abramsky, D. M. Gabbay, and T. S. E. Maibaum, editors, *Handbook of Logic in Computer Science, Vol II*. Oxford University Press, 1989. (to appear).

[WPP77] D. H. D. Warren, L. M. Pereira, and F. Pereira. Prolog - the language and its implementation compared with Lisp. *SIGPLAN Notices*, 12(8), 1977.

Constructivity Issues in Graph Algorithms

Michael R. Fellows
Department of Computer Science
University of Victoria
Victoria, British Columbia V8W 3P6, Canada

Michael A. Langston
Department of Computer Science
University of Tennessee
Knoxville, Tennessee 37996-1301, U.S.A.

Abstract. Decision problems have traditionally been classified as either "tractable" or "intractable," depending on whether polynomial-time decision algorithms exist to solve them. Until recently, one could expect proofs of tractability to be constructive. This comfortable situation is altered, however, by the new and inherently *non*constructive developments in the theory of well-partially-ordered sets. In this paper, we survey some of the main results and open questions related to this topic.

1. Overview

Theoretical computer science has focused a great deal of scrutiny on the complexity class $\mathcal{P}$, those problems for which there *exist* polynomial-time decision algorithms. Despite the potential nonconstructivity permitted by this definition, most researchers would have until recently scoffed at the notion of anything less than a fully constructive proof that a problem is in $\mathcal{P}$. Prior to the developments we survey in this paper, the known methods for proving membership in $\mathcal{P}$ have been constructive. That is, a proof of polynomial-time decidability itself provided "positive" evidence in the form of the promised polynomial-time algorithm. When a polynomial-time algorithm has been known to exist, attention has generally been riveted only on making it as efficient as possible, rather than on trying to find the algorithm.

Surprising recent advances in graph theory, however, radically change this situation. Membership in $\mathcal{P}$ can now be proved for a host of challenging problems, by means of nonconstructive arguments that establish only the existence of a small-degree polynomial-time algorithm. Many of these problems were not previously known to be in $\mathcal{P}$ at all. Some were not previously known even to be decidable. Others were only known to be in

$\mathcal{P}$ by way of exhaustive dynamic programming methods with running times bounded by large degree polynomials.

These developments clearly challenge the established view that equates tractability with polynomial-time decidability (knowledge only of the existence of an algorithm may be of little value). In this paper, we shall review some of the main results on this subject, and report on new techniques for dealing with these powerful tools.

2. Background

Except where explicitly noted otherwise, all graphs we consider are finite and undirected. A graph H is less than or equal to a graph G in the *minor* order, written $H \leq_m G$, if and only if a graph isomorphic to H can be obtained from a subgraph of G by contracting edges.

A family F of graphs is said to be *closed* under the minor ordering if the facts that G is in F and that $H \leq_m G$ together imply that H must be in F. The *obstruction set* for a family F of graphs is the set of graphs in the complement of F that are minimal in the minor ordering. Therefore, if F is closed under the minor ordering, it has the following characterization: G is in F if and only if there is no H in the obstruction set for F such that $H \leq_m G$.

Note that the relation $\leq_m$ defines a partial ordering on graphs. A partially-ordered set $(X, \leq)$ is said to be *well-partially-ordered* if (1) every subset of X has finitely many minimal elements and (2) X contains no infinite descending chain $x_1 \geq x_2 \geq x_3 \geq \ldots$ of distinct elements. Condition (2) of course holds for any collection of graphs under $\leq_m$. The question of whether condition (1) always holds as well has recently been answered in the affirmative by the deep work on the subject by Robertson and Seymour.

Theorem 1. [RS5] *Graphs are well-partially-ordered by $\leq_m$.*

Thus, a closed family must have a finite obstruction set. Furthermore, Robertson and Seymour have established that minor containment for each obstruction can be tested in small-degree polynomial time.

Theorem 2. [RS4] *For every fixed graph H, the problem that takes as input a graph G and* *determines whether $H \leq_m G$ is solvable in $O(n^3)$ time.*

Progressive strengthenings of Theorem 2 have been obtained in the case where H is planar. The best current result is a time bound of $O(n \log n)$ for minor testing for every fixed planar graph H due to Reed [Re]. The results of Reed have the further consequence that any minor closed family of graphs that does not contain some planar graph can be recognized in time $O(n \log n)$.

Theorems 1 and 2 guarantee the existence of a polynomial-time decision algorithm for any minor-closed family of graphs, but do not provide any details of what that algorithm might be. It has been shown that Theorem 1 is independent of constructive axiomatic systems and, indeed, any proof of Theorem 1 must use impredicative methods [FRS].

In fact, an easy reduction from the halting problem tells us that *no* proof of Theorem 1 can be entirely constructive in the most general sense.

Theorem 3. [FL4] *There is no algorithm to compute, from a finite description of a minor-closed family F of graphs as represented by a Turing machine that accepts precisely the graphs in F, the set of obstructions for F.*

Theorems 1 and 2 are nonconstructive in another respect as well. Even if the promised decision algorithm were somehow made available, there is absolutely no guarantee that it will be of any use in solving the *search* version [GJ] of the problem at hand.

A graph H is less than or equal to a graph G in the *immersion* order, written $H \leq_i G$, if and only if a graph isomorphic to H can be obtained from G by a series of these two operations: taking a subgraph and lifting. (A pair of adjacent edges uv and vw, with $u \neq v \neq w$, is *lifted* by deleting the edges uv and vw and adding the edge uw.) The relation $\leq_i$, like $\leq_m$, defines a partial ordering on graphs for which we then have the associated notions of closure and obstruction sets. Analogs of the above theorems hold for the immersion order as well.

3. Sample Applications

The minor and immersion well-qausiorders provide new paradigms for proving polynomial-time decidability. The range of problems amenable to this approach is remarkable.

Many computational problems of practical significance are fixed-parameter versions of problems that are $\mathcal{NP}$-hard in general. For example, in the general gate matrix layout problem, we are given an $n \times m$ Boolean matrix M and an integer k, and are asked whether we can permute the columns of M so that, if in each row we change to $*$ every 0 lying between the row's leftmost and rightmost 1, then no column contains more than k 1s and $*$s. This problem was previously only known to be decidable in $O(2^n)$ time for each fixed $k > 2$.

Theorem 4. [FL1] + [Re] *For any fixed k, gate matrix layout can be decided in $O(n \log n)$ time.*

Others problems have no associated (fixed) parameter. For example, in the knotlessness problem, we are given a graph G, and are asked whether G can be embedded in 3-space so that all of its cycles are unknotted. This problem was not previously known to be decidable.

Theorem 5. [FL2] *Knotlessness can be decided in $O(n^3)$ time.*

A number of other results are summarized in the table below, where n denotes the number of vertices in an input graph and k denotes the appropriate fixed parameter where relevant.

Problem	Previous Upper Bound	New Result	Reference
emulation	open	$O(n^3)$	[FL2]
topological bandwidth*	$O(n^k)$	$O(n \log n)$	[FL2] + [Re]
linklessness	open	$O(n^3)$	[FL2]
crossing number*	open	$O(n^3)$	[FL2]
min cut linear arrangement	$O(n^{k-1})$	$O(n \log n)$	[FL3] + [Re]
search number	$O(n^{2k^2+4k+8})$	$O(n \log n)$	[FL3] + [Re]
load factor	open	$O(n \log n)$	[FL3] + [Re]
max leaf spanning tree	$O(n^{2k+1})$	$O(n \log n)$	[FL3] + [Re]

* input restricted to graphs of maximum degree three

4. Constructivization Strategies

We now describe a constructivization technique that can be employed in the vast majority of known applications. The surprising result is that we can, in very general circumstances, *constructively produce* a low-degree polynomial-time algorithm for search (and hence decision) without having to discover the finite obstruction set.

Decision algorithms based on finite obstruction sets do not decide by solving associated search problems. The situation may be modeled in terms of relations. Associated with a relation $\Pi \subseteq \Sigma^* \times \Sigma^*$ are a number of basic computational problems.

checking the problem of determining, for input (x, y), whether $(x, y) \in \Pi$,

domain decision the problem of determining, for input x, whether there exists a y such that $(x, y) \in \Pi$, and

search the problem of computing a search function for Π, where such a function $f \colon \Sigma^* \to \Sigma^* \cup \{\bot\}$ satisfies

(1) $f(x) = y$ implies that (x, y) is in Π and

(2) $f(x) = \bot \notin \Sigma$ implies that there exists no y for which (x, y) is in Π.

An oracle algorithm that computes a search function for Π with oracle language domain$(\Pi) = \{\, x \mid$ there exists a y for which (x,y) is in $\Pi\}$ is termed a *self-reduction* [BFL]. The *overhead* of such an oracle algorithm is simply its time complexity, where each oracle invocation is charged with a unit-time cost.

A partial order $(R, \leq)$ is *uniformly enumerable* if there is a recursive enumeration $(r_0, r_1, r_2, \ldots)$ of R with the property that $r_i \leq r_j$ implies $i \leq j$. Under a uniform enumeration of the elements of domain(Π), we say that a self-reduction algorithm for Π is *uniform* if, on input r_j, the oracle for domain(Π) is consulted concerning r_i only for $i \leq j$. An oracle algorithm is *honest* if, on inputs of size n, its oracle is consulted concerning only instances of size $O(n)$. An oracle algorithm A with overhead bounded by $T(n)$ is *robust* (with respect to $T(n)$) if A is guaranteed to halt within $T(n)$ steps for any oracle language.

Theorem 6. [FL4] *Let $F = $ domain(Π) be a closed family in a uniformly enumerable well-*

partial-order, and suppose the following are known:

(1) *an algorithm that solves the checking problem for* Π *in* $O(T_1(n))$ *time,*

(2) *order tests that require* $O(T_2(n))$ *time,*

(3) *a uniform self-reduction algorithm (its time bound is immaterial), and*

(4) *an honest robust self-reduction algorithm that requires* $O(T_3(n))$ *overhead.*

Then an algorithm requiring $O(\max\{T_1(n), T_2(n) \cdot T_3(n)\})$ *time is known that solves the search problem for* Π.

Most of the known applications can be made constructive with the above theorem. Gate matrix layout is a good example. Checking a candidate solution is easily performed in $O(n)$ time. $O(n \log n)$ order tests are known. Uniform self-reduction is achievable (although it is rather cumbersome). Honest robust self-reduction can be accomplished with $O(n)$ overhead [Bo]. Thus, for any fixed k, the search and decision versions of gate matrix layout can be solved in $O(n^2 \log n)$ time with a *known* algorithm.

A diagonalization originally noted by Levin can be extended in a way that admits its use in these same circumstances.

Theorem 7. [FL4] *Let* $F = domain(\Pi)$ *be a closed family in a uniformly enumerable well-partial-order, and suppose the following are known:*

(1) *an algorithm that solves the checking problem for* Π *in* $O(T_1(n))$ *time,*

(2) *order tests that require* $O(T_2(n))$ *time, and*

(3) *a uniform self-reduction algorithm (its time bound is immaterial).*

Then an algorithm requiring $O(\max\{T_0(n) + T_1(n) \cdot \log T_0(n), T_2(n)\})$ *time is known that solves the search problem for* Π, *where* $T_0(n)$ *denotes the time complexity of any algorithm solving this search problem.*

We emphasize that constructivization techniques such as these do not depend on knowing, nor do they provide a means for computing, the relevant obstruction sets.

Consider a constructive analog of $\mathcal{P} \stackrel{?}{=} \mathcal{NP}$. By the statement $\mathcal{P}$ *is constructively equal to* $\mathcal{NP}$ we mean that an algorithm is known that computes, from the index of a nondeterministic polynomial-time Turing machine M that recognizes a set X and a bound on the running time of M, the index of a deterministic polynomial-time Turing machine that recognizes X. A set X is *constructively $\mathcal{NP}$-hard* if a polynomial-time many-to-one reduction with a known polynomial time bound from SAT to X is known. Every problem presently known to be $\mathcal{NP}$-hard is constructively $\mathcal{NP}$-hard as well, simply because the relevant reductions and their running times are constructively known.

Could arguments based on well-partially-ordered sets be used to provide an inherently nonconstructive resolution to $\mathcal{P} \stackrel{?}{=} \mathcal{NP}$? It turns out that these tools and known problem-reduction schemes could not.

Theorem 8. [FL4] *Let F denote a closed family in a uniformly enumerable well-partial-order. If it is constructively $\mathcal{NP}$-hard to determine membership in F, then $\mathcal{P}$ is constructively equal to $\mathcal{NP}$.*

5. Perspective

Of course, dealing with nonconstructivity is certainly not really new in computer science as a whole (although it is very much overlooked!). In formal language theory, for example, a number of "folk theorems" concerning nonconstructive proofs are known [Wi]. To illustrate, a nonconstructive argument suffices to prove:

Theorem 9. *Regular languages are closed under quotients with arbitrary languages.*

More to the point, however, a straightforward application of Rice's Theorem yields:

Theorem 10. *No constructive proof of Theorem 9 is possible.*

New developments continue apace. A proof of Higman's Lemma in a constructive formal system has recently been announced [Mu] as well as a similar result for Kruskal's Theorem [Gu]. (These results, however, seem to have more to do with formal systems and proof theory, and so far do not seem be useful for producing known algorithms for the various concrete applications of well-quasiordering.) Techniques that exploit the notion of bounded tree width now make obstruction sets computable (at least in principle) in many cases [FL5]. For particularly important problems, obstruction sets have even been isolated with the aid of massive computational resources [Ki].

The limitations of traditional set-based definitions of complexity classes are at last becoming recognized. In [AFLM], for example, a relational approach is used instead, in which a decision problem is paired with positive evidence. In this setting, a number of results can be proved that remain conjectural within the classical theory.

It will be interesting to see what impact these and related developments have on the field as the existence of nonconstructive algorithmic tools becomes more widely known and understood.

Acknowledgement. This research is supported in part by the National Science Foundation under grant MIP–8919312 and by the Office of Naval Research under contracts N00014–88–K–0343 and N00014–88–K–0456.

References

[AFLM] K. Abrahamson, M. R. Fellows, M. A. Langston and B. Moret, "Constructive Complexity," *Discrete Applied Mathematics*, to appear.

[Bo] H. H. Bodlaender, "Improved Self-Reduction Algorithms for Graphs with Bounded Treewidth," to appear.

[BFL] D. J. Brown, M. R. Fellows and M. A. Langston, "Polynomial-Time Self-Reducibility: Theoretical Motivations and Practical Results," *Int'l J. of Computer Mathematics* 31 (1989), 1–9.

[FL1] M. R. Fellows and M. A. Langston, "Nonconstructive Advances in Polynomial-Time Complexity," *Information Processing Letters* 26 (1987), 157–162.

[FL2] —————, "Nonconstructive Tools for Proving Polynomial-Time Decidability," *J. of the ACM* 35 (1988), 727–739.

[FL3] —————, "Layout Permutation Problems and Well-Partially-Ordered Sets," *Proc. 5th MIT Conf. on Advanced Research in VLSI* (1988), 315–327.

[FL4] —————, "On Search, Decision and the Efficiency of Polynomial-Time Algorithms," *Proc. 21st ACM Symp. on Theory of Computing* (1989), 501–512.

[FL5] —————, "An Analogue of the Myhill-Nerode Theorem and Its Use in Computing Finite-Basis Characterizations," *Proc. 30th IEEE Symposium on Foundations of Computer Science*, (1989), 520–525.

FRS] H. Friedman, N. Robertson and P. D. Seymour, "The Metamathematics of the Graph Minor Theorem," in Applications of Logic to Combinatorics, American Math. Soc., Providence, RI, to appear.

GJ] M. R. Garey and D. S. Johnson, Computers and Intractability: A Guide to the Theory of $\mathcal{NP}$-Completeness, Freeman, San Francisco, CA, 1979.

Gu] A. Gupta, Ph.d. dissertation, University of Toronto, Department of Computer Science, 1991.

Ki] N. G. Kinnersley, "Obstruction Set Isolation for Layout Permutation Problems," Ph.D. Thesis, Department of Computer Science, Washington State University, 1989.

Mu] C. Murthy, "Extracting Constructive Content from Classical Proofs," Ph.D. Thesis, Department of Computer Science, Cornell University, 1990.

Re] B. Reed, to appear. (Presented at the AMS Summer Workshop on Graph Minors, Seattle, June 1991.)

RS1] N. Robertson and P. D. Seymour, "Graph Minors IV. Tree-Width and Well-Quasi-Ordering," J. Combinatorial Theory Series B, to appear.

RS2] _____, "Graph Minors V. Excluding a Planar Graph," J. Combinatorial Theory Series B 41 (1986), 92–114.

RS3] _____, "Graph Minors X. Obstructions to Tree-Decomposition," to appear.

RS4] _____, "Graph Minors XIII. The Disjoint Paths Problem," to appear.

RS5] _____, "Graph Minors XVI. Wagner's Conjecture," to appear.

Wi] K. Winklmann, private communication.

CONSTRUCTIVE TOPOLOGY AND COMBINATORICS

THIERRY COQUAND

Abstract

We present a method to extract constructive proofs from classical arguments proved by topogical means. Typically, this method will apply to the nonconstructive use of compactness in combinatorics, often in the form of the use of König's lemma (which says that a finitely branching tree that is infinite has an infinite branch.) The method consists roughly of working with the corresponding point-free version of the topological argument, which can be proven constructively using only as primitive the notion of inductive definition. We illustrate here this method on the classical "minimal bad sequence" argument used by Nash-Williams in his proof of Kruskal's theorem. The proofs we get by this method are well-suited for mechanisation in interactive proof systems that allow the user to introduce inductively defined notions, such as NuPrl, or Martin-Löf set theory.

Introduction and description of the method

The idea of point-free topology seems to go back to Whitehead (around 1910) and has been presented by Russell (see Russell 1914, 1936 and Wiener 1914). The problem is to relate the mathematical notion of points and what is given to us as observable by our sense data. The point-free approach is then a kind of reverse of the usual set-theoretical approach: instead of defining an observable as a collection of points, the notion of observable is taken as primitive, and a point is defined as a collection of observable. It is hoped then that it will be easier to relate a mathematical result to its potential application, since its formulation is using the concrete notion of observable. The main application of point-free topology for Russell seems indeed to have been an analysis of the use of mathematics in physics. For instance, the notion of instants of time is analysed in terms of the notion of events. One problem that Russell mentions is the need of using the axiom of choice in order to construct points.

This idea of point-free topology has been revived recently in the framework of the theory of locale, see Johnstone 1981, and Vickers 1989. It is instructive to see how the problem of the use of the axiom of choice in the construction of points of a locale is mentionned by Russell is analysed in this framework. As described by Johnstone 1981, what happens is that it is actually in many ways more convenient to work directly with locales, and simply to forget about points (Johnstone traces this idea back to Isbell).

Typeset by $\mathcal{A}_{\mathcal{M}}\mathcal{S}$-TeX

The situation is particularly interesting from a constructive point of view. The point-free approach to topology can be rightly described as constructive topology, and many constructions that need the axiom of choice or highly nonconstructive arguments in terms of points get a direct constructive proof when stated in a point-free way. A typical example is the theorem of Tychonoff, which is classically equivalent to the axiom of choice, but whose point-free version is provable without this axiom, see Johnstone 1981.

In this paper, we want to present point-free topology as a way to get constructive contents from classical proofs. An early application of this method appears in De Bruijn and Van der Meiden 1967, where sufficiently large closed ideals are used instead of maximal ideals in order to give a version of Wiener's proof on absolutely convergent series which does not use the axiom of choice. As noticed by the authors, we get by this point-free approach a constructive development entirely parallel to the classical approach, such that it is possible at every stage to reach the corresponding stage in the usual theory by an application of the excluded middle and the axiom of choice.

The point-free approach has been used also in domain theory, see Martin-Löf 83, Abramsky 87, leading to a generalisation of the predicate transformer method of Dijkstra. In Martin-Löf 83, it is particularly clear that the point-free spaces are naturally described in the framework of inductive definitions. This seems to be a general phenomenon, which is particularly interesting for mechanisation of proofs, that both the basic notions and proofs of point-free topology get a natural formulation with inductive definitions. For instance, in Coquand 91, it is shown that Tychonoff's theorem has a direct inductive point-free proof.

In this paper, we present yet another application of the point-free approach. It concerns combinatoric statements, that are proved using topological argument. Here, we present a point-free analysis of a topological formulation of the classical "minimal bad sequence" argument used by Nash-Williams 1963 in his proof of Kruskal's theorem. The result follows the general pattern: while the classical statement is highly nonconstructive and needs the axiom of choice in its proof, its point-free version gets a direct inductive proof.

1. An open induction principle

Let X be a set, and $W(X)$ be the set of finite words over X, $S(X)$ the set of infinite words (or streams) over X. We can consider $W(X)$ as a transition system, letting $u \to v$ mean that v can be written ux, $x \in X$. We say that a subset $B \subseteq W(X)$ is **monotone** iff $u \in B$ and $u \to v$ imply $v \in B$. If $B \subseteq W(X)$ is monotone we let $U(B) \subseteq S(X)$ be the set of infinite words $\alpha_1 \alpha_2 \ldots$ such that there exists p such that $\alpha_1 \ldots \alpha_p \in B$. We say that $U \subseteq S(X)$ is **open** iff it can be written $U = U(B)$ for one monotone $B \subseteq W(X)$.

For instance, let $\prec$ be a binary relation on X, and let $U \subseteq S(X)$ be the set of infinite words $\alpha_1 \alpha_2 \ldots$ such that there exist i, j such that $i < j$ and $\alpha_i \prec \alpha_j$. Then U is open. To say that $\prec$ is a well-quasi-order on X is the same as to say that $U = S(X)$.

We recall that a relation $<$ on X is **well-founded** iff the following induction principle holds: if $M \subseteq X$ is such that $x \in M$ whenever $y \in M$ for all $y \in X$ such that $y < x$, then M is equal to X. This implies that there are no infinite decreasing sequences. The converse holds intuitionistically only if we consider arbitrary choice sequences.

Classically, we use rather the following equivalent definition, that X is well-founded iff all non-empty subset of X has a minimal element for the relation $<$.

We suppose that $<$ a well-founded relation over a set X. We extend $<$ alphabetically on $W(X)$ by letting $x_1 \ldots x_p < y_1 \ldots y_q$ mean that $p \le q$ and $x_1 = y_1, \ldots, x_{p-1} = y_{p-1}$, $x_p < y_p$. Notice that if $u < v$, then $u < vx$ for all $x \in X$.

We extend next the relation $<$ to $S(X)$: we let $\alpha < \beta$ between $\alpha = \alpha_1 \alpha_2 \ldots$ and $\beta = \beta_1 \beta_2 \ldots$ mean that there exists p such that $\alpha_1 \ldots \alpha_p < \beta_1 \ldots \beta_p$. The relation $<$ on infinite words is not well-founded in general. For instance, if X is the set $\{0,1\}$ with the ordering $0 < 1$, we have the infinite decreasing sequence $1111\cdots > 0111\cdots > 0011\cdots > 0001\ldots$.

However, we can still express the following induction principle, where we restrict ourselves to open properties.

Theorem. *If $U \subseteq S(X)$ is open, and $\alpha \in U$ whenever $\beta \in U$ for all $\beta < \alpha$, then U is equal to $S(X)$.*

This fact can be seen as a topological formulation of Nash-Williams minimal bad sequence arguments, and has been noticed by Raoult 1988.

The proof given in Raoult 1988 is however based on a general argument that uses Zorn's lemma. It is also possible to prove the theorem directly, following the minimal bad sequence argument, by reasoning classically and using the axiom of choice.

Here is a sketch of the classical reasoning. Notice how difficult it would be to formalise this argument completely.

If U is not equal to $S(X)$ there exists $\alpha \in S(X)$ not in U. The set of $x \in X$ such that there exists $\alpha \in S(X)$ not in U satisfying $\alpha_1 = x$ is thus non empty. Since $<$ is well-founded, there exists $x_1 \in X$ minimal such that there exists $\alpha \in S(X)$ not in U satisfying $\alpha_1 = x$. In the same way, we can find $x_2 \in X$ minimal such that there exists $\alpha \in S(X)$ not in U satisfying $\alpha_1 = x_1$ and $\alpha_2 = x_2$. Using the axiom of dependent choice, we build an infinite word $x_1 x_2 x_3 \ldots$ such that, for each k, there exists $\alpha \in S(X)$ not in U extending $x_1 \ldots x_k$. Since U is open, this infinite word $\alpha = x_1 x_2 \ldots$ is not in U. Furthermore, we have by construction that $\beta < \alpha$ implies $\beta \in U$. This contradicts the hypothesis on U.

We show in the next section that the point-free version of this theorem can be proved directly constructively, using only as primitive the notion of inductive definition. This is yet another example of a general fact, already noticed in the case of Tychonoff's theorem, that a point-free version of a topological theorem can be proved directly constructively.

2. A point-free formulation

The point-free approach will consider only "observable elements" that is, only finite words. We will see now an infinite words $\alpha_1\alpha_2\ldots$ as a description of a sequence of successive choices $\alpha_1 \to \alpha_1\alpha_2 \to \alpha_1\alpha_2\alpha_3 \to \ldots$

A subset Q of $W(X)$ is **hereditary** iff $u \in Q$ whenever all v such that $u \to v$ are in Q. More generally, if $E \subseteq W(X)$, a subset $Q \subseteq W(X)$ is **hereditary on** E iff $u \in Q$ whenever $u \in E$ and $v \in Q$ if $u \to v$, $v \in E$. If $B \subseteq W(X)$, the hereditary closure E_B of B is the intersection of all hereditary subsets of $W(X)$ that contain B. This is a set-theoretical description of E_B. We can also consider E_B as an instance of an inductive definition, see Martin-Löf 71, and define E_B inductively by the introduction rule that $B \subseteq E_B$ and that E_B is hereditary. The operation from B to E_B correspons to the "eventually" operator used in temporal logic.

Notice that $v \in E_B$ mean classically that any infinite words that extend v is in $U(B)$.

It can be shown directly that E_B is monotone if B is monotone.

We can express in a point-free way when an open $U(B) \subseteq S(X)$ is the set of all infinite words. We say that a monotone $B \subseteq W(X)$ is a **bar** iff $E_B = W(X)$. For any subset $B \subseteq W(X)$, it can be checked classically, using the axiom of dependent choice, that B is a bar iff $U(B) = S(X)$.

More generally, if $B \subseteq W(X)$ is monotone and $E \subseteq W(X)$, we say that B is a **bar on** E iff the hereditary closure of $B \cap E$ in E is E, that is, if $F \subseteq E$ is such that $E \cap B \subseteq F$, and $x \in F$ whenever $x \in E$ and $y \in F$ if $x \to y$ and $y \in E$, then $E = F$.

Intuitively, a monotone $B \subseteq W(X)$ is a bar iff for any sequence of successive choice $u_1 \to u_2 \to u_3 \ldots$, we have eventually $u_p \in B$. Likewise, a monotone $B \subseteq W(X)$ is a bar on $E \subseteq W(X)$ iff for any sequence of successive choice $u_1 \to u_2 \to u_3 \ldots$, such that all u_k are in E, we have eventually $u_p \in B$.

In order to find a point-free equivalent of the open induction principle, let us first express it in intuitive terms that mention only observable terms. We start with a monotone subset $B \subseteq W(X)$ and we want to find a sufficient condition that implies that B is a bar. The open induction principle says that, in order for this to hold, it is enough to have $\alpha \in U(B)$ whenever $\beta < \alpha$ implies $\beta \in U(B)$. The condition that $\beta < \alpha$ implies $\beta \in U(B)$ can be interpreted as follow: for all p, if $v < \alpha_1\ldots\alpha_p$ and $\beta \in S(X)$ extends v, then $\beta \in U(B)$. This can still be expressed by saying that $v < \alpha_1\ldots\alpha_p$ implies $v \in E_B$. We then get by this analysis a statement that involves only observable terms. This is the following sufficient condition for B to be a bar. If we make a sequence of successive choices $\alpha_1 \to \alpha_1\alpha_2 \to \ldots$ in such a way that at each steps, $v < \alpha_1\ldots\alpha_k$ implies $v \in E_B$, then eventually $\alpha_1\ldots\alpha_p \in B$.

Let us state now formally the result of our analysis. Let $M(B)$ be the set of words $u \in W(X)$ such that $v < u$ implies $v \in E_B$.

Proposition. E_B *is hereditary on* $M(B)$.

Proof. Let $w \in M(B)$ be such that $wx \in E_B$ if $wx \in M(B)$. We show then $wx \in E_B$ for all $x \in X$ by well-founded induction on x. Indeed, if $wy \in E_B$ for all $y < x$, then $u \in E_B$ for all u such that $u < wx$, because in this case either $u < w$, which implies $u \in E_B$ because $w \in M(B)$, or $u = wy$ for some $y < x$, which implies $u \in E_B$ by hypothesis. Hence $wx \in M(B)$. This implies by hypothesis $wx \in E_B$. $\square$

Corollary. *If B is a bar on $M(B)$, then B is a bar.*

Proof. If B is a bar on $M(B)$, then by the proposition, the empty word is in E_B. $\square$

The content of this corollary is classically the same as the one of the open induction principle, hence the same as the one of the "minimal bad sequence" argument. It would be possible from it to deduce a purely inductive proof of Kruskal theorem. Observe that this statement refers only to observable terms, and the proof uses only the notion of inductive definition.

5. Conclusion

A general remark is how a (classically minor) change in the formulation of a mathematical result can change its logical strength. The open induction principle for instance is classically directly equivalent to its point-free version. But its point-free version is proved directly whereas the version with points need the excluded middle and the axiom of choice to be proved. A similar remark appears already in Lorenzen 1958, where it is shown how a point-free version of Cantor-Bendixson is directly expressed, whereas the usual formulation needs very strong logical principle.

It may be interesting to analyse how this remark applies to statements analysed in reverse mathematics, typically to statements that are proved equivalent to weak König's lemma.

It would be also quite interesting to apply this point-free method in more complicated combinatorial statements proved by topological means. For instance, the topological proof of Van der Waerden theorem about arithmetical progression seems to call for a point-free theory of a continuous action of a group on a compact space.

REFERENCES

1. Brouwer. L. E. J., *Uber definitionsbereiche von Funktionen*, Math. Ann. 96 (1927), 60-75.

2. Martin-Löf, P., *Notes on Constructive Mathematics*, Almqvist & Wiksell, 1968.

3. Martin-Löf, P., *Hauptsatz for the Intuitionistic Theory of Iterated Inductive Definitions*, Proceedings of the Second Scandinavian Logic Symposium, (1971), 179 - 216, J.E. Fenstad, editor.

4. De Bruijn, N. G. and Van Der Meiden, W., *Notes on Gelfand's Theory*, Indagationes 31 (1968), 467-464.

5. Nash-Williams, C., *On well-quasi-ordering finite trees*, Proc. Cambridge Phil. Soc. 59, (1963), 833-835.

6. Vickers, S., *Topology via Logic*, Cambridge Tracts in Theoretical Computer Science 5, 1989.

7. Johnstone, P.J., *Stone Spaces*, Cambridge Studies in Advanced Mathematics, 1981.

8. Coquand, Th., *An Intuitionistic Proof of Tychonoff's Theorem*, submitted to the Journal of Symbolic Logic (1991).

9. Martin-Löf, P., *Domain interpretation of type theory*, Workshop on the Semantics of Programming Languages, Chalmers (1983).

10. Abramsky, S., *Domain Theory in Logical Form*, Annals of Pure and Applied Logic (1991).

11. Lorenzen, P., *Logical Reflection and Formalism*, Journal of Symbolic Logic (1958).

12. Russell, B., *On order in time*, in Logic and Knowledges, essays 1901-1950, R.C. Marsh, editor, 1936.

13. Wiener, N., *A Contribution to the Theory of Relative Position*, Proc. Camb. Phil. Soc., Vol. 17 (1914).

14. Raoult, J.C., *An open induction principle*, INRIA Report (1988).

15. Russell, B., *Our Knowledge of the External World,* Cambridge University Press, 1914.

Implementing Constructive Real Analysis (Preliminary Report)

Jawahar Chirimar
Department of Computer and Information Science
University of Pennsylvania
chirimar@saul.upenn.cis.edu

Douglas J. Howe*
Department of Computer Science
Cornell University
howe@cs.cornell.edu

Abstract

In this paper we present the results of an investigation into the use of the Nuprl proof development system to implement higher constructive mathematics. As a first step in exploring the issues involved, we have developed a basis for formalizing substantial parts of real analysis. More specifically, we have: developed type-theoretic representations of concepts from Bishop's treatment of constructive mathematics that allow reasonably direct formalizations; used Nuprl's facility for sound extension of its inference system to implement automated reasoners for analysis; and tested these ideas in a formalization of rational and real arithmetic and of a proof of the completeness theorem for the reals (every Cauchy sequence converges).

1 Introduction

The seminal work of Errett Bishop on real analysis [1] demonstrated that much of higher mathematics can be rebuilt on constructive grounds. In Bishop's approach to mathematics, logical connectives are interpreted constructively so that, for example, knowing an existence property requires knowing a particular witness. Reasoning is limited to principles preserving this interpretation. Mathematics built in this way

---

*Both authors were supported, in part, by ONR contract N00014-88-K-0409 and NSF grant CCR-8616552.

has the property that proofs always supply the computational content of theorems. Thus, a proof that every member of some class of real-valued functions has a root must give a procedure for finding a root.

Formal logics for constructive reasoning are faithful to the idea that proofs give rise to programs. For each such logic, there is a procedure that maps formal proofs to the implicit programs. This gives a way of *implementing* mathematics: if a formal argument is carried out on a computer, then a program can be automatically *extracted*. If the computer checks that the proof is correct, then we will have a high degree of confidence that the program is correct.

In this paper we present the results of an investigation into the use of the Nuprl proof development system to implement higher constructive mathematics. This is a largely unexplored area. Most existing theorem-proving or proof-checking systems are not applicable since they are based on classical logics. The logic of Boyer and Moore's system is constructive, but it is not sufficiently expressive to deal with, for example, real analysis. Several interactive proof systems feature more expressive logics for constructive mathematics, logics such as the Calculus of Constructions [3], some of the type theories of Martin-Löf (for example [10]), and Hayashi's PX [6] which is based on Feferman's T_0 [4]. While many of these systems have been extensively used in a variety of problem domains, before our work there was very little investigation of their use as a basis for higher constructive mathematics. Since our work, there has been another substantial effort in this area. This is the work of Jones [8] which involves the formalization of some real analysis in the LEGO system of Pollack, using an extension of the Calculus of Constructions. Jones deals with roughly the same level of mathematics, but the formal proofs are far from complete.

The mathematics we focus on is constructive real analysis since it is a well-developed area and is rich in interesting computational content. There are two main problem areas in formalizing real analysis: choosing formal representations of mathematical concepts, and developing the automated reasoning facilities to make formal proof construction feasible. It is difficult to design a proof development system which supports both of these areas. Generally, the richer and more expressive the logic, the harder it is to find uniform theorem-proving procedures.

Nuprl has one of the most expressive logics of computation. This level of expressiveness seems to be required to deal with analysis; even with it, the representational problem is a significant one. The automation problem is more difficult; for example, there are no known analogues for Nuprl of the general purpose theorem-proving machinery that one finds in systems based on, say, resolution or term-rewriting.

As a first step in exploring these issues we have developed a basis for formalizing substantial parts of real analysis (for example, calculus). More specifically, we have

- developed type-theoretic representations of concepts from Bishop's mathematics that allow reasonably direct formalizations,

- used Nuprl's facility for sound extension of its inference system to implement automated reasoners that provide a reasonable level of assistance without user direction, and

- tested these ideas in an actual Nuprl formalization of rational and real arithmetic and of a proof of the completeness theorem for the reals (every Cauchy sequence converges).

Other theorems that are at the same level of difficulty as the completeness theorem, given the library of facts we have constructed, include the intermediate value theorem[1] (a function which is negative at 0 and positive at 1 has a root), and the Baire category theorem: in a complete metric space, a countable intersection of dense open sets is dense.

Because of time limitations, some of the lemmas about elementary arithmetic were left unproven (but assumed proven). We estimate that the proof effort for arithmetic is about 90% complete. Except for these general facts about arithmetic, our formal proof of the completeness of the reals is complete.

One of the main difficulties in representing Bishop's mathematics in type theory is dealing with "hidden" computational information. For example, consider sequences of real numbers. A sequence *converges* if it has a limit. Constructively, if we are given a convergent sequence then we must also be given the limit; thus we can think of a convergent sequence as a pair consisting of the sequence and its limit. However, in our mathematics we can deal with it as just being the sequence; the limit is implicitly there, and we can refer to it when necessary, as in the expression $\lim_{n \to \infty} x_n$. This device allows Bishop's mathematics to appear very similar to conventional mathematics. However, in a formalization this "hidden" information must be made explicit. Another representational difficulty has to do with constructive set theory.

The rest of the paper is organized as follows. First we give an example of a constructive proof which illustrates some of the issues that must be confronted when formalizing. Next we give a brief account of Nuprl's type theory. In Section 4 we discuss some of the representational problems that arise when formalizing in type theory, describe how we dealt with them, and give some details on how we formalized some of the basic concepts in real analysis. In Section 5 we describe some aspects of the proof of the main theorem, including the methods used to automate reasoning.

2 Example

We consider a simple, and somewhat contrived, theorem about finite sets. We start by chosing a computationally useful representation of finite sets. If A is a set, then define $\mathcal{F}(A)$, the set of finite sets over A, to be A *list*, the set of lists whose

---

[1]Part of a Nuprl formalization of this has already been done by Fred Brown, an undergraduate at Cornell who worked under the direction of the second author. This proof was developed far enough for the system to extract its computational content (a root finding program).

members come from A. In conventional mathematics we would additionally take equivalence classes to get a set with the right equality. However, equivalence classes are not computationally meaningful in general, so instead we will explicitly define an equality relation for $\mathcal{F}(A)$ and use it in place of equality over a quotient. Functions over $\mathcal{F}(A)$ must respect this equality.

For $s \in \mathcal{F}(A)$ and $x \in A$, define $x \in s$ if there is an i such that $s[i] = x$, where $s[i]$ is the ith member of the list s. Define $s \subseteq s'$ if for every $x \in s$, $x \in s'$. The equality relation for $\mathcal{F}(A)$ is defined by $s = s'$ if $s \subseteq s'$ and $s' \subseteq s$. We will also need the following definitions:

$$
\begin{aligned}
nonempty(s) &= \exists x.\, x \in s \\
s \cup s' &= s@s' \\
\mathcal{F}^+(A) &= \{\, s : \mathcal{F}(A) \mid nonempty(s) \,\},
\end{aligned}
$$

where $s@s'$ is the concatenation of the lists s and s'.

The theorem is as follows. Suppose

1. $f \in \mathcal{F}(A) \to \mathcal{F}(A)$,

2. $f(s \cup s') = f(s) \cup f(s')$ for all s and s', and

3. $nonempty(f(\{x\}))$ for all x.

Then for all $s \in \mathcal{F}^+(A)$, $nonempty(f(s))$.

The proof is as follows. Suppose

The proof is trivial: since $s \in \mathcal{F}^+(A)$, there exists $x \in s$, and we have

$$
f(s) = f(\{x\} \cup s) = f(\{x\}) \cup f(s)
$$

and $nonempty(f(\{x\}))$.

Since this proof is constructive, it should implicitly give us a procedure for computing the "witnessing" information in the conclusion, $nonempty(f(s))$, of the theorem. By definition of $nonempty$ and $\in$, this information should consist of a $y \in A$ and an i such that $f(s)[i] = y$. We now look at how a procedure to compute these values can be derived from our proof.

Assume the hypotheses 1–3 of the theorem, and suppose we are given an $s \in \mathcal{F}^+(A)$. Let $x \in s$ be the element guaranteed by the definition of $\mathcal{F}^+(A)$. The final witness is computed as follows.

1. Hypothesis 3 says that from x we can compute $y \in f(\{x\})$ and the position j of y in the list $f(\{x\})$.

2. By the definition of $\cup$, y also occurs at position j in $f(\{x\}) \cup f(s)$.

3. Because of the definitions of $=$ and $\subseteq$, Hypothesis 2 gives a procedure for computing, from y and j, a j' such that y occurs at position j' in $f(\{x\} \cup s)$.

4. Hypothesis 1 gives a procedure that computes, from witnessing information for $\{x\} \cup s = s$, witnessing information for $f(\{x\} \cup s) = f(s)$. This in turn gives a procedure which, given y and j', produces a j'' such that y occurs at position j'' in $f(s)$.

This procedure draws on witnessing information from various sources. First, membership in a subset is coupled with evidence: the assumption $s \in \mathcal{F}^+(A)$ provides x. Second, as in 3 and 4 above, equalities provide procedures which give a correspondence between equal sets. Finally, in 4 the fact that a function respects equality gives a procedure mapping equality witnesses to equality witnesses.

3 The Type Theory of Nuprl

We give a brief semantic account of Nuprl's type theory. For a more detailed account, and for the associated inference rules, see the Nuprl book [2].

The type theory consists of a programming language together with a type system over it. The programming language consists of a set *Term* of closed terms, together with a binary relation $\cdot \Downarrow \cdot$ over *Term*. A term t *evaluates* to t' if $t \Downarrow t'$. *Term* contains many of the constructs one finds in functional programming languages, and in particular it contains the untyped λ-calculus. We will introduce specific constructors as needed below.

Term also contains terms which serve to name types. These terms are the members of the set *Type* $\subset$ *Term*. For each $T \in$ *Type* there is an associated set $X_T \subset$ *Term* and equivalence relation $r_T \in X_T \times X_T$. The elements of X_T are the *members* of the type T, and r_T is the *equality* of T. We will write $t \in T$ for $t \in X_T$ and $t = t' \in T$ for $(t, t') \in r_T$.

Membership in types respects evaluation, in the sense that if $t \Downarrow t'$, then $t \in T$ if and only if $t' \in T$. Furthermore, if one of t, t' is in T, then $t = t' \in T$. As a simple example, consider the type of integers. This type is named by the constant *Int* $\in$ *Type*, and $t \in$ *Int* if and only if there is an integer constant n such that $t \Downarrow n$. Two members t and t' of *Int* are equal in *Int* exactly if they evaluate to the same constant. Note that, in distinction to other typed programming languages, being a member of a type says nothing about the "well-formedness" of components. A term containing arbitrary junk may be a member of a type, as long as a value of the right kind is produced by evaluation.

The type theory contains no logic *per se*; instead, logic is encoded in *Type* using the *propositions-as-types* correspondence. We only outline this correspondence here. See [2] for a treatment of this general idea that is tailored to Nuprl.

The basic idea is that we represent a logical proposition P by a type $T_P \in$ *Type* such that P is true exactly if T_P has a member. This member can be thought of as the computational content, or realizer, of P. Conversely, we can view any type as a proposition by viewing truth as non-emptiness and falsity by emptiness. For each of the connectives of first-order logic, there is a corresponding type constructor. This correspondence is as follows.

1. $x : A \to B$ corresponds to $(\forall x : A)B$

2. $x : A \times B$ corresponds to $(\exists x : A)B$

3. $A \to B$ corresponds to $A \Rightarrow B$

4. $A \times B$ corresponds to $A \land B$

5. $A \mid B$ corresponds to $A \lor B$

6. ϕ corresponds to $\perp$

7. $a = b \in A$ corresponds to $a = b \in A$

More common notations for the first two types are $\Pi\, x{:}\, A\,.\, B$ for 1 and $\Sigma\, x{:}\, A\,.\, B$ for 2.

A rough semantics for these types is as follows. We only deal with membership (ignoring equality). We give the form that the value of a term has when it is in the corresponding type above.

1. $\lambda x.\, b$, where for every $a \in A$, $b[a/x] \in B[a/x]$.

2. $\langle a, b \rangle$, where $a \in A$ and $b \in B[a/x]$.

3. $\lambda x.\, b$, where for every $a \in A$, $b[a/x] \in B$.

4. $\langle a, b \rangle$, where $a \in A$ and $b \in B$.

5. $inl(a)$, where $a \in A$, or $inr(b)$, where $b \in B$.

6.

7. *axiom*, where a and b are equal members of A (if a and b are unequal, then $a = b \in A$ has no members).

Thus 1 and 2 are "dependent" generalizations of function space and cartesian product respectively, 5 is a disjoint union type, 6 is the empty type, and 7 is a type whose purpose is precisely to encode the proposition that two members of a type are equal.

There are also *subtypes* and *quotient types*. The subtype type $\{\, x{:}A \mid B\,\}$ restricts A to members for which B, viewed as a proposition, is true. It satisfies

$$t \in \{\, x{:}A \mid B\,\} \;\Leftrightarrow\; t \in A \;\&\; B[t/x] \neq \emptyset.$$

Using the subtype constructor we can define the *squash* operator on types

$$\downarrow A \equiv \{\, x\colon Unit \,|\, A \,\}$$

where *Unit* is some fixed one-element type. Thus $\downarrow A$, viewed as a proposition, is true exactly if A is, but the members have been "squashed" into one element. This allows us to express the truth of A without specifying its computational content. The quotient type A/E has the same members as A but has a new equality relation given by E. It satisfies

$$t \in A/E \quad \Leftrightarrow \quad t \in A$$
$$t = t' \in A/E \quad \Leftrightarrow \quad E(x,y) \neq \emptyset.$$

Finally, there are types of types. In particular, there is a cumulative hierarchy of *universes* $U_1, U_2, \ldots$, where each U_i results from closing under the type constructors listed above together with the universes U_j for $j < i$.

4 Representing Analysis in Type Theory

Two fundamental notions in BCM (Bishop's constructive mathematics) are *set* and *operation*. Operations are naturally represented as members of function types in Nuprl. It is not so clear what to do with sets, however. According to Bishop, to define a set we must do two things:

- define how to construct a member of the set, and

- specify when two members are to be considered equal.

The type language of Nuprl gives sufficient means to specify membership, and it might seem natural to use the quotient type constructor to introduce the equality relation for a set. However, as pointed out in [2], this does not work, failing in particular for the set of real numbers. The problem is that there are operations we will want to define which do not respect the equality of real numbers. An example is the operation κ which computes an integer upper bound for a real number. If the set of real numbers were defined as a quotient $R/=$, then we would not have $\kappa \in R/= \rightarrow Int$, since this requires functionality.

The quotient type also seems attractive when one considers the *axiom of choice*. Consider the following formulation AC, for sets S and S'.

$$(\forall x\colon S.\, \exists y\colon S'.\, P(x,y)) \;\Rightarrow\; (\exists f\colon S \rightarrow S'.\, \forall x\colon S.\, P(x,f(x)))$$

In BCM, this is either true or false[2], depending on the interpretation of $S \rightarrow S'$. If it is interpreted as the set of operations, not necessarily equality-respecting, that

---

[2] i.e., there is a recursion-theoretic refutation

map members of S to S', then AC is true since it merely captures the constructive meaning of the quantifiers $\forall$ and $\exists$. If members of $S \to S'$ must also respect equality, then AC is false, since we cannot take an arbitrary method that computes y from x and turn it into a functional one. However, in type theory, if we take $\to$ to mean function-type and use quotients for S and S', then AC is again true, since the $\forall$ quantifier is represented as a function type. So, although this might seem to be a useful feature of using quotients, it is likely of little utility in formalizing BCM since the corresponding property in BCM is false.

So, instead of forming quotients, we will keep equalities separate from the types representing collections of members. The function type constructor of Nuprl will be used for operations, and a function *set* constructor will be defined that restricts operations to those respecting equality.

A second issue related to sets is the computational content of equality propositions. As we saw in the example of Section 2, it is possible for equality reasoning in a constructive proof to contribute to the computational content of the proof. However, in the "elementary" parts of BCM (not including, in particular, measure theory) equality has no interesting computational content. We capture this by insisting that equality relations of sets be *self-realizing*, in the sense that a witness for the equality of members x and y can be determined from x and y, in the case that they are equal. This restriction somewhat simplifies our representation, allowing, for example, functions over sets to be represented as functions in type theory (instead of as functions together with a method of transforming equality witnesses).

There is a third representational issue related to sets that we avoid in this development, and that is how to represent subsets. As the example in Section 2 shows, membership in a subset can be computationally significant. This precludes using Nuprl's subtype constructor to represent subsets. Finding the right representation of subsets is one of the most difficult problems in formalizing BCM in type theory. *In principle* there is no problem, since Bishop's treatment of subsets, where a subset is a set together with an injection into the containing set, can be directly formalized in type theory. The problem is that there appears to be no good way to "abstract away from", or suppress, the injection information when formalizing this way. This kind of suppression is essential to making BCM appear as conventional mathematics, and the inability to do it in a formal account will likely necessitate a large amount of clutter.

We now give some of the details of our representation of real analysis in type theory. We show the actual Nuprl terms representing various mathematical concepts. We show the terms essentially as they would appear to a user of the Nuprl system (this includes the use of "special symbols" such as $\forall$).

We take a set to consist of a type A in U_1 together with an equivalence relation of type $A \times A \to U_1$. The predicate which asserts that a relation is an equivalence has type

```
A:U1 -> r:((A#A)->U1) -> U1
```

where we use the Nuprl syntax # for × and -> for →, and is defined to be

$$\lambda \ A \ r. \ refl\{A:U1\}(r) \ \& \ sym\{A:U1\}(r) \ \& \ trans\{A:U1\}(r)$$

where refl, etc, have the obvious definitions. We are using here the definitions of the logical connectives that were given in the previous section.

The type of all sets is

$$A:U1 \ \# \ r:((A\#A)\text{->}U1) \ \# \ eq\_reln\{A:U1\}(r)$$
$$\# \ \forall x,y:A. \ \downarrow(r(x,y)) \ => \ r(x,y)$$

Thus a set is a 4-tuple, consisting of a type, an equality, a witness for the fact that the relation is an equality, and a witness for the self-realizability of equality (using the squash operator defined in the previous section). We define $|S|$ to be the type component of the set S, and define x=y in S to be the application of the relation component to the pair <x,y>.

It is straightforward to define the product and function space constructors for sets. The type component of the product set is given by

$$\lambda \ S1 \ S2. \ |S1|\# \ |S2|$$

and the relation component by

$$\lambda \ S1 \ S2. \ \lambda \ x,y. \ x.1 = y.1 \ in \ S1 \ \& \ x.2 = y.2 \ in \ S2$$

For the function space, the type component is given by

$$\lambda \ S1 \ S2. \ \{ \ f:|S1|\text{->}|S2| \ |$$
$$\forall x,y:|S1|. \ x = y \ in \ S1 \ => \ f(x) = f(y) \ in \ S2 \ \}$$

and the relation component by

$$\lambda \ S1 \ S2. \ \lambda \ f,g. \ \forall x:|S1|. \ f(x) = g(x) \ in \ S2.$$

It is straightforward to define the set Q of rational numbers and the associated operations. Using these we define the type of real numbers, just as in Bishop's book [1], as

$$\{ \ x:|Q|sequence \ | \ \forall i,j:N+. \ |x(i)-x(j)|\leq(1/i)+(1/j) \ \},$$

where N+ is the type of positive natural numbers. The equality on real numbers is

$$\lambda \ a,b. \ \forall n:N+. \ |a(n)-b(n)|\leq(2/n)$$

The formalization of the required basic arithmetic proceeds as in Bishop's book.

Beyond basic arithmetic, there are only a few definitions required to state the main theorem. First, we need a definition of when a sequence converges. For a a sequence of real numbers we define

```
a↓ == ∃b:|R|. ∀i:N+. ∃Ni:N+.  ∀m:N+. Ni≤m => |a(m)-b|≤(1/i)·
```

where `1/i·` is the injection of the rational `1/i` into the reals. Note that by this definition, a sequence converges just if it has a limit; constructively, this means that if we are given a convergent sequence, we are also given its limit. Thus the usual limit operation merely accesses this information. We define

```
lim(p) == p.1
```

where `p.1` is the first component of the pair `p`, and with this definition we have the theorem

```
>> ∀a:|R|sequence. ∀p:a↓. lim(p) in |R|
```

(`>>` is Nuprl's syntax for `⊢`). This is an example where a reference to witnessing information, which is implicit in the informal account, must be made explicit. A more natural formulation of `lim` would take as an argument a member of the subset of convergent sequences, but it was not worth developing a theory of subsets just for this one case.

A sequence a is *Cauchy* if

```
∃M:N+sequence. ∀k:N+.  ∀m,n:N+.
    M(k)≤m => M(k)≤n => |a(m)-a(n)|≤(1/k)·
```

Finally, we can state the main theorem.

```
>> ∀a:|R|sequence. Cauchy(a) => a↓
```

5 Lemmas and Proofs

The entire Nuprl library of definitions, lemmas and proofs leading up to the completeness theorem took about 3/4 of a semester of part-time work on the part of the first author (who was an undergraduate at the time). About 3/4 of this time was spent proving elementary facts about rational arithmetic. The amount of rational arithmetic developed was roughly comparable to what is contained in the book "Grundlagen" that was formalized in the AUTOMATH system [9].

```
1. a: R_type
2. b: R_type
3. ∀n:N. |a(n)-b(n)| ≤ 2/n
4. j: N+
>> ∃Nj:N+. ∀m:N+. Nj≤m => |a(m)-b(m)| ≤ 1/j

BY (ITerm '2*j'...)

1* >> 0<2*j

2* 5. m: N+
   6. 2*j ≤ m
   7. 1/j < |a(m)-b(m)|
   >> False
```

Figure 1: A proof step.

Because of time limitations, some of the lemmas about elementary arithmetic were left unproven (but assumed proven). We estimate that the proof effort for arithmetic is about 90% complete.

Except for these general facts about arithmetic, our formal proof of the completeness of the reals is complete. It involves roughly 150 user-specified inference steps. Just what is involved in a "step" is difficult to characterize, so we will just give two examples of rather typical simple steps.

The steps we look at are from a proof of a theorem that gives a useful characterization of equality of real numbers. The theorem is

```
>> ∀a,b:R_type.
     R_eq(a,b) <=>
     ∀j:N+. ∃Nj:N+. ∀m:N+. Nj≤m => |a(m)-b(m)| ≤ 1/j
```

where R_eq stands for real equality.

In Figure 1 is the main step in proving the left-to-right implication in the theorem. The *goal* in this step is to prove ∃Nj... under the hypotheses numbered 1 to 4. In other words, assuming that a and b are equal (hypothesis 3), and given a positive integer j, show that for sufficiently large m, the rational approximations a(m) and b(m) are within 1/j of each other. Because of hypothesis 3, "sufficiently large" here can be taken to be 2*j. In the figure, the text following the word BY was entered by the user and invokes a program, called a *tactic* and written in the ML programming language [5], that makes progress toward a proof of the goal by supplying the term 2*j for Nj and attempting to prove any simple proof obligations that arise. The two

```
* top 2 1 1 1 1
1. a: |Q| sequence
2. ∀i,j:N+. |a(i)-a(j)| ≤ 1/i+1/j
3. b: |Q| sequence
4. ∀i,j:N+. |b(i)-b(j)| ≤ 1/i+1/j
5. ∀j:N+. ∃Nj:N+. ∀m:N+. Nj≤m => |a(m)-b(m)| ≤ 1/j
6. n: N+
7. c: N+
8. 2/n+1/c < |a(n)-b(n)|
>> False

BY (InstantiateHyp ['3*c'] 5 ...)

1* >> 0<3*c

2* 9. ∃Nj:N+. ∀m:N+. Nj≤m => |a(m)-b(m)| ≤ 1/(3*c)
   >> False
```

Figure 2: Another proof step.

subgoals, appearing below the tactic, are generated by the system, and it suffices to prove these in order to prove the goal. The two subgoals share the same hypotheses as the goal, although their display is suppressed by the system, and the second subgoal adds new hypotheses 5 to 7. The first subgoal is to prove 2*j is positive, and the second is to show that $|a(m)-b(m)|{\leq}1/j$, or more specifically, because of the way we defined $\leq$ in terms of <, to show that $1/j<|a(m)-b(m)|$ is contradictory.

In Figure 2 is a step from the proof of the right-to-left implication. This step is after we invoked a lemma which reduced the proof to showing that

$$|a(n)-b(n)| \leq 2/n + 1/c$$

for arbitrary c. We prove this by instantiating assumption 5, taking j to be 3*c.

A collection of tactics developed as a general package for Nuprl was employed in the development of our library. This collection contains basic tactics to instantiate lemmas, normalize terms, expand definitions, backchain, and guess many of the parameters required by the inference rules. A special tactic called the *autotactic* was used to prove simple facts about integer arithmetic, typechecking goals (to show a term belongs to some type), and simple propositional facts.

In spite of this basic collection, proving the repetitive membership goals and equalities and inequalities was still extremely time consuming and tedious, as was instantiating many of the lemmas one by one. So we developed some special tactics which were applicable in the specific setting of the analysis library.

The approach we took to equality proofs was a modification to a tactic originally conceived by Brown. The idea is to rewrite the two sides of an equality to normal forms and then compare these whether these are equal or not. This requires some lemmas to be proved about the commutativity and associativity of the operations involved such as addition and multiplication. We also wrote tactics to automatically apply functionality facts (that is, operators' respect for equality).

The approach to inequalities was rather novel. Here we devised several different strategies. One of the typical applications is replacing equals by equals and chaining through inequalities or equalities as appropriate. Obviously a tactic which would chain through a series of lemmas would be much better than having to instantiate all of these by hand. So we have a rather elaborate list of lemmas about inequalities of rationals which are used by a tactic. We also implemented a brute force search procedure which tries all possibilities until the proof is complete. This is rather time intensive, but it possible to sharpen the tactic by having special cases and possibly more redundancy in the lemma library.

In Nuprl's type theory, it is in general undecidable whether a term has a given type. Thus, typechecking is not automatic (and efficient) as in other type theories, and most inference rules generate subgoals that require certain parts of a goal to be well-typed. Because of the number of such goals that arise, proving them, even when, as is usually the case, they can be handled automatically, is time-consuming.

We solved this problem by keeping track of membership proofs we construct. When trying to prove a membership goal we reduce the hypotheses to a canonical form and prune the list of hypotheses to a certain minimal one. Now in the case of membership goals the list is fortunately identical when the same membership goal turns up the next time. So we install the proof in a hash table the first time, and when the same goal recurs we just paste in the stored original proof. We extended this idea to allow hashing of any proofs. These two rather simple tricks turned out to be very effective. The saving of non-membership goals was especially useful for simple inequalities that frequently arose when we were working with sequences of reals and rationals.

6 Conclusion

Clearly we have just begun to explore the implementation of real analysis in Nuprl. There are representational problems, such as the treatment of subsets, which must be dealt with. More importantly, we need to design new tactics. For example, we need tactics for simplification and for aiding reasoning about the monotonicity of operators with respect to inequalities. As reasoning about equalities and inequalities is so pervasive in analysis at this level, we expect that *reflection*, of the kind developed by Howe [7], will be useful.

References

[1] E. Bishop. *Foundations of Constructive Analysis.* McGraw-Hill, New York, 1967.

[2] R. L. Constable, et al. *Implementing Mathematics with the Nuprl Proof Development System.* Prentice-Hall, Englewood Cliffs, New Jersey, 1986.

[3] T. Coquand and G. Huet. The calculus of constructions. *Information and Computation*, 76:95–120, 1988.

[4] S. Feferman. A language and axioms for explicit mathematics. In Dold, A. and B. Eckmann, editor, *Algebra and Logic*, volume 450 of *Lecture Notes in Mathematics*, pages 87–139. Springer-Verlag, 1975.

[5] M. J. Gordon, R. Milner, and C. P. Wadsworth. *Edinburgh LCF: A Mechanized Logic of Computation*, volume 78 of *Lecture Notes in Computer Science*. Springer-Verlag, 1979.

[6] S. Hayashi and H. Nakano. *PX: A Computational Logic.* Foundations of Computing. MIT Press, Cambridge, MA, 1988.

[7] D. J. Howe. Computational metatheory in Nuprl. *CADE-9*, pages 238–257, May 1988.

[8] C. Jones. Completing the rationals and metric spaces in LEGO. In *Proceedings of the Second B.R.A. Workshop on Logical Frameworks*, Edinburgh, UK, May 1991. (To appear.).

[9] L. S. Jutting. *Checking Landau's "Grundlagen" in the AUTOMATH system.* PhD thesis, Eindhoven University, 1977.

[10] P. Martin-Löf. Constructive mathematics and computer programming. In *Sixth International Congress for Logic, Methodology, and Philosophy of Science*, pages 153–175, Amsterdam, 1982. North Holland.

Examples of semicomputable sets of real and complex numbers

J.V. Tucker[1]

Department of Mathematics and Computer Science,
University College of Swansea, Swansea SA2 8PP, Wales

J.I. Zucker[2]

Department of Computer Science and Systems,
McMaster University, Hamilton, Ontario L8S 4K1, Canada

Abstract. We investigate the concept of semicomputability of relations on abstract structures. We consider three possible definitions of this concept, which all reduce to the classical notion of recursive enumerability over the natural numbers. By working in the algebra of the reals, with and without order, we find examples of sets which distinguish between these three notions. We also find interesting examples of sets of real and complex numbers which are semicomputable but not computable.

0 Introduction

The theory of computability over classes of many-sorted abstract algebras generalises the theory of computable functions and relations on the natural numbers. In principle, the theory has many uses in computer science, logic and mathematics because it can be applied to analyse *finite computation* on *any* algebraic structure A, whether A represents the data types of some software or hardware; is a model of a first order theory; or is a specific group, ring, field, etc.

In this computability theory many equivalent models of computation are used to define computable functions on an algebra A (uniformly for all A in a class K of algebras). Some of these models are generalisations to A (and K) of sequential deterministic models of computation on the natural numbers. Here we use parallel deterministic functional and imperative models.

The basic computability theory has been developed primarily to tackle problems in programming language theory, starting with the work of E. Engeler in 1965. An early study of recursion theoretic aspects is Friedman [1971]; early studies of algebraic applications are Herman and Isard [1970], Engeler [1975] and Tucker [1980]. A fairly detailed picture can be obtained from Shepherdson [1985] and Tucker and Zucker [1988].

---

[1] Research supported by SERC Research Grant GR/F 59070, by MRC Research Grant SPG 9017859, and by an academic travel grant from the British Council.

[2] Research supported by a grant from the Science & Engineering Research Board of McMaster University, by a grant from the Natural Sciences and Engineering Research Council of Canada, and by an academic travel grant from the British Council.

Typeset by $\mathcal{A}\mathcal{M}\mathcal{S}$-TEX

Definitions of a semicomputable set or relation R on an algebra A generalise those of a recursively enumerable set of natural numbers, and play a leading role in a specification theory that complements the computation theory; in the basic proof theory of first order theories; and in applications, such as dynamical systems on spaces $\mathbf{R}^n$ of real and complex numbers. Recent relevant papers illustrating these areas of application respectively are: Tucker and Zucker [1989, 1991b]; Tucker, Wainer and Zucker [1990] and Tucker and Zucker [1991a]; and Blum, Shub and Smale [1989].

In this note we study concepts to do with semicomputability in the case of algebras of real and complex numbers. It is one of a series taking up the theory of computation on many-sorted algebras which we established in Tucker and Zucker [1988]. Hopefully this paper will illustrate the versatility of the general theory when applied to a specific situation—here, the real and complex numbers.

To compute on a many-sorted algebra A we first extend A by adding the sets of natural numbers, N, booleans, B, and finite sequences, or arrays, A^*, of elements of A, together with appropriate operations to make a structure A^* with reduct A. We will consider functions computed by induction schemes, or (equivalently) by 'while' programs, over A^*. (See Tucker and Zucker [1988, 1989, 1991a, b, c] for details.)

We consider the effectivity of a set or relation R on A by three ideas:

(i) R is *simply semicomputable*, in the sense that R is the domain of a partial computable function on A^*.

(ii) R is *projectively semicomputable*, in the sense that R is the projection over A of the domain of a partial computable function on A^*.

(iii) R is *projectively star-semicomputable*, in the sense that R is the projection over A^* of the domain of a partial computable function on A^*.

These definitions generalise the two main characterisations of recursive enumerability on the natural numbers. (The other main characterisation, using ranges of computable functions, is equivalent to the use of projections.)

We will prove that the above three notions of semicomputability are distinct by giving relations on the field R of real numbers, and considering their semicomputability relative either to the structure $\mathcal{R}$ of real numbers without order, or to the structure $\mathcal{R}_<$ of the ordered real field. We will show that:

(a) the order relation on R is (ii) but not (i) in $\mathcal{R}$; and

(b) a certain real closed subfield of R is (iii) but not (ii) in $\mathcal{R}_<$.

In Section 1 we summarise ideas about many-sorted algebras and their extension by booleans, natural numbers and finite sequences; and we consider algebras of real and complex numbers. We then review, in Section 2, computability over such structures, and in Section 3, the various notions of semicomputability. In Section 4 we give new versions of a theorem of E. Engeler that are invaluable tools in what follows. In Sections 5 and 6 we give the examples which distinguish the various notions of semicomputability, and in Section 7 we reconsider an example of a semicomputable, non-computable set of reals described in Blum, Shub and Smale [1989]. The effective content of their work can be obtained from the general theory. The paper concludes with Section 8, in which we consider the significance of our computability theory for constructivity.

We thank the referees for helpful suggestions.

1 Abstract structures

In this section we review briefly some concepts defined and discussed in Tucker and Zucker [1988, 1989, 1991a, b, c].

1.1 Standard signatures and structures

A *standard signature* Σ specifies (1) a finite set of *sorts: algebraic sorts* $1, \ldots, r$ (for some $r \geq 0$), and the *numerical sort* N and *boolean sort* B; and (2) finitely many *function symbols* F, each having a *type* $(i_1, \ldots, i_m; i)$, where $m \geq 0$ is the arity of F, $i_1, \ldots, i_m$ are the *domain sorts* and i is the *range sort* (including the case $m = 0$ for *constant symbols*). These include symbols for certain *standard operations* associated with the sorts N and B: (*a*) *arithmetical operations*, namely the constant 0, successor operation S and order relation '<' on the natural numbers; and (*b*) *boolean operations*, including a complete set of propositional connectives, the constants true and false, and an *equality* operator eq_i at some sorts i, including (at least) $i = \mathsf{N}$ and $i = \mathsf{B}$. We call those sorts i with the equality operator eq_i, *equality sorts*.

Relations are interpreted as boolean-valued functions.

We make one further assumption on Σ:

INSTANTIATION ASSUMPTION. *Each sort of Σ is* instantiated, *i.e. there is a closed term of each sort.*

A Σ-structure A has, for each sort i of Σ, a carrier set A_i, and for each function symbol F of type $(i_1, \ldots, i_m; i)$, a function $F^A : A_{i_1} \times \cdots \times A_{i_m} \longrightarrow A_i$. A structure A is *standard* if $A_\mathsf{N} = \mathsf{N}$, the set of natural numbers, $A_\mathsf{B} = \mathsf{B} = \{\mathsf{tt}, \mathsf{f}\}$, the set of truth values, and the standard operations have their *standard interpretations* on N and B, so that, in particular, the equality symbol is interpreted as identity on each equality sort.

We only consider standard signatures and structures. Note however that any many-sorted structure can be *standardized* by the adjunction of the sets N and B, together with their standard operations.

1.2 Strictly standard signatures and structures: Three examples

We consider a notion stricter than standardness, namely *strict standardness*.

A standard signature Σ is *strictly standard* if the *only operations* with range sort N are the *standard operations* listed in §1.1. The structure A is *strictly standard* if its signature is.

REMARK. Any standardized structure is automatically strictly standard!

We present four examples of such structures which will be the focus of attention in the rest of this paper.

(1) The *field of reals* $\mathcal{R} = (\mathsf{R},\mathsf{N},\mathsf{B},\dots)$, with constants 0 and 1, the operations $+$, $\times$ and $-$, and equality on R, standardized by the adjunction of the sets N and B, together with their standard operations;

(2) the *ordered field of reals* $\mathcal{R}_<$ formed from $\mathcal{R}$ by adjoining the order relation $<$: $\mathsf{R}^2 \to \mathsf{B}$.

(3) the *field of complex numbers* $\mathcal{C} = (\mathsf{C},\mathsf{R},\mathsf{N},\mathsf{B},\dots)$, with constants 0, 1 and i, the operations $+$, $\times$ and $-$, and equality on C, and including the structure $\mathcal{R}$, as well as the operations re, im$: \mathsf{C} \to \mathsf{R}$ and their inverse $\pi : \mathsf{R}^2 \to \mathsf{C}$.

(4) the *field of complex numbers with order* $\mathcal{C}_<$ formed from $\mathcal{C}$ by again adjoining the order relation '$<$' on R.

Let $\Sigma(\mathcal{R})$, $\Sigma(\mathcal{R}_<)$, $\Sigma(\mathcal{C})$ and $\Sigma(\mathcal{C}_<)$ be their signatures respectively.

Note, by the above Remark, that all four structures are strictly standard. Further, $\mathcal{R}$ is a $\Sigma(\mathcal{R})$-reduct of $\mathcal{R}_<$ and $\mathcal{C}$, which are, in turn, $\Sigma(\mathcal{R}_<)$- and $\Sigma(\mathcal{C})$-reducts of $\mathcal{C}_<$.

1.3 The unspecified value u; Structures A^u of signature Σ^u

Now fix a (not necessarily strictly) standard signature Σ, and fix a particular standard Σ-structure A. We will expand A in two stages (in this subsection and the next).

For each sort i let u_i be a new object, representing an "unspecified value", and let $A_i^u = A_i \cup \{u_i\}$. For each function symbol F of Σ of type $(i_1,\dots,i_m; i)$, extend its interpretation F^A on A to a function $F^{A,u} : A_{i_1}^u \times \cdots \times A_{i_m}^u \longrightarrow A_i^u$ by *strictness*— i.e., the value is defined as u whenever any argument is u. Then the structure A^u, with signature Σ^u, contains: (i) the original carriers A_i of sort i, and functions F^A on them; (ii) the new carriers A_i^u of sort i^u, and functions $F^{A,u}$ on them; (iii) a constant unspec$_i$ of type i^u to denote u_i as a distinguished element of A_i^u; (iv) a boolean-valued function Unspec$_i$ of type $(i; \mathsf{B})$, the characteristic function of u_i; (v) an *embedding function* i$_i$ of type $(i; i^u)$ to denote the embedding of A_i into A_i^u, and the *inverse* function j$_i$ of type $(i^u; i)$, which maps u_i to the denotation of some closed term in A_i (this being possible by the Instantiation Assumption) for each sort i; and finally (vi) an *equality* operator on A_i^u for each equality sort i.

1.4 Structures A^* of signature Σ^*

Define, for each sort i, the carrier A_i^* to be the set of pairs $a^* = (\xi, l)$ where $\xi : \mathsf{N} \to A_i^u$, $l \in \mathsf{N}$ and for all $n \geq l$, $\xi(n) = u_i$. So l is a witness to the "finiteness" of ξ, or an "effective upper bound" for a^*. The elements of A_i^* have "starred sort" i^*, and can be considered as finite sequences or *arrays*.

The resulting structures A^* have signature Σ^*, which extends Σ^u by including, for each sort i, the new "starred sorts" i^* as well as i^u, and also symbols for functions to return the *length* of an array; to *change the length* of an array; to return the *value* $a^*[n]$ of an array a^* at an index n; to *update* an array at an index by a given value (of the correct sort); and also a constant for the *null array* of each sort; and the *equality* operator on A_i^* for each *equality sort* i.

(The justification for this is that if a sort i has computable equality, then clearly so has the sort i^*, since it amounts to testing equality of finitely many pairs of objects of sort i, up to a computable length.)

Note that A is a Σ-reduct of A^u and of A^*, and A^u is a Σ^u-reduct of A^*.

The reason for the use of starred sorts in the present work (as in Tucker and Zucker [1988, 1991a, b, c]) is the lack of effective coding of finite sequences within abstract structures in general.

1.5 Projections

We collect some definitions and notation.

(1) If $\vec{k} = k_1, \ldots, k_m$ $(m \geq 0)$ is a list of sorts then $A[\vec{k}]$ denotes $A_{k_1} \times \cdots \times A_{k_m}$.

(2) A function on A of *type* $(\vec{k}; l)$ is a partial function from $A[\vec{k}]$ to A_l (by "function" we will always mean *partial function*), and a relation on A of *type* $\vec{k}$ is a subset of $A[\vec{k}]$.

(3) A relation on A of type $\vec{k}$ is *algebraic* if none of its arguments is of sort N, i.e., for $i = 1, \ldots, n$, $k_i \neq \mathsf{N}$.

2 Computable functions

In this section, we again review work discussed in Tucker and Zucker [1988, 1989, 1991a, b, c]. We consider two approaches to the definition of classes of computable functions: by *induction schemes* and by *imperative programs*.

2.1 Definability of functions by induction schemes

We use *induction schemes* over Σ to define functions over A. These generalize the schemes for partial recursive functions over N in Kleene [1952]. We define two classes of functions.

2.1.1 A model with bounded memory: Inductive computability

The class of inductively computable functions is generated by schemes for the operations of Σ as *initial functions and constants, projection, composition, definition by cases, simultaneous primitive recursion on N and the μ or least number operator*.

Notice the last two schemes use the standardness of the structures, *i.e.* the carrier N.

2.1.2 A model with unbounded memory: Star-inductive computability

A function on A is star-inductively computable ("star-computable" for short) if it is defined by an induction scheme over Σ^*, interpreted on A^* (*i.e.*, using starred sorts in its definition).

2.2 Definability of functions by input/output programs

An *i/o-program* over Σ is defined to be a triple $[S, \vec{v}, w]$, consisting of a deterministic program S in some programming language over Σ, together with a list of *input variables* $\vec{v}$ and an *output variable* w (of sorts $\vec{k}$ and l, say). Such a triple defines (in an obvious way) a function on A of type $(\vec{k}; l)$.

Note that there may also be *auxiliary variables* in S (distinct from the input and output variables), which we assume to be completely uninitialized.

2.2.1 A model with bounded memory: 'while'-computability

Now consider a 'while' programming language (without arrays) over Σ. A function on A is 'while'-*computable* if it is computable by a 'while' program on A.

This approach to the definition of computable functions is equivalent to that in §2.1.1:

THEOREM. *Let f be a function on A. Then*

$$f \text{ is 'while'-computable} \iff f \text{ is inductively computable.}$$

2.2.2 A model with unbounded memory: 'while'-star computability:

A function on A is 'while'-*star-computable* if it is computable by a 'while' program on A^*. (Of course, the input and output variables will be of unstarred sorts; only the auxiliary variables may be of starred sorts.)

Again, this notion coincides with that in §2.1.2, by the above Theorem applied to A^*:

THEOREM. *Let f be a function on A. Then*

$$f \text{ is 'while'-star-computable} \iff f \text{ is star-inductively computable.}$$

2.3 Generalized Church-Turing Thesis for deterministic computation

In Chapter 4 of Tucker and Zucker [1988], (a class equivalent to) the class of star inductively computable functions was examined as a possible formalization of "effective calculability" over abstract data types. Many formulations were found to be equivalent, and this and other considerations led to the postulation of a *Generalized Church-Turing Thesis*, which (in the present context) can be formulated as follows:

Computability of functions on structures by deterministic algorithms can be formalized by star-inductive (or equivalently, 'while'-star) computability.

2.4 Two facts about inductive computability

PROPOSITION 1. *(a) If f is an inductively computable function on A of type $(\vec{k}; l)$ and $\vec{x} \in A[\vec{k}]$, then $f(\vec{x})$ (if defined) lies in the Σ-substructure of A generated by $\vec{x}$.*

(b) If f is a star-computable function on A of type $(\vec{k}; l)$ and $\vec{x} \in A[\vec{k}]$, then $f(\vec{x})$ (if defined) lies in the Σ^*-substructure of A generated by $\vec{x}$.

This yields the following negative computability results.

COROLLARIES. *(1) The square root function is not star-computable in $\mathcal{R}$ or $\mathcal{R}_<$.*

(2) The mod function $(z \mapsto |z|)$ is not star-computable in $\mathcal{C}$ or $\mathcal{C}_<$.

PROOF: (1) The subset of R generated from the empty set by the constants and operations of $\Sigma(\mathcal{R})^*$ (or $\Sigma(\mathcal{R}_<)^*$) is the set Z of integers. But $\sqrt{2}$ is not in this set.

(2) The subset of R generated from the empty set by the constants and operations of $\Sigma(\mathcal{C})^*$ (or $\Sigma(\mathcal{C}_<)^*$) is again Z. But again, $|1 + i| = \sqrt{2}$ is not in this set. $\square$

Note that the mod function *would* be computable in $\mathcal{C}$ if we adjoined the square root function to the structure $\mathcal{R}$ (as a reduct of $\mathcal{C}$).

For computability in ordered "euclidean" fields incorporating the square root operation, see Engeler [1974].

The following proposition will be used later.

PROPOSITION 2. *(a) Term evaluation on A is star-inductively computable on A.*

(b) If A is one of the structures $\mathcal{R}$, $\mathcal{R}_<$, $\mathcal{C}$ or $\mathcal{C}_<$, then term evaluation on A is actually inductively computable.

More precisely: Fix a list of variables $\vec{w}$ of sorts $\vec{k} = k_1, ..., k_m$. Let *Term*$(\vec{w})$ be the class of terms over Σ with variables among $\vec{w}$ only, and let $\ulcorner t \urcorner \in \mathsf{N}$ denote the Gödel number of the term $t \in$ *Term*$(\vec{w})$. Then for $i = 1, ..., m$, the map

$$te_A^i : \mathsf{N} \times A[\vec{k}] \to A_{k_i},$$

where $te_A^i(\ulcorner t \urcorner, \vec{a})$ is the value of t when $\vec{w}$ is evaluated as $\vec{a}$, is *star-inductively computable* on A, or actually *inductively computable* if A is one of the structures named in part (b).

3 Notions of Semicomputability

We consider *semicomputability* of relations, intended to generalize to A the notion of *recursive enumerability* over N. Again, there are two approaches to the definition of classes of semicomputable relations: by *induction schemes* and by *imperative programs*.

3.1 Definability of relations by induction schemes

3.1.1 Computable relations

A relation on A is defined to be *(inductively) computable*, or *star-computable*, if its (boolean-valued) characteristic function is computable or star-computable (respectively).

Now we consider different notions of *semicomputability*, all coinciding in the classical case over N. Let R be a relation on A.

3.1.2 Semicomputable relations

DEFINITIONS. (1) R is *(simply) semicomputable* iff R is the domain of an inductively computable function.

(2) R is *star-semicomputable* iff R is the domain of a star-computable function.

For these notions we have versions of Post's Theorem:

PROPOSITIONS. *(1) A set on A is computable iff both it and its complement are semicomputable.*

(2) A set on A is star-computable iff both it and its complement are star-semicomputable.

3.1.3 Projectively semicomputable relations

DEFINITIONS. (1) R is *projectively semicomputable* iff R is a *projection* of a semicomputable relation on A (ie, "existentially quantifying out" certain of the arguments).

(2) R is *projectively star-semicomputable* iff R is a *projection* on A of a semicomputable relation on A^*.

In general, projective semicomputability is weaker than semicomputability, as we will see (§6). However, in the special case of existential quantification over N, we have equivalence:

PROPOSITION. *Suppose $R \subseteq A[\vec{k}, \mathsf{N}]$ is semicomputable. Then so is its projection $\{\vec{x} \mid \exists z\, R(\vec{x}, z)\}$ on $A[\vec{k}]$.*

PROOF: Briefly, we can effectively "search" for the existentially quantified z by means of the μ operator. $\square$

Further, projective star-semicomputability is in general weaker than projective semicomputability, as we will see (§7).

We will give a "structural" characterization for star-semicomputable relations, among those which are algebraic on strictly standard structures (Corollary 1 of Engeler's Lemma, Version 2, in Section 4).

3.2 Definability of relations by imperative programs

3.2.1 Definability by input programs

An *i-program (input program)* is defined to be a pair $[S, \vec{v}]$ consisting of a deterministic program S in some programing language over Σ, together with a list of *input variables* $\vec{v}$ (but no output variables). Such a pair defines a relation on A, namely the *halting set* of $[S, \vec{v}]$ on A.

There may also be *auxiliary variables* in S (distinct from the input variables), which we assume to be completely uninitialized.

Let us call a program variable *simple* if it is of some sort in Σ, and *starred* if it is of some starred sort in Σ^*.

We will assume that S is a 'while' program over A or A^*, *i.e.*, without or with starred auxiliary variables. However, we will always assume that the *input variables* $\vec{v}$ *are simple*, so that in all cases S defines a relation on A.

DEFINITIONS. Let R be a relation on A, defined by $[S, \vec{v}]$.

(1) R is 'while' *definable* by $[S, \vec{v}]$ if S is a 'while' program over A.

(2) R is 'while'-*star definable* by $[S, \vec{v}]$ if S is a 'while' program over A^*.

As simple consequences of the theorems in §2.2, we have:

COROLLARY. *Let R be a relation on A.*

(1) R is 'while' definable $\iff$ R is simply semicomputable.

(2) R is 'while'-star definable $\iff$ R is star-semicomputable.

3.2.2 Definability by input/search programs

Now we introduce a new feature: definability with the possibility of arbitrary initialization of *search variables*.

An *i/s-program (input program with search variables)* is defined to be a triple $[S, \vec{v}, \vec{z}]$ consisting of a deterministic program S, together with a list of *input variables* $\vec{v}$ and *search variables* $\vec{z}$. The relation defined by such a triple on A is the set $\vec{x}$ of tuples of elements of A, such that when $\vec{v}$ is initialized to $\vec{x}$ then *for some (non-deterministic) initialization of $\vec{z}$, S halts*.

Again, there may also be *auxiliary variables* in S (distinct from the input and search variables), which we assume to be completely uninitialized.

DEFINITIONS. Let R be a relation on A, defined by $[S, \vec{v}, \vec{z}]$.

(1) R is 'while' *definable with initialization* by $[S, \vec{v}, \vec{z}]$ if S is a 'while' program over A.

(2) R is 'while'-*star definable with initialization* by $[S, \vec{v}, \vec{z}]$ if S is a 'while' program over A^* (so that the search and auxiliary variables may be starred).

As further simple consequences of the theorems in §2.2, we have:

COROLLARY. *Let R be a relation on A.*

(1) *R is 'while' definable with initialization $\iff$ R is projectively semicomputable.*

(2) *R is 'while'-star definable with initialization $\iff$ R is projectively star-semicomputable.*

3.3 Generalized Church-Turing Thesis for nondeterministic specification

In Tucker and Zucker [1991b], we showed that projective star-semicomputability is also equivalent to definability by a Horn program over Σ^* ("Horn-star definability"). This, and other considerations, led us to formulate a *Generalized Church-Turing Thesis for non-deterministic specifications* of relations:

Definability or specifiability of relations by effective specifications or non-deterministic algorithms can be formalized by projective star semicomputability (or, equivalently, 'while'-star definability with initialization).

4 Computation trees and Engeler's Lemma

One can define, for any 'while' program S over Σ, and vector $\vec{v}$ of program variables such that $var(S) \subseteq \vec{v}$, the *computation tree* $T[S, \vec{v}]$, which is like an "unfolded flow chart" of S. (Full details are given in Tucker and Zucker [199c].)

For each leaf λ of this tree, there is a *boolean term* $B[S, \vec{v}, \lambda]$ (*i.e.*, a program term over Σ of sort B) with free variables among $\vec{v}$, which expresses the conjunction of results of all the successive tests, that (the current values of) the variables $\vec{v}$ must satisfy in order for the computation to "follow" the finite path from the root of the tree to λ.

Then if $\langle \lambda_j \mid j \geq 0 \rangle$ is an *effective enumeration* of the leaves of $T[S, \vec{v}]$, the *halting predicate* $Halt[S, \vec{v}]$ of S with respect to $\vec{v}$ is the infinite disjunction

$$\bigvee_{j=0}^{\infty} B[S, \vec{v}, \lambda_j]$$

which expresses that execution of S eventually *halts*, if started in the initial state (represented by) $\vec{v}$.

Furthermore, the predicate $B[S, \vec{v}, \lambda_j]$ is *effective* in S, $\vec{v}$ and j. This means that there is a partial recursive function β of three arguments such that $\beta(\ulcorner S \urcorner, \ulcorner \vec{v} \urcorner, j)$ is the Gödel number of $B[S, \vec{v}, \lambda_j]$. So S halts on the initialization $\vec{x} \in A[\vec{k}]$ iff for some j,

$$te_A(\beta(\ulcorner S \urcorner, \ulcorner \vec{v} \urcorner, j), \vec{x}) \downarrow \text{tt},$$

where te_A is the *term evaluation function* on A (Proposition 2 in §2.4)

This give us (by §3.2.1, Corollary 1) the following form of *Engeler's Lemma* (Engeler [1968]):

THEOREM 1 (ENGELER'S LEMMA, VERSION 1). *Let R be a semicomputable relation on a Σ-structure A. Then R can be expressed as an effective (infinite) disjunction of boolean terms over Σ.*

Actually, we need a stronger version of Engeler's Lemma, applied to 'while'-star programs, which we now formulate.

THEOREM 2 (ENGELER'S LEMMA, VERSION 2). *Let R be an algebraic relation on a strictly standard Σ-structure A. Suppose R is star-semicomputable on A. Then R can be expressed as an effective (infinite) disjunction of boolean terms over Σ.*

(Algebraic relations were defined in §1.5(3).) The proof is given in Tucker and Zucker [1991c]. In the converse direction, we have:

PROPOSITION. *Let R be a relation on the Σ-structure A.*

(a) If R can be expressed as an effective (infinite) disjunction of boolean terms over Σ, then R is star-semicomputable.

(b) If, further, A is $\mathcal{R}$, $\mathcal{R}_<$, $\mathcal{C}$ or $\mathcal{C}_<$, then R is semicomputable.

PROOF: For (a), use the Proposition in §3.1.3 and Proposition 2(a) in §2.4. For (b), use Proposition 2(b) instead of 2(a). $\square$

Combining Engeler's Lemma (Version 2) with this result, gives the following "structural" characterization of star-semicomputable relations, among those which are algebraic on strictly standard structures.

COROLLARY 1. *Suppose R is an algebraic relation on a strictly standard Σ-structure. Then R can be expressed as an effective (infinite) disjunction of boolean terms over Σ iff R is star-semicomputable.*

COROLLARY 2. *Suppose A is $\mathcal{R}$ or $\mathcal{R}_<$, and $R \subseteq \mathrm{R}^n$; or A is $\mathcal{C}$ or $\mathcal{C}_<$, and $R \subseteq \mathrm{C}^n$. Then the following are equivalent:*

(i) R can be expressed as an effective (infinite) disjunction of boolean terms over Σ,

(ii) R is semicomputable over A,

(iii) R is star-semicomputable over A.

In the remaining three sections, we will apply Engeler's Lemma (Version 2), and especially Corollary 2, to the structures $\mathcal{R}$, $\mathcal{R}_<$, $\mathcal{C}$ and $\mathcal{C}_<$.

5 The field of reals; A projectively semicomputable set which is not star-semicomputable

5.1 Application of Engeler's Lemma

As an application of Engeler's Lemma (Version 2) to $\mathcal{R}$, we have:

THEOREM. *Any subset S of $\mathbf{R}^n$ is star-semicomputable in $\mathcal{R}$ iff it can be expressed as an effective (infinite) disjunction of boolean terms*

$$\vec{x} \in S \iff \bigvee_i B_i(\vec{x}) \tag{1}$$

where each $B_i(\vec{x})$ is a finite conjunction of equations and negations of equations of the form

$$p(\vec{x}) = 0 \quad \text{and} \quad q(\vec{x}) \neq 0, \tag{2}$$

where p and q are polynomials in $\vec{x}$ with coefficients in $\mathbf{Z}$.

Let us define a point $(\alpha_1, \ldots, \alpha_n) \in \mathbf{R}^n$ to be (i) *algebraic* if it is the root of a polynomial in n variables with coefficients in $\mathbf{Z}$; and (ii) *transcendental* if it is not algebraic, or, equivalently, if for each $i = 1, \ldots, n$, α_i is transcendental (in the usual sense) over $\mathbf{Q}(\alpha_1, \ldots, \alpha_{i-1})$.

The following Corollary was stated for $n = 1$ in Herman and Isard [1970].

COROLLARY 1. *If $S \subseteq \mathbf{R}^n$ is star-semicomputable in $\mathcal{R}$, and contains a transcendental point $\vec{\alpha}$, then S contains some open neighbourhood of $\vec{\alpha}$.*

PROOF: In the notation of (1): $\vec{\alpha}$ satisfies $B_i(\vec{x})$ for some i. Then (for this i) $B_i(\vec{x})$ cannot contain any equations (as in (2)) since $\vec{\alpha}$ is transcendental, and so it must contain negations of equations only. The result follows from the continuity of polynomial functions. $\square$

An immediate consequence of this is

COROLLARY 2. *Any subset of $\mathbf{R}^n$ which is both dense and co-dense in $\mathbf{R}^n$ cannot be star-computable in $\mathcal{R}$.*

EXAMPLES. The following sets of points in $\mathbf{R}^n$ are easily seen to be semicomputable in $\mathcal{R}$, but are not computable, or even star-computable, by Corollary 2: (i) the set of points with *rational coordinates*; (ii) the set of points with *algebraic coordinates*; (iii) the set of *algebraic points*; (iv) the *complement of the Cantor set* in $[0,1]$ (with $n = 1$).

COROLLARY 3. *If $S \subseteq \mathbf{R}$ is star-semicomputable in $\mathcal{R}$, then S is either countable or co-finite in $\mathbf{R}$.*

PROOF: By the *Fundamental Theorem of Algebra*, each polynomial equation with coefficients in $\mathbf{Z}$ has *at most finitely many roots* in $\mathbf{R}$. Hence, regarding the disjunction in (1) again, there are two cases:

Case 1. For some i, $B_i(x)$ contains *only negations of equations*. Then (for this i) $B_i(x)$ holds for *all but finitely many* $x \in$ R. Hence S is *co-finite* in R.

Case 2. For all i, $B_i(x)$ contains *at least one equation*. Then (for all i) $B_i(x)$ holds for *at most finitely many* $x \in$ R. Hence S is *countable*. $\square$

5.2 An example of a projectively semicomputable set which is not star-semicomputable.

Consider the relation $R(x, y) \equiv_{df} x = y^2$ on R. Then R is semicomputable over $\mathcal{R}$, but the projection of R on the first argument:

$$\{x \mid \exists y(x = y^2)\}$$

i.e., the set of all *non-negative reals*, is not semicomputable, or even star-semicomputable. This follows immediately from Corollary 3 in §5.1.

6 The ordered field of reals; A projectively star-semicomputable set which is not projectively semicomputable

6.1 Mathematical background

First we collect some definitions and lemmas. (Background information on algebraic geometry can be found, *e.g.*, in Shafarevich [1975] or Chapter 12 of Bröcker and Lander [1975].) For the application in §6.3, we relativize our concepts to an arbitrary subset D of R.

DEFINITION 1. (*a*) An interval in R (open, half-open or closed) is *algebraic in D* iff its end-points are.

(*b*) A *patch* in R is a finite union of points and intervals.

(*c*) A *D-algebraic patch* in R is a finite union of points and intervals algebraic in D.

DEFINITION 2. A set in $\mathbb{R}^n$ is *D-semialgebraic* iff it can be defined by the conjunction of finitely many equations and inequalities of the form

$$p(\vec{x}) = 0 \quad \text{and} \quad q(\vec{x}) > 0,$$

where p and q are polynomials in $\vec{x}$ with coefficients in $\mathbb{Z}[D]$.

(We will drop the 'D' when it denotes the empty set.) We need three lemmas.

LEMMA 1. *A semialgebraic set in $\mathbb{R}^n$ has a finite number of connected components.*

(See Becker [1986]. This will be used in §7.) It follows that a semialgebraic subset of R is a patch. However, for $n = 1$ we need a stronger result:

LEMMA 2. *A subset of R is D-semialgebraic iff it is a D-algebraic patch.*

LEMMA 3. *A projection of a D-semialgebraic set in $\mathbb{R}^n$ on $\mathbb{R}^m$ ($m < n$) is again D-semialgebraic.*

This follows from Tarski's quantifier-elimination theorem for real closed fields (see, e.g., Chapter 4 of Kreisel and Krivine [1971]). From this and Lemma 2:

COROLLARY. *A projection of a D-semialgebraic set in $\mathbb{R}^n$ on $\mathbb{R}$ is a D-algebraic patch.*

6.2 Application of Engeler's Lemma

As an application of Engeler's Lemma (Version 2) to $\mathcal{R}_<$, we have:

THEOREM. *Any subset S of $\mathbb{R}^n$ is star-semicomputable in $\mathcal{R}_<$ iff it can be expressed as an effective (infinite) disjunction of boolean terms*

$$\vec{x} \in S \iff \bigvee_i B_i(\vec{x}) \tag{1}$$

where each $B_i(\vec{x})$ is semialgebraic.

Corollaries 1 and 2 from §5.1 hold for $\mathcal{R}_<$ as well as $\mathcal{R}$, leading to the same examples of $\mathcal{R}_<$-semicomputable, but not (star-)computable, subsets of $\mathbb{R}^n$.

The following Corollary, however, points out a contrast with $\mathcal{R}$.

COROLLARY 3. *In the structure $\mathcal{R}_<$, the following three notions coincide for subsets of $\mathbb{R}^n$:*

(i) simple semicomputability,

(ii) star-semicomputability,

(iii) projective semicomputability.

This follows from Corollary 2 of Engeler's Lemma in Section 4, and Lemma 3 of §6.1. (This Corollary fails in the structure $\mathcal{R}$, since in that structure, Lemma 3 of §6.1, depending on Tarski's quantifier-elimination theorem, does not hold.)

However, the above three notions of semicomputability differ in $\mathcal{R}_<$ from our fourth notion: *projective star-semicomputability*, as we will see next. But first we need:

COROLLARY 4. *A star-semicomputable subset of $\mathbb{R}^n$ consists of countably many connected components.*

This follows from the above Theorem and Lemma 1 of §6.1.

6.3 An example of a projectively star-semicomputable set which is not projectively semicomputable

First we must enrich the structure $\mathcal{R}_<$. Let $E = \{e_0, e_1, e_2, \ldots\}$ be a sequence of reals such that for all i,

$$e_i \text{ is transcendental over } Q(e_0, \ldots, e_{i-1}). \tag{1}$$

NOTATION. (1) $\mathcal{R}_{<,E}$ is $\mathcal{R}_<$ augmented by the set E as a separate sort E, with the embedding $j : E \hookrightarrow \mathsf{R}$ in the signature.

(2) $\bar{E} \subset \mathsf{R}$ is the real algebraic closure of $Q(E)$.

It is easy to see that $\bar{E}$ is projectively star-semicomputable in $\mathcal{R}_{<,E}$. (In fact, $\bar{E}$ is the projection on R of a semicomputable relation on $\mathsf{R} \times E^*$.) We will now show that, on the other hand, $\bar{E}$ is not projectively semicomputable in $\mathcal{R}_{<,E}$.

THEOREM. Let $F \subseteq \bar{E}$ be projectively semicomputable in $\mathcal{R}_{<,E}$. Then $F \neq \bar{E}$. Specifically, suppose for some inductively computable function φ:

$$F = \{x \in \mathsf{R} \mid \exists y_1^{\mathsf{R}} \ldots y_r^{\mathsf{R}} z_1^{\mathsf{E}} \ldots z_s^{\mathsf{E}} u_1^{\mathsf{N}} \ldots u_k^{\mathsf{N}} v_1^{\mathsf{B}} \ldots v_l^{\mathsf{B}} \; \varphi(x, \vec{y}, \vec{z}, \vec{u}, \vec{v}) \downarrow\} \tag{2}$$

(with existential quantification over all 4 sorts in $\mathcal{R}_{<,E}$). Then for all $x \in F$, x is algebraic over some subset of E of cardinality s (= the number of quantifiers of sort E in (2)).

The rest of this section is a sketch of the proof.

LEMMA 1. (In the notation of the Theorem:) F can be represented as a countable union of the form $F = \bigcup_{i=0}^{\infty} F_i$, where

$$F_i = \{x \mid \exists y_1^{\mathsf{R}} \ldots y_r^{\mathsf{R}} z_1^{\mathsf{E}} \ldots z_s^{\mathsf{E}} B_i(x, \vec{y}, \vec{z})\}$$

and B_i is a finite conjunction of equations and inequalities of the form

$$p(x, \vec{y}, \vec{z}) = 0 \quad \text{and} \quad q(x, \vec{y}, \vec{z}) > 0$$

where p and q are polynomials in $x, \vec{y}, \vec{z}$ with coefficients in Z.

PROOF: Apply Engeler's Lemma (Version 1) to $\mathcal{R}_{<,E}$. Also replace existential quantification over N and B by countable disjunctions. $\square$

LEMMA 2. (In the notation of Lemma 1:) For any s-tuple $\vec{e}$ of elements of E, put

$$F_i[\vec{e}] =_{df} \{x \mid \exists y_1^{\mathsf{R}}, \ldots, y_r^{\mathsf{R}} B_i(x, \vec{y}, \vec{e})\}.$$

Then for all $\vec{e}$, $F_i[\vec{e}]$ is a (finite) set of points, all algebraic in $\vec{e}$.

PROOF: Note that $F_i[\vec{e}]$ is a projection on R of an $\vec{e}$-semialgebraic set in R^{r+1}. Hence, by the Corollary in §6.1, it is an $\vec{e}$-algebraic patch. Since by assumption

$$F_i[\vec{e}] \subseteq F \subseteq \bar{E},$$

$F_i[\vec{e}]$ is countable, and hence cannot contain any (non-degenerate) interval. The result follows from the definition of $\vec{e}$-algebraic patch. □

Since F is the union of $F_i[\vec{e}]$ over all i, and all s-tuples $\vec{e}$ from E, the Theorem follows from Lemma 2 and the following

LEMMA 3. *For all n, there exists a real which is algebraic over E but not over any subset of E of cardinality n.*

This follows from the construction (1) of E.

REMARK. We have shown that $\bar{E}$ (although a projection on R of a semicomputable relation on $R \times E^*$) is not a projection of a semicomputable relation in $\mathcal{R}_{<,E}$. In fact, we can see (using Version 2 instead of Version 1 of Engeler's Lemma) that $\bar{E}$ is not even a projection of a *star*-semicomputable relation on $R^n \times E^m$ (for any $n, m > 0$). Thus to define $\bar{E}$, we must project off the *starred sort* E^*, or (in other words) existentially quantify over a *finite, but unbounded* sequence of elements of E.

7 The field of complex numbers; A semicomputable set which is not computable

7.1 Relating computability in the complex and real structures

We consider the structures $\mathcal{C}$ and $\mathcal{C}_<$.

NOTATION. If $S \subseteq C^n$, we write

$$\hat{S} =_{df} \{(\mathrm{re}(z_1), \mathrm{im}(z_1), \ldots, \mathrm{re}(z_n), \mathrm{im}(z_n)) \mid (z_1, \ldots, z_n) \in C^n\} \subseteq R^{2n}.$$

REDUCTION LEMMA. *Let $S \subseteq C^n$.*

(a) S is semicomputable in $\mathcal{C}$ iff $\hat{S}$ is semicomputable in $\mathcal{R}$.

(b) S is star-semicomputable in $\mathcal{C}$ iff $\hat{S}$ is star-semicomputable in $\mathcal{R}$.

(c) S is semicomputable in $\mathcal{C}_<$ iff $\hat{S}$ is semicomputable in $\mathcal{R}_<$.

(d) S is star-semicomputable in $\mathcal{C}_<$ iff $\hat{S}$ is star-semicomputable in $\mathcal{R}_<$.

This lemma will enable us to reduce problems of semicomputability in the structures $\mathcal{C}$ or $\mathcal{C}_<$ to those in the corresponding real structures. For example, from this Lemma and Corollary 3 in §6.2 we have:

COROLLARY. *In $\mathcal{C}_<$, the notions of semicomputability, star-semicomputability and projective semicomputability all coincide for subsets of C^n.*

Note that the Reduction Lemma would not be true if we included the mod function $(z \mapsto |z|)$ in $\mathcal{C}$ or $\mathcal{C}_<$, by Corollary 2 in §2.4.

7.2 Julia sets

We reconsider an example from Blum, Shub and Smale [1989], and show how it follows from our general theory of semicomputability.

We work from now on in $\mathcal{C}_<$.

Let $g : C \to C$ be a function. For $z \in C$, the *orbit of g at z* is the set

$$\mathbf{orb}(g,z) = \{g^n(z) \mid n = 0, 1, 2, \ldots\}.$$

Let

$$U(g) = \{z \in C \mid \mathbf{orb}(g,z) \text{ is unbounded}\}$$

and

$$F(g) = \{z \in C \mid \mathbf{orb}(g,z) \text{ is bounded}\}$$
$$= C - U(g).$$

The set $F(g)$ is the *filled Julia set* of g; the boundary $J(g)$ of $F(g)$ is the *Julia set* of g.

For any $r \in R$ define

$$V_r(g) = \{z \in C \mid \exists n(|g^n(z)| > r)\}.$$

Clearly, $U(g) \subseteq V_r(g)$ for all r.

LEMMA. For $g(z) = z^2 - c$, with $|c| > 4$, we have $U(g)$ is semicomputable but not computable. Thus, $F(g)$ is co-semicomputable but not semicomputable.

PROOF: Assume for now that $|c| \geq 1$, and choose $r \geq 1 + |c|$. Then for $|z| > r$,

$$|g(z)| = |z^2 - c| \geq |z|^2 - |c| \geq \tfrac{3}{2}|z|.$$

Hence for all n,

$$|g^n(z)| \geq (\tfrac{3}{2})^n |z|.$$

and so

$$g^n(z) \to \infty \quad \text{as} \quad n \to \infty.$$

Hence for such r, $V_r(g) \subseteq U(g)$, and so

$$U(g) = V_r(g) = \{z \in C \mid \exists n(|g^n(z)| > r)\}.$$

Now choose $r = 1 + |c|^2 \geq 1 + |c|$, for example. (Note that although the function $z \mapsto |z|$ is not computable, the function $z \mapsto |z|^2 = \mathrm{re}(z)^2 + \mathrm{im}(z)^2$ is.)

To show that $U(g)$ is semicomputable is routine; for example, with the i-program

```
input a : complex
var   b : complex
begin
    b := a;
    while |b|² ≤ (1 + |c|²)²  do  b := b² − c  od
end
```

hich converges precisely on $U(g)$.

To conclude the proof we must show that $F(g)$ is not semicomputable. Suppose was, then (by Corollary 4 in §6.2 and the Reduction Lemma) it would consist of ɔuntably many connected components. But if we choose $|c| > 4$ we can show that $F(g)$ a Cantor set, and so we have a contradiction. $\square$

Conclusion: Significance for constructive methods

n conclusion, we may ask: since our computability theory is carried out in a classical amework, what is its applicability to constructive methods? An answer may be: the ɯme as the applicability of classical recursion theory to constructivity. For example dentifying for now "computability" with "constructivity") we have certain negative ɛsults. Thus the square root function on the reals (with our signature) is not computable (see §2.4). As another example, this theory shows that certain sets, such as the lled Julia set of §7.2, are not computable, even though approximations to these can be fectively displayed.

Let us consider more carefully the justification for our identification of computability ɩccording to our theory) with constructivity.

One point to note is that our computation schemes are appropriate for constructiv-y in that, unlike many recursion theories over abstract structures, there is no (non-ɔnstructive) search operator over arbitrary domains.

Another point is that (obviously) the class of computable functions and relations on structure depends on the class of initial functions, i.e., the signature of the structure.

To illustrate the latter point, consider, in particular, the ordered field $\mathcal{R}_<$ of reals. ɯr approach is algebraic, i.e., the reals are taken as "primitive" entities, with the ɛlations '=' and '<' in the signature, and hence certainly computable. Now in most ʾeatments of constructivity over the reals, the reals are taken not as primitive, but as ɔnstructed from the rationals; for example, as equivalence classes of Cauchy sequences. ʒee, e.g., Troelstra and Van Dalen [1988, Chapter 5], and the papers by Chirimar and owe and by Weihrauch in these Proceedings.) In such an approach, the equality and ʾder relations on the reals are clearly not computable. Note also that with equality the signature, it is possible to define computable but discontinuous functions on the ɛals! Such a function would not be constructive in the traditional sense.

On the other hand, let us return to our example above of the square root. The square root function *is* constructive, in the sense that Newton's method (say) gives a construction which effectively maps a stream of rationals defining a given positive real, to another stream of rationals defining its square root.

This does not invalidate our approach, which is valuable for its purposes — in the present case, to find examples in the fields of real and complex numbers so as to distinguish between different notions of semicomputability. However it will be important to attempt a comparison, or reconciliation, between these two approaches; or, to put the matter another way, to analyse carefully the roles of the equality and order relations in theories of computability over the reals.

Finally, let us note another, more positive application of classical recursion theory to constructive mathematics: namely, in the use of *realizability interpretations* for constructive formal systems. It would be interesting to see if such an approach would similarly be useful in the framework of our computability theory for *formal deductive systems* over abstract data types. A beginning of an investigation of such formal systems is given in Tucker, Wainer and Zucker [1990] and Tucker and Zucker [1991a], but without (so far) the use of realizability interpretations.

References

E. Becker (1986), *On the real spectrum of a ring and its application to semialgebraic geometry*, Bull. Amer. Math. Soc. (N.S.) **15**, 19–60.

L. Blum, M. Shub and S. Smale (1989), *On a theory of computation and complexity over the real numbers: NP-completeness, recursive functions and universal machines*, Bull. Amer. Math. Soc. (N.S.) **21**, 1–46.

T. Bröcker and L.C. Lander (1975), *Differentiable Germs and Catastrophes*, London Math. Soc.

E. Engeler (1968), *Formal Languages: Automata and Structures*, Markham Publ. Co.

E. Engeler (1975), *On the solvability of algorithmic problems*, Logic Colloquium '73 (H.E. Rose and J.C. Shepherdson, eds.), North-Holland, pp. 231–251.

H. Friedman (1971), *Algorithmic procedures, generalised Turing algorithms and elementary recursion theories*, Logic Colloquium '69 (R.O. Gandy and C.M.E. Yates eds.), North-Holland, pp. 361–389.

G.T. Herman and S.D. Isard (1970), *Computability over arbitrary fields*, J. London Math. Soc. (2) **2**, 73–79.

S.C. Kleene (1952), *Introduction to Metamathematics*, North-Holland.

A. Kreczmar (1977), *Programmability in fields*, Fundamenta Informaticae **1**, 195–230.

G. Kreisel and J.L. Krivine (1971), *Elements of Mathematical Logic*, North-Holland.

I.R. Shafarevich (1977), *Basic Algebraic Geometry* (transl. K.A. Hirsch), Springer-Verlag.

J. Shepherdson (1985), *Algorithmic procedures, generalized Turing algorithms, and elementary recursion theory*, Harvey Friedman's Research on the Foundations of Mathematics (L.A. Harrington, M.D. Morley, A. Ščedrov and S.G. Simpson, eds.), North-Holland, pp. 309–315.

A.S. Troelstra and D. van Dalen (1988), *Constructivism in Mathematics: An Introduction, Vol. I*, North-Holland.

J.V. Tucker (1980), *Computing in algebraic systems*, Recursion Theory, Its Generalisations and Applications (F.R. Drake and S.S. Wainer, eds.), LMS/Cambridge University Press.

J. V. Tucker, S.S. Wainer and J. I. Zucker (1990), *Provable computable functions on abstract data types*, Proceedings of the 17th International Colloquium on Automata, Languages and Programming, Warwick University, England, July 1990 (M.S. Paterson, ed.), Lecture Notes in Computer Science 443, Springer-Verlag, pp. 745–760.

J.V. Tucker and J.I. Zucker (1988), *Program Correctness over Abstract Data Types, with Error-State Semantics*, CWI Monograph Series, North-Holland and the Centre for Mathematics and Computer Science (CWI), Amsterdam.

J.V. Tucker and J.I. Zucker (1989), *Horn programs and semicomputable relations on abstract structures*, Proceedings of the 16th International Colloquium on Automata, Languages and Programming, Stresa, Italy, July 1989, Lecture Notes in Computer Science **372**, Springer-Verlag, pp. 745–760.

J.V. Tucker and J.I. Zucker (1991a), *Provable computable selection functions on abstract structures*, Leeds Proof Theory 1990 (P. Aczel, H. Simmons and S.S. Wainer, eds.), Cambridge University Press (to appear).

J.V. Tucker and J.I. Zucker (1991b), *Deterministic and nondeterministic computation, and Horn programs, on abstract data structures*, J. Logic Programming (to appear).

J.V. Tucker and J.I. Zucker (1991c), *Specifications, relations and selection functions on abstract data types*, Department of Computer Science & Systems, McMaster University, Technical Report 91-14.

Bringing Mathematics Education
Into the Algorithmic Age

Newcomb Greenleaf
Department of Computer Science
Columbia University, New York, N. Y. 10027
`newcomb@cs.columbia.edu`

Abstract

We began from the observation that most of our students find *algorithms* easy and natural and *proofs* difficult and obscure, and are totally unaware of the close relationship between algorithms and proofs. This observation led to the hypothesis that part of the problem lay in the fact that the students had been born into the algorithmic age, which their mathematics courses had largely yet to enter. This paper explores various ways in which mathematics courses can be made more algorithmic, both in style and in content. Particular attention will be paid to the term *non-computable function*, which will be seen as oxymoronic. An algorithmic explanation will be developed, particularly for the *busy beaver function*. We shall also give an algorithmic analysis of *Cantor's diagonal method*.

1. Learning Mathematics in the Algorithmic Age

The observations that follow are based on my own experience in teaching a variety of courses involving both theory and programming at Columbia University, on the comments (and complaints) of several of my colleagues about our students, and on numerous anecdotal reports from teachers at other institutions.

Our good students are very good algorithmists. They find the concept of an algorithm a natural one and delight in understanding the subtleties of intricate algorithms. When they write programs, they take pride not just in the correctness and robustness of their code, but also in its intelligibility and aesthetic appearance.

With a few exceptions, these same students are not good mathematicians. After four semesters of calculus they have only the foggiest notions of mathematical analysis. They find the notion of mathematical proof uninteresting and unintelligible. While they have seen proofs by mathematical induction in a variety of courses, they are still generally unable to do even the simplest proofs by mathematical induction.

In view of this dichotomy, it is not surprising that they see no close connection between algorithms and proofs. They are capable of giving cogent informal arguments for the correctness of their programs, but they are generally skeptical and rather fearful of the subject of program verification (which, indeed, is little practiced at Columbia). Almost never have they been exposed to the idea that mathematics can be done algorithmically or constructively. They find this idea both intriguing and astonishing.

It seems natural to conjecture that the problem lies, in part, in the failure of instruction in mathematics to adjust and develop to living in the age of algorithms. If mathematics were presented more algorithmically, then it could build on the students' understanding of algorithms, and perhaps students would take to it with more ease and enthusiasm.

After a brief discussion of the algorithm concept and the way in which it has entered into the center of our thought, I will present several suggestions on ways in which mathematics instruction can be made more algorithmic, with some focus on the first course in computer theory.

2. The Algorithm Concept

The name "the Euclidean algorithm" may give the misleading impression that the concept of an algorithm has been securely and explicitly rooted in our mathematical thought for many centuries. In fact, the word **algorithm** acquired its modern meaning only half a century ago. Unabridged dictionaries of the 1930's define *algorithm* only as an "erroneous refashioning" of *algorism*. (*Algorists* used the ten digits for computing, instead of an abacus). This art was brought to Europe by the Latin translation of a text by al-Khwarizmi, the eponymous ninth century mathematician of Baghdad. According to the new **Oxford English Dictionary** of 1989, the first modern use of the word was in the classic 1938 number theory text of Hardy and Wright, where the nomenclature "the Euclidean algorithm" was introduced. While there are doubtless earlier occurrences missed by the OED, the term could not then have been in common parlance. Now it has become so central that Knuth has proposed *algorithmics* as the best name for the discipline of computer science [20].

I shall not attempt an extended discussion of the nature of algorithms. It is a subject to which philosophers have yet to pay sufficient attention. Like all foundational terms, the algorithm concept is difficult to define. A standard discussion occurs in Knuth's *Fundamental Algorithms* [19], pp. 1-9, where algorithms are characterized in terms of five properties: *finiteness*, *definiteness*, *input*, *output*, and *effectiveness*. Many interesting insights about the algorithm concept can be found in the papers presented at the symposium honoring al-Khwarizmi, held near the place of his birth on its 1200-th anniversary [15].

Associated with the algorithm concept is a powerful new language of *algorithms and data structures*. Since Euclid, mathematicians have used algorithms, but only recently have systematic languages for algorithms been developed. And only *very* recently has an evident quorum of mathematicians, through their programming experience, become fluent in higher-level algorithmic languages. A Pascal-like pseudo-code has become a new *lingua franca* and is used in many recent texts on discrete mathematics. There has been much discussion of the proper mathematics prerequisites for computing courses. Soon we may expect to see mathematics courses with a computer science prerequisite, since students master the language of algorithms through learning to program [17].

Of course, we have had the languages of Turing machines and recursive functions for half a century. While these precise formalisms have made their mark on our thought, they are at the level of assembly or machine language, and, as Martin-Lof and others have shown us, mathematics might best be regarded as a *very high level programming language* [22].

There is a complex symbiotic relationship between *algorithm* and *proof*. In computer science proofs are used to verify algorithms. Indeed, any algorithm must be supported by some form of proof to be believed. Such proofs often consist of a very informal argument buttressed by testing of special cases, but many workers in program verification argue that a program should be a *proof that can be compiled* [2, 10].

On the other hand, mathematicians often use algorithms in their proofs, and many proofs are totally algorithmic, in that the triple

[assumption, proof, conclusion]

can be understood in terms of

[input data, algorithm, output data].

Such proofs are often known as *constructive*, a term which provokes endless unfortunate arguments about ontology.

An interesting discussion of the relation between algorithms and proofs occurs in Knuth's review [20] of Bishop's *Foundations of Constructive Analysis* [4]:

> The interesting thing about this book is that it reads essentially like ordinary mathematics, yet it is entirely algorithmic in nature if you look between the lines.

Knuth goes on to analyze the algorithmic nature of Bishop's proof of the Stone-Weierstrass theorem in great detail, even translating it into pseudocode, and noting that:

> I want to stress that the proof is essentially an algorithm; the algorithm takes any constructively given compact set X and continuous function f and tolerance ε as input, and it outputs a polynomial that approximates f to within ε on all points of X. Furthermore the algorithm operates on algorithms, since f is given by an algorithm of a certain type, and since real numbers are essentially algorithms themselves.

It seems that we are seeing the emergence of a new concept, of which proof and algorithm are but two aspects. Michael Beeson recently put it nicely (in the context of a discussion of Prolog), [3]:

> The flow of information seems now to be logic, now to be computation. Like waves
> and particles, logic and computation are metaphors for different aspects of some
> underlying unity.

have no good candidate for a name for this underlying unity (neither *verified algorithm* nor *constructive proof* does it justice). It should be remarked that the coming together of the concepts of algorithm and proof creates two tensions. The pull to make proofs more like algorithms is the subject of this article. I hope to discuss the reverse pull, which manifests in Prolog and logic programming generally, in a subsequent paper.

For our purposes, this superficial discussion of the relationship between algorithms and proofs will do. The important point is that the explicit notion of an algorithm has become central in mathematical thought only recently. While our students are at home with algorithms and algorithmic languages, instruction in mathematics is only beginning to adjust to the new reality. In particular, almost never do mathematics texts take advantage of the new algorithmic literacy to explain proofs in terms of algorithms.

3. Algorithmic Style

Through our experience in writing computer programs, we have gradually developed a distinctive and effective style for presenting algorithms. While this style has much in common with mathematics, it departs radically from traditional mathematical style in some ways. This is strikingly evident in the different approaches to names and symbols. Mathematicians have generally used single characters for symbols. There is good reason for this, since it leads to brevity and allows complex formulas to be written concisely. But this style demands of the reader that she remember the meanings of all of the symbols, which do not have mnemonic names, and my students generally are unable and/or unwilling to read material written in this style.

Computing also puts a high value on brevity. Good programming style dictates that procedures be fairly short, but this is not generally achieved by using single-character symbols. Rather, symbols are generally words or word fragments, chosen for mnemonic value. Brevity is achieved through procedural abstraction, by giving each procedure a relatively simple task, and using many procedures in a single program.

Good programming style also pays careful attention to the layout of the program on the page or screen. I regularly teach the Scheme dialect of LISP to novice programmers. A great deal of credit for the success of this enterprise goes to the pretty-printing mechanism of the editor which lays out the programs in a way which both illustrates the structure and is aesthetically pleasing. Here is the first substantial procedure which my students see, adapted from [1], which uses Newton's method to compute the square root of a number x. The simple helping procedures `square` and `average` are defined elsewhere.

```
(define (sqrt x tolerance)
  (define (sqrt-iter guess)
    (define (good-enough?)
      (< (abs (- (square guess) x)) tolerance))
    (define (improve)
      (average guess (/ x guess)))
    (if (good-enough?)
        guess
        (sqrt-iter (improve))))
  (sqrt-iter 1))
```

While this is a challenging procedure for the second week of a first course, the combination of mnemonic names and pretty-printing (along with the simple syntax of Scheme) makes it accessible to students. Now consider how inaccessible the same procedure becomes when mnemonic names are replaced by single letter symbols:

```
(define (s x t)
  (define (i g)
    (define (g?)
      (< (abs (- (q g) x)) t))
    (define (p)
      (a g (/ x g)))
    (if (g?)
        g
        (i (p))))
  (i 1))
```

Had the students been presented with the latter version, which mimics the mathematical style of conciseness, their reaction would have been quite different. Instead of loving the course, they might have dropped it.

A glaring example of the failure of mathematics instruction to make the subject algorithmic occurs in the elementary differential calculus, which is still generally presented as a large collection of derivative formulas. Of course, these formulas are intended to be used as the base cases and recursive operations of a grand recursive *derivative algorithm*, which, for want of a proper language, is not made explicit (and therefore never really verified).

The derivative algorithms can be cast in two different forms. If the chain rule is made an explicit recursive operation, then there will be a large number of base case formulas, such as:

$$D(\sin x) = \cos x$$

If the chain rule is built into most derivative formulas, then the sine formula will appear as a recursive operation, corresponding to the formula

$$D(\sin u) = (\cos u) * Du$$

and there will be two base cases which tell us that the derivative of a constant is 0 and the derivative of the identity function is 1. The student of calculus is expected not merely to learn the various formulas, but primarily to understand the operation of the recursive algorithm, which never appears explicitly.

We can hope that tomorrow's calculus texts will express the derivative algorithm explicitly in a (formal or informal) algorithmic language, as is already done in programming texts such as [1]. Here is a simple implementation of the derivative algorithm in the Scheme programming language (any programming language supporting recursion would do), adapted from [1], which takes the derivative of an expression exp with respect to var. Note that the first two cases are the only base cases. The various constructors, selectors, and predicates, such as make-sum, multiplicand, and constant? are defined elsewhere. This simple version handles only sums, products and sines:

```
(define (deriv exp var)
  (cond ((constant? exp) 0)
        ((variable? exp)
         (if (same-variable? exp var) 1 0))
        ((sum? exp)
         (make-sum (deriv (addend exp) var)
                   (deriv (augend exp) var)))
        ((product? exp)
         (make-sum
          (make-product (multiplier exp)
                        (deriv (multiplicand exp) var))
          (make-product (deriv (multiplier exp) var)
                        (multiplicand exp))))
        ((sine? exp)
         (make-product (deriv (arg exp) var)
                       (make-cosine (arg exp))))
        (else
         (error "Unknown expression type -- DERIV" exp))))
```

When the derivative algorithm assumes its rightful place as the *primary explicit structure* of differential calculus, there is a welcome gain in clarity, and also a subtle but profound shift of meaning. The proofs of the various derivative formulas are now part of the *verification* of the algorithm. They prove the *correctness* of an algorithm rather than the *truth* of a theorem. These considerations are amplified in [25].

We can summarize this discussion by enunciating

Four Marks of Algorithmic Style:

- **Mnemonics**: Use mnemonic names for symbols, and achieve brevity through abstraction.

- **Explicitness**: Make the algorithms explicit. Identify input and output and check the effectiveness of the algorithm. When appropriate, pay attention to questions of efficiency and feasibility.

- **Language**: Use an appropriate algorithmic language, which can be pseudocode rather than a formal programming language.

- **Primacy**: Make algorithms the primary structures when appropriate. For instance, the chapter of a calculus text dealing with differentiation could be called

The Derivative Algorithm.

We will illustrate these principles further by considering an example from the first course in computing theory. This course is traditionally called "Computability," a name which I feel is highly inappropriate, for reasons to appear below. A far better title would be "Languages, Grammars, and Machines."

I have taught Computability on numerous occasions at Columbia. The first time I used the excellent traditional text of Lewis and Papadimitriou [21]. While I found the book to be clear and comprehensive, my students found it almost completely unreadable. They would open it only when problems were assigned, and then would skim backwards from the problems to see if they could find a relevant formula. Now this is what students have been doing with mathematics texts for many years, but it is not optimal for learning. I then heard of the text by Daniel Cohen, which takes a very different approach [9]. At first I was skeptical, but after giving it a try became an enthusiastic booster of Cohen's text. True, Cohen takes 800 pages to cover what Lewis and Papadimitriou do in less than half of their 400 pages. But students found the book a joy to read, and in reading it they acquired a much better understanding than they had when I had used a traditional text, and they went on to distinguish themselves in more advanced courses using traditional texts.

The text of Lewis and Papadimitriou is written in a traditional mathematical style and displays none of the marks of algorithmic style. If we examine Cohen's book in light of the four marks, we see that he is very careful to use well-chosen mnemonic names. Indeed, the book contain virtually no traditional-looking mathematical formulas which students new to the subject find so opaque. He always presents his proofs as algorithms and continually points this out. But he does not use any special language for algorithms, preferring to present them in lengthy English descriptions. Of course, this allows the instructor to ask students to summarize various algorithms in pseudocode. And while the algorithms are explicit, they are embedded within the proofs of theorems, which remain the primary structures. So Cohen exhibits two of the four marks to a high degree.

To highlight the difference between Cohen's text and that of Lewis and Papadimitriou, I will present the beginning of a crucial proof from each text. The proof involves showing that any for deterministic finite automaton (DFA) there exists an equivalent regular expression. Lewis and Papadimitriou start like this:

> Let $M = (K, \Sigma, \delta, s, F)$ be a deterministic finite automaton; we need to show that there is a regular language R such that $R = L(M)$. We represent $L(M)$ as the union of many (but a finite number of) simple languages. Let $K = \{q_1, \cdots, q_n\}$, $s = q_1$. For $i, j = 1, \cdots, n$ and $k = 1, \cdots, n+1$, we let $R(i, j, k)$ be the set of all strings in Σ^* which drive M from q_i to q_j without passing through any state numbered k or greater.

The proof is off to an economical start. It will be completed in less than a page. But because of the density of non-mnemonic notation, and the complete lack of an algorithmic framework, few beginning students will even attempt to read it, let alone come to understand it.

In marked contrast, Cohen starts off as follows:

> The proof of this part will be by constructive algorithm. This means that we present a procedure that starts out with a transition graph and ends up with a regular expression that defines the same language. To be acceptable as a method of proof, any algorithm must satisfy two criteria. It must work for every conceivable TG, and it must guarantee to finish its job in a finite time (in a finite number of steps). For the purposes of theorem-proving alone, it does not have to be a good algorithm (quick, least storage used, etc.). It just has to work in every case.

This is hardly an economical start! But while the proof proper has yet to begin, a proper foundation has been laid. The proof itself will occupy twelve or so pages (with many illustrations). But my students read it with enthusiasm. I often get reports that they have stayed up half the night reading it and running hand simulations of the algorithm. And they retain from their reading the ability to execute the algorithm on simple machines.

4. Are All Functions Computable?

I remarked above that the name "Computability" seems to be inappropriate for the first course in computing theory. Traditionally, a high point of this course has been the demonstration of the existence of "non-computable functions." By accepting uncritically the traditional mathematical account of these phenomena, we introduce radically non-algorithmic ideas into the heart of our curriculum.

Indeed, the very concept of a non-computable function is problematic. First we explain functions to students in computational terms. A *function* from X to Y takes an element of X as input and delivers an element of Y as output. Then, a few pages or weeks later, we turn around and announce to the students that, in fact, almost all functions are non-computable! But what is a non-computable function? But if a function does not correspond to an algorithm, what can it be? There is in this context no higher court corresponding to the set theory of logical mathematics. Indeed, one thing we can say is that, for our students, the term *non-computable function* is an oxymoron which undermines their algorithmic understanding of functions.[1] Since there are evident advantages of simplicity and unity in defining functions in terms of algorithms, we shall take the stand that functions are, by definition, computable, and then test those phenomena which are standardly taken as evidence for the existence of non-computable functions, to see if we need to yield any ground. Our strategy will be the following. Given a putative function f, we do not ask

Is f computable?

---

[1] Following Umberto Eco we might imagine a **College of Oxymoronics** in which a course in *Non-Computable Functions* would take its place in the curriculum along side such offerings as *Tradition in Revolution, Democratic Oligarchy, Parmenidean Dynamics, Heraclitean Statics, Boolean Eristic*, and *Tautological Dialectics* [14].

but rather

What are the **proper data types** for the domain and range of f?

The latter question may have more than one natural answer, and we can consider both restricted and expanded domain/range pairs. Only if you attempt to pair an expanded domain for f with a restricted range will you come to the conclusion that f is "non-computable."

The usual argument given for the existence of non-computable functions involves Cantor's diagonal algorithm. Let us look at a typical example of how this argument is usually presented. As we will see, it is, from an algorithmic standpoint, badly flawed. I chose the following excerpt from a review by Ian Stewart not because it is unusual, but because it states the received view so well [27].

> A real number is computable if its binary expansion can be the output of a computer program. It is a theorem--at first sight surprising, but true, and even easy--that 'almost all' real numbers are not computable. For example, Turing used an argument that goes back to Georg Cantor to prove the existence of at least one non-computable real number.

> The proof is based on the idea that all possible computer programs, each of which must be represented by a finite sequence of digits, can be listed in order. To do this, interpret the program's defining sequence as a whole number, expressed in binary, and arrange these numbers in increasing numerical order. Now assign to each program in this list its output data, a real number expressed in binary. Run down the diagonal of this table of numbers, changing the nth digit in the nth number. The new diagonal is a number that is not on the list, which therefore corresponds to the output of no computer program whatsoever.

Note that Stewart expresses himself in a rather algorithmic fashion, using words which imply feasible actions, like *arrange, assign, run down,* and *change.* But in fact what is proposed is completely non-algorithmic. We can indeed arrange all possible computer programs producing binary sequences in an infinite list (more precisely, we can write a program which will generate as much of this list as desired), but these programs will be partial functions which may produce a binary sequence but may, at some of the places in the sequence, compute forever without producing a 0 or a 1. We cannot change the value in such a case because there is no value to change. But we cannot be sure that there is no value, because perhaps a value will be found if we just let the program run for a longer time. The undecidability of the halting problem bars the way.

But perhaps Stewart means only that we should arrange in a list only those programs which correspond to *total* functions. Then the diagonal algorithm will work without a hitch, but a problem remains: without an oracle we have no way of extracting from the list of all programs those which correspond to total functions, and hence cannot arrange the latter in a list. Again the halting problem halts our progress.

The pioneering researches of Turing, Church, and others showed that the functions defined by Turing machines (or equivalent formalisms) are typically partial, and their domains are typically undecidable (because of the undecidability of the *halting problem*). They also concluded that functions are typically non-computable, on the grounds that Turing machines can be enumerated, while functions cannot. Later, specific examples of non-computable functions were found, most notably the *busy beaver function*. We will present another interpretation of the busy beaver phenomenon, based on careful attention to the data types of domain and range, in which the busy beaver function is indeed computable. Then we will consider the Cantor diagonal algorithm and questions of cardinality.

Note that I use the simple and intuitive terms *decidable* and *enumerable* instead of the more traditional "recursive" and "recursively enumerable." A set of integers is *decidable* if there is an algorithm or Turing machine for deciding membership, and *enumerable* if there is an algorithm or Turing machine for listing its members.

5. The Busy Beaver Function

The busy beaver phenomenon concerns Turing machines (TMs) whose tape alphabet consists of a single non-blank symbol, say "*". A *beaver* is a TM which, when started on a blank tape, halts and computes an integer, known as its *productivity*. Several conventions are commonly used for what counts as the computation of an integer. The most restrictive requires that the machine halt on the leftmost * of a contiguous block on an otherwise blank tape. The less restrictive require only that the machine halt and take as productivity either the number of *'s on the tape, the number of steps of the computation, or some other measure of the complexity of the computation. A *k*-state beaver is *busy* if, among all TMs with *k* states, it has greatest productivity. It does not matter which convention is taken, beavers turn out to be extremely busy. Already Rado had proved the following [26]:

> **Rado's Theorem.** Let f be any (total) Turing-computable function. Then there is an integer n such that for all integers $k \geq n$ there is a k-state beaver with productivity greater than $f(k)$.

An extremely careful proof is given in Chapter 4 of the text [7]. If we define the *busy beaver function bb* by taking $bb(k)$ to be the maximum productivity of any k-state beaver, then the theorem shows that *bb* grows faster than any Turing-computable function. Hence, under the Church-Turing Thesis, it might appear that *bb* is a non-computable function.

But Rado's Theorem gives no hint of the extraordinary complexity of computations performed by extremely small machines. While k-state busy beavers have been found for $k \leq 4$, computer searches are continually finding busier and busier 5-state beavers. Recently a 5-state machine which halts with 4,098 symbols on the tape after running for 23,554,760 steps has been announced! Note that these are 5-tuple machines, which simultaneously print and move. Other authors, like [7], work with 4-tuple machines which can either move or print (but not both at once). A 5-tuple machine with 5 states will generally convert to a 4-tuple machine with 8 or 9 states. For further discussion of busy beavers, we particularly recommend Brady's fascinating article [8] and the entertaining account in the *Scientific American* column of Dewdney [13].

The busy beaver function *bb* becomes computable when its domain and range are properly defined. When the domain is taken to be **N**, the range will be the set of "weak integers," a superset of **N** which we shall define shortly. Rado's Theorem then demonstrates that *bb* grows faster than any *integer-valued* function.

To determine the proper data type for $bb(k)$, simply attempt to compute it by brute force. First list all k-state machines (the number of such machines is staggeringly large, so this is, of course, feasible only for very small k. Then run all of the k-state machines in parallel. Whenever one of them halts, determine its productivity and include it in the output. We obtain an enumerable set of integers, of cardinality bounded by the (very large) number of k-state TMs. But the process of generating this enumerable set will not halt, even though the set has bounded cardinality. Many of the TMs will never halt, and while we can weed out many obvious non-halters, there will be others whose status we will be unable to decide, so we will have to keep them running. We are led to the following definition.

Definition. A **weak integer** X is an enumerable set of positive integers which contains at least one and at most B elements, for some integer B.

Intuitively, a weak integer X represents an approximation from below, and every element $x \in X$ establishes a lower bound. It is crucial to understand that while we are given a bound B on the number of elements in a weak integer X, we do not necessarily have any bound on the values of these elements. For we do not generally have access to the entire list, but only to the algorithm or TM which enumerates it. So, in concrete terms, a weak integer is an algorithm or Turing machine which enumerates a set of integers of bounded cardinality. Given k, the function bb produces such a TM $bb(k)$ by the finite process outlined above.

The situation here is analogous to that of real numbers. When we speak of a real number x, we generally have in mind a Cauchy sequence of rationals. But when an algorithm produces a real number x which is not rational, what it actually delivers as x is an algorithm for computing as much of the Cauchy sequence as we may wish to see [20]. Similarly, when we speak of a weak integer X, we have in mind an enumerable set of integers. But when an algorithm produces a weak integer which is not an integer, what it actually delivers is an algorithm or Turing machine for enumerating as much of that set as we may wish to see.

Keeping in mind that weak integers approximate from below, it is natural to define equality and order on the collection of all weak integers as follows. For weak integers X and Y:

- $X \le Y$ means $(\forall x \in X)\,(\exists y \in Y)\,(x \le y)$

- $X = Y$ means $(X \le Y) \wedge (Y \le X)$

- $X < Y$ means $(\exists\, y \in Y)\, (\forall\, x \in X)\, (x < y)$

By associating each integer n with the singleton set $\{n\}$ the integers become a subset of the weak integers. Clearly a weak integer X equals an integer x if and only if x is the maximum element of X.

Hence there are really two busy beaver functions. Rather than extend the range to the set of weak integers, we could shrink the domain to **D**, the set of integers at which *bb* takes integer values (**D** contains at least $\{1..4\}$).

We can now deduce from Rado's Theorem:

> **The Busy Beaver Theorem.** Let f be any total Turing-computable function from **N** to **N**. Then there is an integer n such that
>
> $$bb(k) > f(k)$$
>
> for all $k \ge n$. That is, the busy beaver function grows faster than any total *integer-valued* function.

This theorem confirms our intuition that the complexity of a computation is incomparably more sensitively linked to the size of the machine than to the size of the input. There can be no universal total machine which computes all (total) functions from **N** to **N**; no single machine can keep up with a sequence of ever larger machines. Note that, since we do have a universal machine for partial functions, this implies the unsolvability of the halting problem, for if we could decide the halting problem then we could carry out a brute force computation of $bb(k)$ as an integer by running all machines with k states which halt when started on a blank tape.

The theorem also shows that we can obtain faster growing functions by relaxing the data type of the range. Functions to the weak integers can grow faster than functions to the integers. Hence the weak integers cannot be identified with the integers, and this interpretation requires that we use intuitionistic rather than classical logic. Using classical logic, it can be shown that every set of integers of bounded cardinality contains a greatest element, and hence every weak integer equals an integer. The weak integers form an *intuitionistic extension* of **N** [28].

6. The Diagonal Algorithm

We shall consider the Cantor diagonal method (which we will call the diagonal algorithm) in the context of the set $\mathbf{N^N}$ of all integer valued functions. The algorithm takes as input a sequence of such functions $\{F_k(n)\}$ and produces as output a function G different from each F_k. It was Cantor's genius to notice that this is achieved if we simply "go down the diagonal'' and construct G by a simple rule such as

$$G(n) = F_n(n) + 1.$$

As long as the functions F_k are total, this procedure is wholly algorithmic, and the implication is similar to one drawn from busy beavers: there does not exist a universal total machine. A single fixed Turing machine which takes two integer inputs k and n cannot, by fixing one input, imitate the behavior of an arbitrary machine which takes one integer input, when all functions are required to be total (which is hardly surprising, since Cantor also showed that two integers are really no better than one).

The diagonal algorithm is commonly used to point to the existence of non-computable functions in two different ways. It is used to directly construct specific non-computable functions as in the argument of Stewart given above. Used indirectly, it is the source of the theory of infinite cardinal numbers, which seems to imply that almost all functions are non-computable. The direct construction of a non-computable function by the diagonal algorithm is carried out with great care in Chapter 5 of [7]. The basic idea is very simple. The collection of all Turing machines which compute partial functions is arranged in a list $\{F_k\}$ so that $F_k(n)$ represents the value computed by the k-th TM when given input n. Note that while we have a list of all Turing machines which compute partial functions, we have no way of extracting the list of those machines which compute total functions, because of the undecidability of the halting problem. Because the functions are partial, we must modify the diagonal algorithm:

$$G(n) = \begin{cases} 1 & \text{if } F_n(n) \text{ is undefined} \\ F_n(n) + 1 & \text{otherwise} \end{cases}$$

Certainly G is a total function which is distinct from every Turing computable function, and it is very tempting to say that $G(n)$ is an integer, since it is an integer if $F_n(n)$ is defined or is undefined. Using classical logic, we could conclude that logically $G(n)$ *must* be an integer. On the other hand, it is certainly not an integer in any computational sense, since we have no general way of finding its value. (Indeed, here lurks another proof of the undecidability of the halting problem). So the question arises, what is the data type of $G(n)$? Again, we must describe the proper superset of N. {*Warning*: this definition will seem artificial and even paradoxical to those unused to intuitionistic logic. But the artificiality really resides in the application of the diagonal algorithm to a sequence of partial functions.}

> **Definition.** A **singleton integer** is a set X of integers satisfying:
> - X contains at most one integer (i.e. if $x \in X$ and $y \in X$, then $x = y$),
> - X is non-empty (in the sense that it is contradictory that $X = \varnothing$).
> A similar definition is found in [11].

The function G, defined above by the diagonal algorithm, is a (computable) function from N to the set of singleton integers. If we identify each integer m with the singleton set $\{m\}$, then the singleton integers become an (intuitionistic) extension of N. If we require G to take integer values, then the domain must be restricted to the set of integers n for which $F_n(n)$ is either defined or undefined. This subset, while it has empty complement, cannot be identified with N (again a paradoxical situation for those unused to the algorithmic niceties of intuitionistic logic). We should emphasize that, unlike weak integers, singleton integers are not in general enumerable. Indeed, an enumerable singleton integer is very close to being an ordinary integer and can be proved equal to an ordinary integer if *Markov's principle* is assumed (see [5], p. 63). While Markov's principle is often considered constructive or very weakly non-constructive, it has the effect of erasing the algorithmically significant distinction between algorithms which are merely known not to run forever and those for which there exists an upper bound on runtime.

It is often felt that the existence of non-computable functions shows that mathematics necessarily transcends the algorithmic. I have tried to show that this is not so, that the phenomena standardly connected with non-computability can better be understood in purely algorithmic terms. The usual approach simply lumps together the busy beaver function and the function output by the diagonal algorithm as "non-computable functions." Our approach distinguishes them in terms of the very different data types of their natural ranges.

7. Cardinality as shape

Cantor argued that the diagonal algorithm showed that the set N^N contained *more* elements than the set N of natural numbers. It is this last step which introduces into mathematics a supposed universe of non-algorithmic functions. We might well wonder how so simple an algorithm could transcend the computable.[2] Cantor did indeed show that there is a fundamental difference between the sets N^N and N, but this difference can be understood not as a quantitative difference, but as a difference of quality or *structure*. Rather than call the set N^N uncountable, it might better be called *productive*, because there are very powerful methods for producing elements of N^N, in particular for producing an element outside of any given sequence in N^N [16].

> **Definition.** A set X is **productive** if, for every function $f: N \rightarrow X$ there exists an $x \in X$ outside of $f(N)$.

> **Cantor's Theorem.** The set N^N is productive.

This use of the term *productive*, taken from recursive function theory, grounds our

---

[2]This point was perhaps first made by Wittgenstein, who wrote of the diagonal algorithm: "Our suspicion ought always to be aroused when a proof proves more than its means allow it. Something of this sort might be called a 'puffed-up proof'." ([29], p. 56)

understanding in algorithmic reality rather than idealistic fantasy.[3] Given any sequence of elements of N^N, we can find an element of N^N outside the sequence, not because there are more functions than integers, but because of the structure of N^N.

Of course it follows from the diagonal algorithm that there is no surjection from N to N^N. Let N^N_{par} denote the set of all partial binary functions on N, with intensional equality. Then, under the assumptions of the Church-Turing Thesis, the set of all partial functions can be represented by the set of Turing machines, which can be listed, so there is indeed a bijection from N to N^N_{par}. Since N^N is a subset of N^N_{par}, this might be considered as evidence that N is *larger* than N^N, were one inclined to make a quantitative comparison between them. Cantor, and most mathematicians after him, considered sets as "mere collections of elements," which could differ only in quantity. We do not find this position algorithmically intelligible, since the extra structure of N^N plays an essential role in the diagonal algorithm.

If we want a simple metaphor for differences such as that between N and N^N, *shape* might be better than *size*. It is then perfectly intelligible to maintain that

$$N \subset N^N \subset N^N_{par}$$

where N and N^N_{par} have the same shape, which is distinct from that of N^N.

8. Algorithmic logics

We have seen that if we are to take the stand that all functions are computable, we must distinguish the ordinary integers from such intuitionistic extensions as the weak integers and the singleton integers. This requires that we adopt a logic such as intuitionistic logic which allows us to make such distinctions.

Brouwer claimed that the Law of Excluded Middle was not algorithmically valid and developed intuitionistic logic as a subset of classical logic. However, there is another way to view the difference between the two logics. Since intuitionistic logic allows us to make more distinctions, it must itself contain additional distinctions. Most basically, intuitionistic logic distinguishes between a statement and its double negation, and distinguishes the constructive disjunction $A \vee B$ from the classical disjunction $\neg(\neg A \wedge \neg B)$. It is this richness of distinctions which allows the distinction between the integers and the weak integers or singleton integers.

---

[3]The fantasy here is not the theory of sets, as elegantly elaborated in the concrete confines of Zermelo-Fraenkel set theory, but rather the notion that, somewhere, there are really "more" elements of N^N than of N. It is worth remembering that Cantor lobbied the Vatican to recognize that the higher cardinals pointed the way to God [12, 18].

Looking at algebras rather than sets of truths, we can see classical logic as a *quotient* or collapsing of intuitionistic logic. From the algorithmic standpoint, classical logic does not make enough distinctions. In the logical paradigm for mathematics, the traditional test for correctness is *consistency*, and while this has an appealing simplicity, it turns out not to be sufficient for the needs of algorithms. As Errett Bishop forcefully stated: *meaningful distinctions deserve to be maintained* [6]. Similar ideas were recently expressed in [23]:

A common misconception among mathematicians is to think of intuitionistic mathematics as "mathematics without the law of the excluded middle" (the law asserting that every statement is either true or false. From this point of view, intuitionistic mathematics is a proper subset of ordinary mathematics, and doing your mathematics intuitionistically is like doing it with your hands tied behind your back.

Another more realistic viewpoint is to regard intuitionistic logic, and the mathematics based on that logic, as the logic of sets with some structure, rather than of bare sets. Traditional examples are sets growing in time (as in Kripke semantics), or sets with some recursive structure (as in Kleene's realizability interpretation), or sets continuously varying over some fixed parameter space. Universes of such sets are perfectly suitable for developing mathematics, but one is often *forced* to use intuitionistic logic.

. . .

In the first half of this century, the interest in intuitionistic logic and mathematics was mainly of a philosophical and foundational nature. More recently, it has become apparent that intuitionistic logic or some variant thereof is often the right logic to use in theories of computing.

Of course, from an algorithmic standpoint, there can be no such thing as a "bare" set, except possibly finite sets. Any set carries a structure by virtue of its construction. The weak integers and singleton integers can be considered as further examples of sets "growing in time."

What logic should we teach our students? Traditionally, classical logic has been the only logic used. Indeed, students have often been taught that "to prove something, assume it false and try to derive a contradiction." From an algorithmic standpoint there could hardly be worse general advice. When intuitionistic logic has been mentioned, it has often been in the pejorative category of a "deviant logic." But intuitionistic logic makes it natural to think of proofs as programs, and on this grounds should be preferred. We have found that some sort of natural deduction system provides an excellent way to introduce students to logic with an algorithmic flavor.

Ultimately we would, of course, like our students to be literate in both classical and intuitionistic logic. Experience has shown that those who thinking classically have a very difficult time understanding intuitionistic thought. The law of excluded middle is made part of one's subconscious thought and even questioning it seems strange. On the other hand, those who start with intuitionistic logic have no difficulty in adding excluded middle when appropriate (for example for decidable propositions). This seems to me to be a very strong argument for starting with intuitionistic logic.

An analogous situation holds in programming languages. Ultimately, of course, programmers will want to be able to use non-local control mechanisms, but experience has shown that those who start using GOTO as novice programmers develop bad habits which are very hard to erase. This analogy between local control and intuitionistic logic and non-local control and classical logic actually runs quite deep, as shown in other papers at this symposium [24].

The algorithmic spirit is diverse. We have a multitude of programming languages, programming paradigms, operating systems, etc., and this is recognized as a healthy situation in which the diversity leads to evolutionary progress (even though our students may complain about having to learn more than one programming language).

Mathematics, on the other hand, has been characterized by conformity. Physicists may still be seeking the "grand unified theory," but mathematicians seemed to have found theirs. For the past half century adherence to Zermelo-Frankel set theory and classical logic as the foundations of mathematics has been virtually unquestioned. While this unity has some advantages, it has also led to a stifling of fresh inquiry into the nature of mathematics.

Following the algorithmic spirit, we might hope that in the future mathematics will become more diverse, that students will be presented with a genuine choice among a range of foundational outlooks, and that they will learn, in particular, to view mathematics algorithmically.

References

1. Abelson, H. and Sussman, G. J. *Structure and Interpretation of Computer Programs.* M. I. T. Press, 1984.

2. Bates, J. L. and Constable, R. L. "Proofs as programs". *ACM Transactions on Programming Languages and Systems 7* (1985), 113-136.

3. Beeson, M. Computerizing mathematics: logic and computation. In Herken, R., Ed., *The Universal Turing Machine: a Half-Century Survey*, Oxford University Press, 1988, pp. 191-225.

4. Bishop, E. *Foundations of Constructive Analysis.* McGraw-Hill, 1967. [5] is a new edition.

5. Bishop, E. and Bridges, D. *Constructive Analysis.* Springer-Verlag, 1985. Revised edition of [4].

6. Bishop, E. *Contemporary Mathematics.* Volume 39: Schizophrenia in Contemporary Mathematics. In Rosenblatt, M., Ed., *Errett Bishop: Reflections on Him and His Research,* American Mathematical Society, 1985. Originally distributed as the American Mathematical Society Colloquium Lectures in 1973.

7. Boolos, G. S., and Jeffrey, R. C. *Computability and Logic.* Cambridge University Press, 1980.

8. Brady, A. H. The busy beaver game and the meaning of life. In Herken, R., Ed., *The Universal Turing Machine: a Half-Century Survey,* Oxford University Press, 1988, pp. 259-278.

9. Cohen, D. I. A. *Introduction to Computer Theory.* John Wiley and Sons, 1986.

10. Constable, R. L. "Programs as proofs: a synopsis". *Information Processing Letters 16* (1983), 105-112.

11. Dalen, D. van. Singleton Reals. In *Logic Colloquium '80,* North-Holland, 1982, pp. 83-94.

12. Dauben, J., W. *Georg Cantor: His Mathematics and Philosophy of the Infinite.* Harvard University Press, 1979.

13. Dewdney, A. K. "Computer Recreations". *Scientific American 252* (April, 1984), 20-30. Reprinted in *The Armchair Universe,* Freeman, 1988, 160-171..

14. Eco, Umberto. *Foucault's Pendulum.* Harcourt Brace Jovanovich, 1989.

15. Ershov, A. P., and Knuth, D. E. (Ed.) *Algorithms in Modern Mathematics and Computer Science.* Springer Lecture Notes in Computer Science, 1981.

16. Greenleaf, N. Liberal constructive set theory. In Richman, F., Ed., *Constructive Mathematics,* Springer Lecture Notes in Mathematics, Vol. 873, 1981, pp. 213-240.

17. Greenleaf, N. "Algorithms and proofs: mathematics in the computing curriculum". *ACM SIGCSE Bulletin 21* (1989), 268-272.

18. Hallett, M. *Cantorian Set Theory and Limitation of Size.* Oxford University Press, 1984.

19. Knuth, D. E. *The Art of Computer Programming.* Volume 1:*Fundamental Algorithms.* Addison-Wesley, 1973.

20. Knuth, D. E. Algorithms in modern mathematics and computer science. In *Algorithms in Modern Mathematics and Computer Science,* Springer Lecture Notes in Computer Science, Vol 122, 1981, pp. 82-99. Revised (from ALGOL to Pascal) and reprinted in *American Mathematical Monthly 92* (1985), 170-181.

21. Lewis, H. and Papadimitriou, C. *Elements of the Theory of Computation.* Prentice-Hall, 1981.

22. Martin-Lof, P. Constructive mathematics and computer programming. In *Sixth International Congress for Logic, Methodology, and Philosophy of Science,* North-Holland, 1982. Reprinted in *Mathematical Logic and Programming Languages,* Hoare, C. A. R. and Shepherdson, J. C. (eds.), Prentice-Hall, 1986.

23. Moerdijk, I. "Review of *Mathematical Intuitionism. Introduction to Proof Theory*". *Bull. Amer. Math. Society 22* (1990), 301-304.

24. Murthy, C. Classical Proofs as Programs: How, What, and Why. These Proceedings, 1991.

25. Myers, J. P., Jr. "The central role of mathematical logic in computer science". *ACM SIGCSE Bulletin 22* (1990), 22-26.

26. Rado, T. "On non-computable functions". *Bell Sys. Tech. Journal* (1962), 887-884.

27. Stewart, I. "The ultimate in undecidability". *Nature 332* (10 March 1988), 115-116.

28. Troelstra, A. S. "Intuitionistic extensions of the reals". *Nieuw Arch. Wisk. 28* (1980), 63-113.

29. Wittgenstein, L. *Remarks on the Foundations of Mathematics*. Basil Blackwell, 1956. Translated by G. E. M. Anscombe.

The Type Structure of CAT

J. Paul Myers, Jr. and Ronald E. Prather
Trinity University

Computer science educators have long lamented the fact that so many students show a lack of basic understanding of the fundamental mathematical principles underlying the field. Small wonder, when they are taught that "the integers" end at *maxint,* that fractions are to be rounded off to the nearest decimal, and that recursion is something connected with the notion of "pointers." We may think that we are doing an admirable job with our elementary discrete mathematics course; we may be encouraging any number of students to enroll in substantive courses in logic, algebra, and the like. But so long as we persist in the use of programming languages that are counter to the most fundamental mathematical precepts, obscuring their very nature, we cannot hope that the situation will improve. In this paper, we introduce the type structure of a new programming language, CAT, based in the mathematical theory of categories. Among its many novel features, the language offers an exact arithmetic, a blend of applicative and imperative programming methodologies, strong and consistent typing, a functorial semantics, a proof technique, functions as "first class" objects, and recursive structures without pointers. It is felt that a strong pedagogical basis can be achieved by the early introduction to such a language -- provided of course, that the more esoteric categorical properties are kept "at a distance." The CAT language, even as seen at an informal level (as is our treatment here), nevertheless introduces a number of important constructive notions, in particular that of "provable recursiveness," one that seems to have been overlooked in the literature. In the context developed here, it is seen to form the basis for a whole new programming philosophy.

1. Introduction

In the development of the traditional programming language, we more often find that the applications, the machine model of the computing environment, and the habits and attitudes of the common programmer, in some combination, serve to determine the design philosophy and the overall characteristics of the resulting language. Usually these considerations are mutually competitive and we then obtain something of a compromise in design philosophy and in language features. When, at a later stage, we undertake a study of the semantic interpretation of such languages, it is not surprising that all mathematical machinery adequate to the task is found to be cumbersome, unwieldy, and largely unsatisfactory [1].

We believe that the problem is, in part, a question of "domains," and of the necessary "exception handling" when these computational domains are not well chosen. In pure Lisp [2], we have an unusually simple domain structure, all computational objects are of one type, and accordingly, the semantic interpretation of Lisp is quite straightforward. On the other hand, not everyone is completely happy with such a rudimentary programming environment, however elegant the foundation may be.

In a previous paper [3], one of the authors has proposed an equally simple programming language foundation, drawn from the mathematical theory of categories [4,5], yet having the consequence that more conventional programming structures and techniques might result. Here, it is our task to describe the elements of such a language in somewhat more detail, with particular attention to the underlying type structure, i.e., the nature of the domains to be admitted. Since we eventually arrive at the situation where, as they say [6], "functions are first class objects" (themselves members of some admissible domain), we give an almost equal treatment to these functions, alias morphisms of the category, alias provably recursive programs.

2. The Category of Provable Recursiveness

In the earlier paper [3], it was first suggested that the category (here called **RecSet**) consisting of *recursive sets* as "objects" and (total) *recursive functions* amongst them as "morphisms," might serve as the cornerstone to the CAT language development. Certainly if we are given $f : A \to B$ and $g : B \to C$, the ordinary composition $g \circ f$ is again recursive, and the categorical axioms:

$$\text{associativity} \qquad h \circ (g \circ f) = (h \circ g) \circ f$$
$$\text{identity} \qquad i \circ f = f \quad \text{and} \quad g \circ i = g$$

are clearly satisfied, where $i = i_B$ is the *identity* morphism (recursive function) on B. Moreover, the resulting category **RecSet,** a subcategory of **Set,** has a number of useful properties (e.g., the monomorphisms are the injective morphisms, and dually, the epimorphisms coincide with the surjective morphisms, as in **Set**). But in yet another setting [7], the same author provides a convincing argument that **RecSet** finds its best use as a category for the semantic interpretation of CAT, whereas it falls short of the original purpose owing to the fact that it is not a "cartesian closed" category [5], a consequence of the unsolvability of the halting problem [6].

So instead, we are led to introduce as our programming language foundation, the category **ProRec** of *provable recursiveness,* and ultimately (as in [7]), a semantic functor [5]

$$\text{S: ProRec} \to \text{RecSet}$$

in which the domains of CAT (objects of **ProRec**) and the "provably" recursive functions (morphisms of **ProRec**) are each given a concrete interpretation. Basically, the objects of **ProRec** consist in the smallest class of domains or "types" containing the (six) elementary domains to be presented in the next section, and closed under certain (again, six in number) type constructions, including recursion, as discussed in subsequent sections. And as morphisms, the *provably recursive functions* $f : X \to Y$ among such domains are defined as in the program schemata:

$$\text{function } f(x : X) : Y; \ C \ \{P\}$$

where C is a (most likely compound) command and P is a proof that for each x in the domain X, the execution of C yields a value for f in the domain Y. Both the objects and the morphisms of **ProRec** are subject to a notion of "structural equivalence" [9], causing an identification of suitably equivalent objects (resp. morphisms). But in the case of the latter, we hasten to observe that such equivalence is largely only textual, i.e., many different programs will still compute the same function, and it is only in the semantic interpretation (in **RecSet**) that these functional identifications are made. With this proviso, we may nevertheless introduce a *composition* of morphisms in **ProRec** corresponding to a simple juxtaposition of program texts. After handling certain patching problems in this regard, it can be shown [7] that the categorical axioms are satisfied and that **ProRec** is indeed a category. Denoting the image of the semantic functor by

$$\mathbf{Rec} = S(\mathbf{ProRec})$$

we obtain (again, as shown in [7]) a cartesian closed category **Rec**, a subcategory of **RecSet**, in which the interpretation of CAT finds its proper home.

Finally, we note that the proof mechanism, to be drawn from the predicate calculus, may be thought to be similar to that described in other contexts [10,11], and we do not provide details here. Suffice it to say that (contrary to the implication in the above schemata) the lines of the proof are most likely "spread over" the lines of the accompanying compound command, as is customary. All that is important for our purposes is that it be decidable whether a proof is correct in its determination that a given function is indeed recursive. Note that this is a considerably weaker requirement than in most "proof of correctness" contexts, where the proof must also demonstrate that an a priori given functional specification has been achieved.

We remark that the recursive functions are a classical (and therefore "static") notion; a given function may be there but we "don't know it." The category **Prorec**, on the other hand, is a constructive category. It is subject to growth over time, inasmuch as the library of programs and proofs will effect changes in the "state of knowledge." And there is thus an evolutionary increase in this knowledge [12].

3. Elementary Types

Every discussion of programming language types must eventually propose a class of irreducible types, domains over which all others are built. Usually these are chosen to establish the computational basis as well, an underlying arithmetic, so to speak. In the CAT language, we take the idealistic view that such arithmetics ought to be uncompromising for their faithfulness in representing genuine mathematical systems. With this view in mind we propose the following six *elementary types:*

rational	integer	natural	string	boolean	fuzzy
$\times$	$\times$	$\times$	*	$\wedge$	$\cap$
+	+	+		$\vee$	$\cup$
$-$	$-$			$\neg$	$\sim$
/	div				
	mod				

with built-in arithmetic operations as shown. The first three are self-explanatory, except to emphasize that we intend a full implementation of these countably infinite algebraic systems -- the rational numbers, the integers (with the usual **div** and **mod** replacing division), and the natural numbers. We use the symbols R, I, N as abbreviations herein, and also S, B, F for the other three elementary types, in effect identifying these computational domains with their interpretation in **RecSet** (as recursive sets). In this connection, we assume that $S = A^*$ for some suitable alphabet or character set A (and we should point out that it is to be assumed that there are to be many more string operations on S than the simple concatenation operation listed above), whereas $B = \{false, true\}$ as usual. Finally, we take F as the set of rational numbers in the interval from zero to one, with the hope that we might thereby encourage the computational development of "fuzzy logic" [13], noting that our operations are in this case *min*, *max*, and *reflection* (subtraction from one), respectively, the traditional fuzzy operations.

While we do not speak of "number representations" here, we note that if an implementation of CAT were to use (unbounded -- except for B) linked lists of cells in the representation, it has been shown [14] that all of the above arithmetic operations are, at the worst $O(n^2)$ in their computational complexity, n being the number of cells employed. Indeed, the larger the two integers, the longer the time a multiplication will take. But overall, this figure does not seem to be prohibitive, in exchange for the extraordinary generality of our system. This same figure, incidentally, extends to the operations in the picture below.

We refer to the class of *transfer operations* built into the CAT language for transforming data of one elementary type to another, as follows:

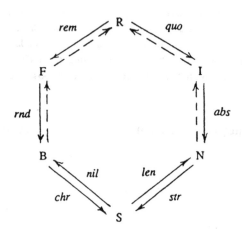

Note that dotted lines signify *implicit* transfer operations, e.g., every natural number can be considered as an integer. Again, these three-letter mnemonics should be self evident as to their meaning, except perhaps for *rnd* (with the value *true* iff applied to a fuzzy number greater than or equal to one-half), and for *quo* and *rem* on the rationals. These latter two operations are to return integer and fractional parts, respectively, indeed the integer quotient and the remainder (technically, a fuzzy number). We should note in this connection, that whereas $F \subseteq R$ in the set-theoretic sense, we intend a completely different representation in these two computational realms. In R we use the numerator-denominator integer pairs, whereas in F we choose the repeating decimal notation, for one reason, to facilitate (with *quo* and *rem*) output of rationals to a more human-readable form.

Any built-in operations beyond these that one might propose should be subject to the condition that they be representable as provably recursive functions (i.e., programs). Otherwise we would frustrate the underlying categorical foundation of CAT. Especially must this be kept in mind regarding the built-in *enumeration facility* that we now describe. Obviously, each of our six elementary types correspond (in **RecSet**) to recursive, and thus recursively enumerable [8] sets. In specific acknowledgement of this fact, we propose that the CAT language provide built-in *first* and *succ* operations, applicable to each of the six elementary types. The intention is clear, e.g., for N, *first* yields zero and *succ* is the ordinary successor function, whereas for I, *succ* will alternate positive and negative in the enumeration. In S we enumerate the members in lexicographical order by length, whereas in B the whole process is trivial (but to remark that $succ(true) = false$ as well as $succ(false) = true$, to ensure *total* recursiveness). In R we use the traditional enumeration along diagonals, again alternating positive and negative, whereas in F we may enumerate along successive rows of the obvious infinitely triangular region. In every case, *first* is a zero-ary operation returning the obvious starting point, while *succ* is unary. But note well that in every instance we are dealing with a provably recursive function, a point of some importance in the sequel.

4. Structured Types

Our category-theoretic foundation [7] serves as the main source in proposing constructs that are appropriate in building new CAT domains (types) from others. This way, there is a good chance that there will be a smooth mathematical underpinning to the semantics of the whole CAT type structure. With this in mind, we note that each of the six "structured" types that we now introduce has an underlying functorial significance in the category **Rec**. That these same structures have, in every instance, a well-recognized computational basis is more than a happy coincidence. Indeed it is a sign that the categories **ProRec** and **Rec** have been chosen wisely. If these structures, further, have at least some similarity to those found in Pascal-like languages [15], so much the better.

Yet it is our feeling that the designers of these languages would have been better advised to have named the structured types in accordance with standard mathematical terminology. Why speak of a "record" when we have a member of a product domain? We feel that it is better to give the domain the name, not the member, i.e., here to speak of "product" rather than "records." Even at the elementary level, it is likely that most students read "n : **integer**" as "n is an integer" rather than "n is a member of the **integer** domain," our preference, as if the colon stood for the set-theoretic epsilon.

This may all seem quite trivial, but it does help to explain the naming conventions used below.

Thus if X, Y and the X_i are previously understood types or domains, we may introduce any of the following *structured types:*

product	**sum** name : type **of**
name$_1$: X_1 ;	value$_1$: X_1 ;
$\cdots$	$\cdots$
name$_n$: X_n	value$_n$: X_n
end	**end**
sets of X	**sequences of** X
exponential Y **of** X	**subdomain** name **of** X

As to their interpretation in **RecSet,** these are the ordinary category-theoretic **product** and **sum,** whereas **sets** and **sequences** refer to the set of all *finite* subsets of X and *finite* sequences from X, respectively. The last pair are more difficult to describe (and accordingly, more interesting).

The **exponential** construct builds the domain of all provably recursive functions from Y to X. *Important:* Not all functions, not even all recursive functions! If X and Y are recursive, the full X^Y consisting of all recursive functions from Y to X is non-recursive. It is undecidable whether $f : Y \rightarrow X$ is (total) recursive. In fact, this is the "halting problem." But our CAT functions carry with them a proof of their total recursiveness, and we suppose that one can indeed decide whether or not such a proof is correct in its claim. As a matter of practicality, we probably would not insist that all programs include such proofs in the actual implementation of CAT. But that does not obviate the need for the programmer to be aware of the responsibility, nor does it argue for the elimination of this requirement in our theoretical treatment of the language. In fact, it is essential, and we will act as though such proofs have indeed been provided.

Equally interesting is the **subdomain** construct, for it references a program, identically named, in its type specification, following a suggestion from [9] (p. 208, Project 12.1), thereby providing a provably recursive decision algorithm for answering the membership question. For example, one may write:

$$\textbf{subdomain even of integer}$$

where "even" is the obvious program

```
function even (n  : integer) : boolean;
  begin
  if n div 2 = 0 then
    even := true
  else
    even := false
  end
```

Note that the (provable) recursiveness ensures that an explicit type-checking mechanism is extended to such subdomain types, in fact, to all of the structured types that we have introduced.

We should not fail to mention, in concluding this section, that each of our structured types are to be provided with the (built-in) canonical morphisms that characterize the constructions in the category-theoretic sense. Thus, the **product** type will include projections onto the coordinate types (the "dot" operator in Pascal [14]), the **sum** will include the canonical injections, and **sets** and **sequences** may similarly be accessed and modified, though we do not discuss the syntactic details here. In effect, such discussions relate to the *expression language* of CAT, whose complete design is still under study. Of special concern here are the functional expressions, those relating to the **exponential** construct, where we intend to provide a comprehensive *applicative* programming capability, to augment the more traditional imperative command structure. However, such discussions would take us quite far from the context of the present work and one should consult [3,7] for more detail.

5. Special Types and Modified Types

The reader may sense that a few of the standard programming language features are so far missing from our type structure, namely "arrays," enumerated types, subrange types, and any provision for recursion in type definitions. Consider the following extract from the proposed *abstract syntax* [9] of CAT [7]:

$$R ::= L \mid I \mid OR$$

$$S ::= L \mid I \mid OS \mid S .. S \mid < \cdots , S , \cdots > \mid [\cdots , S , \cdots]$$

$$T ::= S \mid \textbf{product} \cdots ; I : T ; \cdots \textbf{ end} \mid \textbf{sum } I : S \textbf{ of} \cdots ; R : T ; \cdots \textbf{ end} \mid$$
$$\textbf{sets of } T \mid \textbf{sequences of } T \mid \textbf{exponential } S \textbf{ of } T \mid \textbf{subdomain } I \textbf{ of } T$$

where R = *restricted expressions*, S = *special types*, and T = *types* (the other classes mentioned here -- *literals, identifiers, operators, and commands*, are easily discernible). We thus see that the *enumerated types* come into the picture through the admittance of S into T as a sequential listing of identifiers, i.e.,

$$T = S = < \cdots , S , \cdots > = < \cdots , I , \cdots >$$

whereas the conventional *subrange types*, $S .. S$, are likewise admissible, even though equivalent structures could have been introduced with the **subdomain** construct.

As a further extension of this *special type* facility, note that we have included the provision for introducing products of subrange types, i.e.,

$$T = S = [\cdots , S , \cdots] = [\cdots , S .. S , \cdots]$$

and further, that such a domain is permissible as the first entry of the **exponential** construct. It would thus appear that *arrays* (functions from such products) may in fact be treated as only a special case of the use of the **exponential** construction. Our reply is "yes and no." It must be remembered that the functions of **exponential** S **of** T are programs, software so to speak, whereas the whole notion of an array (though it is a function) is tied up with the hardware, a tabular storage facility. As a compromise, we propose a modification, the use of an adjective (or modifier: **hard**) to effect such a change in meaning, rather than to augment our type structure unnecessarily. In fact, we would argue that it is well to admit that arrays are indeed functions, as in the definition:

type
    matrix = **hard exponential** $[1 .. n, 1 .. n]$ **of rational**;
variable
    A, B : matrix;

Note that such usage does not preclude one also introducing a "soft" version of the same structure, i.e., matrices whose entries are computed by some algorithm.

Finally, there is the matter of recursion. All that we really need is some way to implement the schema:

$$X = A + \mathbf{F}(X)$$

where $\mathbf{F}$ is any monotone $(X \leq Y \rightarrow \mathbf{F}(X) \leq \mathbf{F}(Y))$ functor, as is shown in the following result:

Theorem: If $\mathbf{F}$ is monotone, the recursive equation

$$X = A + \mathbf{F}(X)$$

has the solution:

$$X = \sum_{n=0}^{\infty} X_n$$

where $X_0 = A$, $X_1 = \mathbf{F}(A)$, $X_2 = \mathbf{F}(A + \mathbf{F}(A))$, and so on.

Now the **sum** is already available in our type structure, whereas any of the other type constructs (except for **subdomain**, where the meaning of recursion is problematic at best) will serve well in the role of $\mathbf{F}$.

In fact, we provide slightly more generality in simply modifying the **sum** construct with the adjective: **recursive**, permitting any number of summands and arbitrarily many self-references, e.g.,

type
    binarytree = **recursive sum leaf : boolean of**
        $true : A$;
        $false$: **product**
            left : binarytree ;
            right : binarytree
        end
    end

for binary trees labeled with members of the base type A. Note that there is no mention of "pointers." If they appear at all, it is at a lower linguistic level, perhaps in an implementation of CAT, say in the C language.

6. Concluding Remarks

Consider once again our requirement, imposed by the category **RecSet** in which our semantic interpretation is made, that all of the domains we are using to represent data types be recursive. This is an essential requirement if we are to provide the CAT language with an adequate *type-checking* facility. Our elementary types are recursive. And we are on safe ground in introducing the product and sum, sets and sequences (these latter, owing to the finiteness restrictions imposed). The sub-domain construct returns again a recursive set, in fact, provably so. And finally, the exponential domains are recursive because we limit our functions (sub-programs) to those that are provably recursive.

One may ask, "How restrictive is this limitation?" What kind of functions are we missing? To give some kind of answer, and we feel that it is a most adequate one, first consider the CAT domain

$$t = \textbf{exponential natural of natural}$$

consisting of all provably recursive functions on the natural numbers. This will correspond to a definite recursive set (call it T) in **RecSet** under the semantic functor S (in which "equivalent" programs, those which compute the same mathematical function, are identified). It has been shown [7] in this regard, that any *succ* function for T (while there are definitely ones that are recursive), is not provably recursive. Thus, the set T can be regarded as "productive" in that there are very powerful methods for producing elements of T, even outside of any given sequence [16]. Such a function then plays the same role for provable recursiveness as Ackermann's function [8] does for primitive recursion. In fact, Ackermann's function is obviously provably recursive, so we would thus have established a definite hierarchy:

$$\text{primitive recursive} < \text{provably recursive} < \text{(total) recursive}$$

and it is in this middle ground that the CAT language lives.

REFERENCES

1. Ashcroft, E.A. and W.W. Wedge, Rx for Semantics, *ACM Transactions on Programming Languages and Systems*, 4, 1982.

2. McCarthy, J., Recursive Functions of Symbolic Expressions and their Computation by Machine, *Communications of the ACM, 3, 1960*.

3. Prather, R.E., Proposal for a Categorical Programming Language, *ACM SIGPLAN Notices*, 25, 1990.

4. MacLane, S., *Categories for the Working Mathematician*, Springer-Verlag, 1971.

5. Cohn, P.M., *Universal Algebra*, Harper and Row, 1965.

6. Barr, M. and C. Wells, *Category Theory for Computing Science*, Prentice Hall, 1990.

7. Prather, R.E., Categorical Properties of CAT, submitted to *Fourth Summer Conference on Category Theory and Computer Science*, Paris, 1991.

8. Brainerd, W.S. and L.H. Landweber, *Theory of Computation*, Wiley, 1974.

9. Tennent, R.D., *Principles of Programming Languages*, Prentice Hall, 1981.

10. Reynolds, J.C., *The Craft of Programming*, Prentice Hall, 1981.

11. Constable, R.L. and M. O'Donnell, *A Programming Logic*, Winthrop, 1978.

12. Nerode, A., Some Lectures on Intuitionistic Logic, in *Logic and Computer Science* (ed. P. Odifreddi), Lecture Notes in Mathematics 1429, Springer-Verlag, 1990.

13. Zimmerman, H.S., *Fuzzy Set Theory and its Applications*, Kluwer-Nijhoff, 1985.

14. Prather, R.E., The Arithmetic of CAT, *Unpub. Ms.*, 1989.

15. Jensen, K. and N. Wirth, *Pascal User Manual and Report*, Springer-Verlag, 1974.

16. Greenleaf, N., Algorithms and Proofs: Mathematics in the Computing Curriculum, Proceedings of the 20th Technical Symposium on Computer Science Education, *SIGCSE Bulletin*, 21, 1989.

A Simple and Powerful Approach for Studying Constructivity, Computability, and Complexity

Klaus Weihrauch
Theoretische Informatik, FernUniversität
D – 5800 Hagen
Germany

Abstract

In this contribution a natural and simple as well as general and efficient frame for studying effectivity (*Type 2 theory of effectivity, TTE*) is presented.

TTE is a straightforward "logic free" extension of ordinary computability theory. Three basic kinds of effectivity for functions on Σ^* and Σ^ω are distinguished: continuity, computability, and easy computability (computational complexity).

As the most remarkable property, *continuity* in TTE can be very adequateley interpreted as a basic kind of *constructivity*. Effectivity is transferred from Σ^* and Σ^ω to other sets by *notations*, (where finite words serve as names) and by *representations* (where ω–words serve as names), respectively.

In this contribution the structure of TTE is explained and its applicability is demonstrated by simple examples mainly from analysis. Especially it is shown how the "effectivity gap" between Abstract Analysis and Numerical Analysis can be closed step by step by introducing stronger and stronger effectivity requirements.

It is suggested that the theory outlined in this paper is adequate to introduce effectivity in Analysis into Comupter Science curricula.

1   Introduction

While for denumerable sets we have a far developed computability and complexity theory, in Analysis there is a very abstract mathematical theory on the one hand on a Numerical Analysis for practical computations on the other hand. But there has been no satisfactory theory of effectivity between them.

In the past several partly overlapping, partly competing attemps have been made for closing this "effectivity gap". Ituitionism (Brouwer [1], Heyting [2], Troelstra [3] et al.) reduces classical logic to "constructive" proofs, whereby proofs of pure existence are avoided. Bishop and Bridges [4] show that "constructive proofs in a narrow sense" suffice to develop the important parts of Analysis. Two other approaches are based on recursion theory. The "Russian approach" uses standard numberings of "computable"real numbers etc. and thus transfers computability from IN to sets of "computable" objects in Analysis (Ceitin [5], Markov [6], Kushner [7], Aberth [8]). The "Polish" approach represents real numbers etc. by sequences of rational or natural numbers. Computable operators on sequences of natural numbers yield computable functions on all real numbers, not only on the computable ones (Mazur [9], Grzegorczyk [10], Mostowski [11], Lacombe [12], Klaua [13], Hauck [14]). Also the work of Pour-El and Richards [33] is of this type. Scott [15] suggested to embed real numbers into an interval-cpo and thus to approximate real numbers by chains of intervals. Finally there are some papers concerning computational complexity of real functions (Brent [16], Ko and Friedman [17], Müller [18, 19]). A detailed discussion of Constructive Mathematic is given by M. Beeson[20]. Although each of the approaches has advantages, none of them has been accepted by the majority of Computer Scientists or Mathematicians.

In this contribution a theory of effectivity is presented which is more flexible and powerful than any of the approaches to effectivity in Analysis mentioned above. Among many other applications this "Type 2 theory of effectivity", shortly TTE, admits to close step by step the effectivity gap between Abstract and Numerical Analysis by a concept for constructivity (formally expressed by continuity), a concept for computability, and a concept for computational complexity. TTE includes ordinary recursion theory but does not depend on any kind of restriction of logic (i.e. TTE is "logic–free"). It can be considered as a consequent extension of the Polish Recursive Analysis.

First basic definitions of TTE will be introduced and explained. Then it will be shown by examples, how constructivity, computability and computational complexity in analysis can be investigated in TTE. Finally open questions and suggestions for further studies will be given. More details can be found in papers by Kreitz, Müller, and Weihrauch [18, 19, 21 - 30].

2 The basic effectivity theory

Let Σ be a finite alphabet. We may assume $\{0,1\} \subseteq \Sigma$. We assume that the reader is familiar with basic recursion theory (computability, Turing machines, numbering $\varphi : \mathbb{N} \to P^{(1)}$ of the computable functions satisfying the smn– and utm–theorem, recursiveness on M w.r.t a given numbering $\nu :\subseteq \mathbb{N} \to M$ or notation $\nu :\subseteq \Sigma^* \to M$).

In ordenary ("Type 1") recursion theory computers can, w.l.g., operate only on finite words $w \in \Sigma^*$. In Type 2 Theory, ω–sequences $p \in \Sigma^\omega$ are considered as new primitives, which computers can handle. We fix a standard machine model which is especially suitable for studying computational complexity. A *Type 2 machine* M of kind (k,m) is a Turing machine with k input tapes for words $w \in \Sigma^*$, m one–way input tapes for sequences $p \in \Sigma^\omega$, finitely many work tapes, and one write–only one–way (!) output tape. Two partial functions $f_M :\subseteq (\Sigma^*)^k \times (\Sigma^\omega)^m \to \Sigma^\omega$ and $g_M :\subseteq (\Sigma^*)^k \times (\Sigma^\omega)^m \to \Sigma^*$ can be associated with M. For $z \in (\Sigma^*)^k \times (\Sigma^\omega)^m, p \in \Sigma^\omega$ and $w \in \Sigma^*$ let

$f_M(z) = w$ iff M with input z stops with w on the output tape,
$g_M(z) = p$ iff M with input z computes forever writing p on the
                output tape.

Clearly, f_M and g_M have the following finiteness property FP:
Every finite portion of the output depends only on a finite portion of the input. FP is one of the most elementary conditions for effectivity. Many problems which are not effective intuitively (and which are shown to be not effective in other formal approaches) violate FP in TTE (see below for examples). FP can be expressed as continuity. Consider the discrete topology on Σ^*. For $w \in \Sigma^*$ let $w\Sigma^\omega := \{wp \mid p \in \Sigma^\omega\}$. On Σ^ω consider Cantor's topology τ_C defined by the base $\{w\Sigma^\omega \mid w \in \Sigma^*\}$ of

open sets. FP for f_M and g_M means: f_M and g_M are continuous. Thus, continuity is (in this framework) a very basic kind of effictivity. It is easy to define computable pairings $\Sigma^* \times \Sigma^\omega \to \Sigma^\omega$ and $\Sigma^\omega \times \Sigma^\omega \to \Sigma^\omega$ with computable inverses (notation $< x, p >$ and $< p, q >$ for $x \in \Sigma^*$ and $p, q \in \Sigma^\omega$). As counterparts of the numbering φ there are two "effective" representations

$$\chi : \Sigma^\omega \to [\Sigma^\omega \to \Sigma^*] \text{ and } \psi : \Sigma^\omega \to [\Sigma^\omega \to \Sigma^\omega]$$

where $[\Sigma^\omega \to \Sigma^*] = \{f :\subseteq \Sigma^\omega \to \Sigma^* \mid f$ continuous and $dom(f)$ open$\}$ and $[\Sigma^\omega \to \Sigma^\omega] = \{f :\subseteq \Sigma^\omega \to \Sigma^\omega \mid f$ continuous and $dom(f)$ is $G_\delta\}$. Both, χ and ψ, have computable universal functions and satisfy smn–theorems. From this basis a rich theory of continuity, formally similar to ordinary recursion theory, can be developed. Specialization leads to computability. Especially, the open sets correspond to the recursively enumerable subsets of IN and the closed and open (clopen) sets to the recursive subsets of IN. A subset $X \subseteq \Sigma^\omega$ is called r.e., iff $X = dom(g_M)$ for some Type 2 machine M, and recursive, iff X and $\Sigma^\omega \setminus X$ are r.e.

Example 1

(1) (Brouwer's "limited principle of omniscience") The set $X = \{p \in \Sigma^\omega \mid 0 \in range(p)\}$ is r.e. (hence open) but not closed (hence not recursive).

(2) The function $t : \Sigma^\omega \to \Sigma^*$ with $t(p) = (0$ if $(\forall i)10^{i+1}1$ is a subword of p, 1 otherwise) is not continuous, hence not computable.

(3) If q divides a power of p then there is a computable translation from the p–adic into the q–adic representation of real numbers, otherwise there is no continuous translation.

In this example (and many others) positive answers have the form "there is a computable functions ...", while negative ones have the form "there is no continuous function ..." . In the latter cases non–effectivity does not depend on the definition of computability (Church's thesis) but a more basic principle, namely finiteness (= continuity) is violated.

3 Representations

The uniform way to introduce computability on a denumerable set M is to define an ("effective") notation $\nu :\subseteq \Sigma^* \to M$ (or equivalently a numbering) and to transfer computablility from Σ^* to M by means of ν. By cardinality, sets like 2^ω, $\mathbb{R}$, $O(\mathbb{R}) :=$ open subsets of $\mathbb{R}$ or $K(\mathbb{R}) :=$ compact subsets of $\mathbb{R}$ have no notations. In TTE also infinite sequences $p \in \Sigma^\omega$ are used as names. Such namings are called *representations*. A representation of a set M is a (possibly partial) surjective function $\delta :\subseteq \Sigma^\omega \to M$. A representation transfers topological and computational properties from the set of namens Σ^ω to the represented set. Let $\delta :\subseteq \Sigma^\omega \to M$. and $\delta' :\subseteq \Sigma^\omega \to M'$ be representations. A function $f :\subseteq M \to M'$ is called (δ, δ')–*continuous* $((\delta, \delta')$–*computable*), iff $\forall p \in dom(f\delta) : f\delta(p) = \delta'\Gamma(p)$ for some continuous (computable) function $\Gamma :\subseteq \Sigma^\omega \to \Sigma^\omega$ (correspondingly for (δ, ν)–continuous (computable) where ν is a notation). A subset $X \subseteq M$ is δ–*open* $(\delta$–*r.e.*, δ–*clopen*, δ–*recursive*), iff $\delta^{-1}X = A \cap dom(\delta)$ for some open (r-e-. clopen, recursive) subset $A \subseteq \Sigma^\omega$.
Notice that the set $\tau := \{A \subseteq M \mid A \quad \delta$–open $\}$ is a topology on M (the final topology of δ).
Generalizations to functions $f :\subseteq M_1 \times M_2 \to M'$ $((\delta_1, \delta_2), \delta')$–continuous etc.) and subsets $X \subseteq M_1 \times M_2$ $((\delta_1, \delta_2)$–open etc.) are straightforward.
Reducibility (or translatability) for representations is defined by: $\delta \leq_t \delta'$ $(\delta \leq_c \delta')$ iff $\forall p \in dom(\delta) : \delta(p) = \delta'\Gamma(p)$ for some continuous (computable) $\Gamma :\subseteq \Sigma^\omega \to \Sigma^\omega$.
Deduce equivalence by $\delta \equiv_t \delta' :\Longleftrightarrow (\delta \leq_t \delta')$ and $\delta' \leq_t \delta)$ (correspondigly $\delta \equiv_c \delta'$). The following example shows very clearly why in this context continuity is a general kind of effectivity.

Example 2

Define the enumeration representation En, the complement enumeration representation Enc, and the characterestic function representation Cf of 2^ω by $En(p) := \{i \in \mathbb{N} \mid 10^{i+1}1$ is a subword of $p\}$, $Enc(p) := \mathbb{N} \setminus En(p), Cf(p) := p^{-1}\{1\}$. Then union on 2^ω is $((En, En), En)$–computable and $((Cf, Cf), Cf)$–computable, complementation is not $((En, En), En)$–continuous but $((Cf, Cf)$–computable, $En \not\leq_t Enc, En \not\leq_t Cf, Cf \leq_c En, Cf \leq_c Enc$. $En \not\leq_t$

Cf, e.g., means that enumerations cannot effectively be converted to characteristic functions.

All the proofs are elementary. Notice that formally similar theroems can be proved for the enumeration numbering and the characteristic function numbering of the recursive subsets of IN in ordinary recursion theory.

Example 3

(1) Define the standard representation $\iota :\subseteq \Sigma^\omega \to \omega^\omega$ by $\iota^{-1}(r) :=$ $0^{r(0)}1\, 0^{r(1)}1 \ldots$ (ι^{-1} is a homeomorphism from the Baire Space to a G_δ–subspace of Σ^ω)

(2) Let (M, d) be a separable metric space, let $\nu : $ IN $\to A$ be a total numbering of a dense subset $A \subseteq M$. The derived *normed Cauchy representation* $\delta_C :\subseteq \Sigma^\omega \to M$ is defined by

$$\delta_C(p) = x \text{ iff } p \in dom(\iota) \text{ and } (\forall j > i)d(\nu(\iota p)(i), \nu(\iota p)(j)) < 2^{-\bullet}$$
$$\text{and } x = \lim \nu(\iota p)(i)$$

For example, for $y \in M$ and $i \in$ IN, $\{x \in M \mid d(x,y) < 2^{-i}\}$ is δ_C–open.

(3) The standard representation $\omega_R : \Sigma^\omega \to O(\text{IR})$ of the open subsets of IR is defined by $\omega_R(p) = \cup\{I_k \mid k \in En(p)\}$, where I is a standard numbering of the open intervals on IR with rational end points. For example, union is $((\omega_R, \omega_R), \omega_R)$–computable, and $\{B \in O(\text{IR}) \mid [0;1] \subseteq B\}$ is ω_R–r.e., while $\{B \subseteq O(\text{IR}) \mid (0;1) \subseteq B\}$ is not ω_R–open.

For obtaining a "natural" effectivity theory on a set M, the representation δ must be chosen carefully. Example 2 shows that a set may have various "effective" representations. A general definition of "effective" seems to be difficult, however, there is an important distinguished class of certainly topologically effective representations, the *admissible* representations. By a T_0D–space we shall mean a topological T_0–space with denumerable base.

Definition 4

A representation $\delta :\subseteq \Sigma^\omega \to M$ is *admissibble* , iff (1) and (2) hold.

(1) (M, τ_δ) is a T_0D–space, where τ_δ is the set of δ–open subsets of M.

(2) $\delta' \leq_t \delta$ for any (τ_C, τ_δ)–continuous representation δ' of M.

Theorem 5

(1) For any T_0D–space (M, τ) there is an admissible representation δ of M with $\tau = \tau_\delta$.

(2) Let δ and δ' be admissible representations of M and M' with final topologies τ und τ', respectively. Let $f :\subseteq M \to M'$ be a function. Then f is (τ, τ')–continuous $\Longleftrightarrow$ f is (δ, δ')–continuous.

The majority of spaces used in Analysis and Functional Analysis, especially all the separable metric spaces, are T_0D–spaces. Also the cpo's used for semantics of programming languages are T_0D–spaces. For example the representations En, Enc and Cf from Example 2 are admissible, the first one has Scott's topology the third one Cantor's topology as its final topology. Any normed Cauchy representation of a separable metric space (Example 3) is admissible, where the final topology is that one induced by the metric. Notice that the real line is a separable metric space!

A great number of "effective" numberings, notations, and representations can be introduced by a definition similar to Definition 4, for example the "admissible" numberings $\nu : \mathbb{N} \to P^{(1)} : \nu$ is "admissible" iff (1) the universal function u_ν of ν is computable and (2) $\nu' \leq \nu$ for all numberings ν' of $P^{(1)}$ with computable universal functions. For more examples see [26, 27].

Theorem 5 indicates a uniform way to introduce constructivity, computability, and computational complexity on T_0D–spaces (M, τ) . By Theorem 5, topological continuity is already constructivity in TTE. From the t–equivalence class of t–admissible representations try to select a computationally effective one, which induces the computability theory on M. Finally among the computationally effective representations try to find a computationally "simple" one as a basis for computational complexity on M. Computational complexity will be discussed in Section 5.

In the same way as it is unreasonable to study computational complexity if computability not defined well, it is unreasonable to introduce computability for functions which are not constructively effective (continuous). Both "mistakes" have been made in the past repeatedly. It should be mentioned here that there are representations which are intuitively effective but not admissible according to Definition 4, e.g. $\omega_R(Ex.3), \chi$, and ψ.

4   Constructivity and Computability in Analysis

In this section we apply TTE to some simple questions in Analysis. First, we shall discuss representations of the real numbers. In the past several representations of the real numbers have been proposed (see e.g. Deil [31]). TTE explains clearly why most of them are not natural. Let ν_Q be some standard numbering of the rational numbers, e.g. $\nu_Q < i,j,k >= (i-j)/(k+1)$. Since the rational numbers are dense in the real line (IR, τ_R) we may apply Example 3(1) to ν_Q and the distance $(x,y) \rightarrow |x-y|$ on IR. Let us call the resultant representation δ_C "standard Cauchy representation" of IR. Since δ_C is admissible, by Theorem 5(2) all the functions like addition, multiplication, division, exponentiation, the trigonometric functions, etc. are continuous w.r.t. δ_C, i.e. constructive. By the choice of ν_Q these functions are even computable w.r.t. δ_C. Roughly speaking, there is a Type 2 machine which for any "fast converging" rational Cauchy sequence with limit x determines a "fast converging" rational Cauchy sequence with limit $exp(x)$, etc.

The most familiar representation of the real numbers is the decimal representation δ_D which can be defined as follows: Let ν_z be a standard numbering of the integers.

$$\delta_D(p) = \nu_z(\iota p)(0) + \Sigma((\iota p)(i) \bmod 10) \cdot 10^{-i} \quad (p \in dom(\iota))$$

Although $X \subseteq$ IR is δ_D–open iff $X \in \tau_R$, δ_D is not admissible, especially $\delta_C \leq \delta_D$ is false. As a proof one shows easily, that $x \rightarrow 3x$ is not continuous w.r.t. δ_D. Therefore, the decimal representation is unnatural for topological reasons, i.e. "constructively ineffective".

Also the usual "naive" Cauchy representation δ_N defined by ($\delta_N(p) = x$

iff $(\nu_Q(\iota p)(i))_{i\in\mathbb{N}}$ is a Cauchy sequence with limit x) is topologically bad: $X \subseteq \mathbb{R}$ is δ_N-open iff $X = \emptyset$ or $X = \mathbb{R}$. Informally, no finite initial part of p gives any information about $\delta_N(p)$.

Some further representations of $\mathbb{R}$ drived from Dedekind cuts are discussed in Weihrauch and Kreitz [25, 26]. For studying the real line, mainly admissible representations with final topology τ_R and especially the "computationally admissible" representation δ_C (or any computationally equivalent one) yield "natural" results. We shall apply δ_C-effectivity now.

Proposition 6

No nontrivial property on $\mathbb{R}$ is δ_C-decidable.

Suppose $X \subseteq \mathbb{R}$ is δ_C-decidable. Then X and $\mathbb{R} \setminus X$ are δ_C-open, hence open. Since the real line is topologically connected, X must be empty or equal to $\mathbb{R}$. At most those properties which correspond to open subsets of $\mathbb{R}$ (or $\mathbb{R}^n$) can be δ_C- r.e. . For example, the sets $\{x \mid x > 0\}, \{x \mid x \neq 0\}$ and $\{(x,y) \mid x \neq y\}$ are δ_C- r.e. . Sets which are δ_C- open but not δ_C- r.e. can be defined easily by using some non recursive set or function.

Classically, if $x \cdot y = 0$ then $x = 0$ or $y = 0$ $(x,y \in \mathbb{R})$. Can we determine effectively a factor which is 0?

Proposition 7

There is no continuous function $\Gamma :\subseteq (\Sigma^\omega)^2 \to \Sigma^*$ such that for all $p, q \in dom(\delta_C)$ with $\delta_C(p) \cdot \delta_C(q) = 0$ the following holds: $\Gamma(p, q) \in \{0, 1\}$ exists and $(\Gamma(p, q) = 0 \Rightarrow \delta_C(p) = 0)$ and $(\Gamma(p, q) = 1 \Rightarrow \delta_C(q) = 0)$.

The formal proof is easy. Informally, a function Γ with the above property would yield information which does not only depend on finite initial parts of p and q. Hence Γ cannot be continuous. The next examples concern zeroes of continuous functions. Let $C[0; 1]$ be the set of continuous functions $f : [0; 1] \to \mathbb{R}$. With the max–distance $d(f, g) := max\{|f(x) - g(x)| \mid x \in [0; 1]\}$ $C[0; 1]$ becomes a separable metric space. Examples for dense denumerable subsets are the polynomials with rational coefficients, the trigonometric polynomicals with rational coefficients, and $Pg :=$ the set of finite polygons with rational vertices. Let $\alpha : \mathbb{N} \to Pg$ be some standard numbering of Pg. Let $\delta_\alpha :\subseteq \Sigma^\omega \to C[0;1]$ be the admissible representation derived form α according to Example 3(2). If $\delta_\alpha(p) = f$, then for any $i, B_c(\alpha(i), 2^{-i})$ is a closed ball containing f, where $B_c(g, \varepsilon)$ can be visualized by a band around g with width 2ε. Almost all properties concerning zeroes of continuous functions are not constructive. Let $X_{IV} := \{f \in C[0;1] \mid f(0) \cdot f(1) < 0\}$, $X_{IVD} := \{f \in X_{IV} \mid f^{-1}\{0\}$ is nowhere dense$\}$, $X_1 := \{f \in C[0;1] \mid f$ has exactly one zero $\}$.

Theorem 8

(1) The sets $X_N := \{f \in C[0;1] \mid$ has no zero $\}$ and X_{IV} are δ_α-r.e. but not closed, especially not δ_α-decidable.

(2) There is no continuous function $\Gamma :\subseteq \Sigma^\omega \to \Sigma^\omega$ with $\delta_\alpha(p)\,(\delta_C\Gamma(p)) = 0$ for all $p \in X_{IV}$.

(3) There is a computable function $\Delta :\subseteq \Sigma^\omega \to \Sigma^\omega$ with $\delta_\alpha(p)\,(\delta_C \Delta (p)) = 0$ for all $p \in X_{IVD}$.

(4) There is no $(\delta_\alpha, \delta_C)$-continuous function $F :\subseteq C[0;1] \to \mathbb{R}$ with $f(F(f)) = 0$ for all $f \in X_{IVD}$.

(5) The function $G :\subseteq C[0;1] \to \mathbb{R}$ with $dom(G) = X_1$ and $f(G(f)) = 0$ for all $f \in X_1$. is $(\delta_\alpha, \delta_C)$-computable.

The proofs are not difficult. By (2) the classical intermediate value therorem is not constructive. Under additional conditions it becomes "weakly" computable or even computable (3), (4), (5). Notice especially the difference between (3) and (4)! Similar properties can be proved for positions of maximal values. Notice, however, that the function $M : C[0;1] \to \mathbb{R}$ with $M(f) = max\{f(x) \mid 0 \le x \le 1\}$ is $(\delta_\alpha, \delta_C)$-computable. While determining a zero is not effective in general it is well known that "ε-substitutes" for zeroes can be computed:

Proposition 9

There is a computable function Γ such that $|\delta_\alpha(p)\,(\delta_C\Gamma(0^n 1p))| < 2^{-n}$

for all $n \in \mathbb{N}$ and $p \in dom(\delta_\alpha)$ satisfying $min\{|\delta_\alpha(p)(x)| \, \| \, 0 \leq x \leq 1\} <$ 2^{-n}.

Again, the proof is easy. While integration is a computable operator, differentiation is not even continuous on $C[0;1]$.

Proposition 10

(1) The function $I : C[0;1] \rightarrow C[0;1]$ defined by $I(f)(x) = \int_0^x f(x)dx$ is $(\delta_\alpha, \delta_\alpha)$-computable.

(2) The function $D : \subseteq C[0;1] \rightarrow C[0;1]$ with $dom(D) = \{ f \mid f$ is continuously differentiable $\}$ and $D(f)$ is the derivative of f is not $(\delta_\alpha, \delta_\alpha)$-continuous.

Compactness on the real line is a further instructive example. Consider I and ω_R from Example 3(3) and En and Cf from Example 2.
Let e be a standard numbering of the finite subsets of $\mathbb{N}$ and define $J(n) := \bigcup\{I_k \mid k \in e(n)\}$. We define five representations of the set $K(\mathbb{R})$ of the compact subsets of $\mathbb{R}$: α_1 corresponds to "closed" , α_2 to "closed and bounded", α_3 to "closed, bounded and located" [4], κ_e to the "enumerable" Heine–Borel property and κ_d to the "decidable" Heine–Borel property.

Definition 11

$$dom(\alpha_1) = \{p \mid \mathbb{R} \setminus \omega_R(p) \text{ bounded }\}, \alpha_1(p) := \mathbb{R} \setminus \omega_R(p)$$
$$\text{for } p \in dom(\alpha_1)$$
$$\alpha_2(p) = X \quad \text{iff } (\exists n \in \mathbb{N}, q \in \Sigma^\omega)(p = 0^n 1q, X \subseteq [-n; n]$$
$$\text{and } X = \alpha_1(q))$$
$$\alpha_3(p) = X \quad \text{iff } (\exists q, r \in \Sigma^\omega)(p = < q, r > \text{ and } X = \alpha_2(q)$$
$$\text{and } En(r) = \{k \mid I_k \cap X \neq \emptyset\})$$
$$\kappa_e(p) = X \quad \text{iff } En(p) = \{n \mid X \subseteq J(n)\}$$
$$\kappa_d(p) = X \quad \text{iff } Cf(p) = \{n \mid X \subseteq J(n)\}$$

Theorem 12

(1) $\alpha_3 \leq_c \alpha_2 \leq_c \alpha_1, \alpha_1 \not\leq_t \alpha_2, \alpha_2 \not\leq_t \alpha_3$
(2) $\alpha_2 \equiv_c \kappa_e$
(3) $\alpha_3 \equiv_c \kappa_d$

Properties (2) and (3) are two effective versions of the Heine–Borel Theorem for the real line, by (1) they are different. This example and also Example 2 indicate that representations can be characterized (up to equivalence) by the kind information about named objects which can be continuously (or computably) determined from their names.

5   Computational Complexity

While constructivity (i.e. continuity) and computability can be easily discussed simultaneously, computational complexity requires separate considerations. First results of a complexity theory for operators on the Cantor Space can be found in Weihrauch and Kreitz [28].

Let $\delta :\subseteq \Sigma^\omega \to \mathbb{R}$ be a representation. Let $f :\subseteq \mathbb{R} \to \mathbb{R}$ be a real function and assume that the Type 2 machine M computes f w.r.t. δ, i.e. $f\delta(p) = \delta\, g_M(p)$ for all $p \in dom(f\delta)$. As in ordinary Type 1 complexity theory the complexity should (essentially) not depend on the names $p \in \delta^{-1}\{x\}$ but on $x \in \mathbb{R}$ itself. In general, elements x have too many δ–names such that a uniform complexity bound for all $p \in \delta^{-1}\{x\}$ does not exist. There is, however, a special representation ρ of $\mathbb{R}$ such that $\rho^{-1}\{x\}$ in "small"for any $x \in \mathbb{R}$.

Definition 13 *(modified binary representation)*
Let $T := \{0, 1, -1, :\}$. Define a representation $\delta :\subseteq T^\omega \to \mathbb{R}$ of $\mathbb{R}$ as follows:

$$dom(\rho) := \{a_k a_{k-1} \ldots a_0 : a_{-1} a_{-2} \ldots \mid a_i \in \{0, 1, -1\},$$
$$a_k \neq 0, a_k a_{k-1} \notin \{1 - 1, -11\}\}$$

$$\rho\{a_k a_{k-1} \ldots a_0 : a_{-1} a_{-2} \ldots) := \sum_{i=k}^{-\infty} a_i \cdot 2^i.$$

Thus, ρ is a binary representation with positive and negative digits. The three conditions for $a_k a_{k-1}$ guarantee that the integer part of a name cannot become unnecessarily long. The representation ρ is computationally equivalent to our previous representation δ_C of $\mathbb{R}$. But additionally $\delta^{-1}(X)$ is a compact subset of T^ω for any compact $X \subseteq \mathbb{R}$.

Compact subsets of T^ω are sufficiently "small" such that uniform complexity bounds exist (see eg. Müller [18, 19]).

Let M be a Type 2 machine which computes a function $f :\subseteq \mathbb{R} \to \mathbb{R}$ w.r.t. to the representation ρ, i.e. $f\rho(p) = \rho g_M(p)$ for all $p \in dom(f\rho)$. For any $p \in dom(f\rho)$. define $TIME(M)(p) : \mathbb{N} \to \mathbb{N}$ by $TIME(M)(p)(n) :=$ "the number of steps which M with input p needs for determining the n-th digit after the binary point of $g_M(p)$". Let $X \subseteq dom(f)$ be a compact set. Then there is some uniform bound $t : \mathbb{N} \to \mathbb{N}$ with $(\forall p \in \rho^{-1}X)(\forall n)TIME(M)(p)(n) \leq t(n)$. In this case let us say "M computes f on X in time t". The polynomial time computable functions in Ko and Friedman [17] are the polynomial time computable functions in TTE. Also Brent's [16] definition of complexity is in accord with this one. Let $M(n)$ be an upper time bound for integer multiplication, e.g. $M(n) = n \cdot \log n \cdot \log \log n$. Then the following can be shown (Müller [18], Brent [16]).

Proposition 14

(1) On any compact subset $X \subseteq \mathbb{R}^2$, addition is computable in time $0(n)$.

(2) On any compact subset $X \subseteq \mathbb{R}^2$, multiplication is computable in time $0(M(n))$.

(3) On any compact subset $X \subseteq \mathbb{R}$ with $0 \notin X, x \to 1/x$ is computable in time $0(M(n))$.

(4) Let f be any trigonometric function or the exponential function. Then on any compact subset $X \subseteq dom(f)$, f can be computed in time $0(M(n) \cdot \log n)$

The time complexity of a real number $a \in \mathbb{R}$ (w.r.t. the representation ρ) can be defined as the computational complexity of the constant function $x \to a$. Hierarchy theorems from ordinary Turingmachine complexity theory can be transferred to the complexity of real numbers and of real functions (Müller [18]).

For certain classes of real functions f, the zeroes of f are in the same complexity class as f (Ko and Friedman [17], Müller [18]). If $t : \mathbb{N} \to$

IN is suficiently large (e.g. $\lambda n \cdot n^3 \in 0(t)$) and "regular" then $x \to \int_0^x f(y)dy$ is computable in $0(n^2 \cdot t(n))$ if $f :\subseteq \mathbb{R} \to \mathbb{R}$ is analytic on $[0;1]$ and computable in time $0(t)$ (Müller [19]).

In addition to $TIME(M)$ we define the input lookahead of a machine M by $ILA(M)(p)(n) :=$ "the number of digits after the binary point which M with input p reads for determining the n–th digit after the binary point." Finally we define the dependence of $f : \mathbb{R} \to \mathbb{R}$ at $p \in dom(\rho)$ informally by $DEP(\rho)(p)(n) :=$ "the least number of digits after the binary point of p by which $f\rho(p)$ is determined up to an error of 2^{-n}." If M computes f, then in any case $DEP(\rho)(p)(n) \leq ILA(M)(p)(n)$. The machine M computes "online" iff equality holds. It has been shown (Weihrauch [30]) that in any case ("almost –") online machines can be found but that forcing the online property may increase the computation time more than polynomially iff $P \neq NP$.

6   Conclusion

The short outline of TTE and the examples presented in this text should already explain the character of this approach and demonstrate its wide applicability. Especially in analysis it admits to distinguish five levels of effectivity.

(1) Abstract Analysis: $(\forall x)(\exists y)R(x,y)$

(2) Constructivity: $(\exists \Gamma$ continuous$)$ $(\forall p \in Y)R(p, \Gamma(p))$

(3) Computability: $(\exists \Gamma$ computable$)$ $(\forall p \in Y)R(p, \Gamma(p))$

(4) Complexity: $(\exists \Gamma$ easily computable$)$ $(\forall p \in Y)R(p, \Gamma(p))$

(5) Numerical Analysis: "There is a program operating on floating point numbers such that ..."

Of course, Numerical Analysis is much more comprehensive, and it already provides mainly indirectly many facts concerning constructivity, computability, and computational complexity in TTE. In addition other approaches yield a large number of positive effectivity results in analysis (see e.g. Kushner [7], Bishop and Bridges [4], Beeson [20]).

For most of them corresponding theorems of the form "There is a computable ..." can be proved in TTE. In approaches based on Church's thesis negative results have the form "There is no computable ..." and in approaches based on some constructive logic negative results (if they are formulated at all) are of the form "Property P implies the non constructive principle ...". On the other hand, in TTE most of the fundamental negative results in Analysis have the form "There is no continuous ...". Also results by Pour–El and Richards [33] concerning different topologies can be explained in TTE very naturally.

In Appendix B of his famous book [34] E. Bishop says:
"As written, this book is person–oriented rather than computer–oriented. It would be of great interest to have a computer–oriented version .... A thoughtful computer–oriented version should uncover many interesting phenomena."

TTE is a computer–oriented theory which might formalize Bishop's intentions. By the results obtained so far it seems to be almost straightforward to transform Bishop's book into TTE. It should be noted, however, that such a transformation changes the meanings of definitions and theorems. Furthermore, a mere transformation would not suffice since theorems may split into different versions (e.g. the Heine–Borel–Theorem), negative results had to be proved (degrees of non–effectivity), and computational complexity should be considered. One crucial point in TTE is always the definition of the representations which correspond to definitions of sets and special subsets in Bishop's approach. Roughly speaking, a representation of a set can be charactericed by the information about objects which can be obtained from names effectively.

TTE mainly studies effectivity of properties of the form $(\forall x \in X)(\exists y \in Y)P(x,y)$, where card X, card $Y \leq$ card 2^{ω}. It is, however, easy to add a "concrete existence" operatator $\hat{\exists}$ ("we know"), where $(\hat{\exists}w \in \Sigma^*)$. $P(w)$ is proved by writing down (or by saying how to compute) a word w and proving (classically) $P(w)$.

The results obtained so far show that TTE is a very useful frame for investigating effectivity in Mathematics and especially in Analysis. But many basic questions still have to be settled:

– The Myhill/Shepherdson theorem for computable operators and the Ceitin/Kreisel/Lacombe/Shoenfield/Moschovakis theorem for computable functions on metric spaces obtain a unique simple formulation in TTE.

Let $\delta :\subseteq \Sigma^\omega \to M$ be a representation. Derive a numbering $\nu_\delta :\subseteq \mathbb{N} \to M_C$ of the δ-computable elements of M by $\nu_\delta(i) = x$ iff $\delta\varphi(i) = x$. Let $f : M_C \to M_C$ be a function.

Question: f (δ, δ)-computable $\Longleftrightarrow$ f (ν_δ, ν_δ)-computable?

The case " $\Longrightarrow$ " can be proved easily. For certain admissible representations of metric spaces, " $\Longleftarrow$ " is the CKLSM–Theorem. For sufficiently effective representations of cpo's, " $\Longleftarrow$ " is the Myhill/Shepherdson theorem [35]. No similar results seem to be known about other representations.

- Non constructivity of subsets of Σ^ω can be compared by continuous reduction: $A \leq_w B :\Longleftrightarrow A = f^{-1}B$ for some continuous function f. The resulting degrees are the Wadge–degrees [36, 37]. Several other reducibilities can be defined. Concrete non–effective problems from Analysis can be localized in the corresponding hierarchies [38].

- TTE provides the tools for precisely formulating computational complexity properties in Numerical Analysis. Until now, only the complexity of real functions and numbers is studied. We have theorems of the form: if $f \in C$ then the zero of f (or the integral of $f, \ldots$) is an element of C' where C and C' are complexity classes. But what is the computational complexity of the intergration operators $I : C[0; 1] \to \mathbb{R}$ and $I' : C[0; 1] \to C[0; 1]$? As for the real numbers distinguished representations for $C[0; 1]$ (which exist!) must be considered in order to obtain reasonable definitions. We do not yet know, for which spaces there is a reasonable complexity theory. We know, that on $C[0; 1]$ polynomials can be approximated by rational trigonometric polynomials or by rational polygons and vice versa. We have results about the degrees of polynomials or the numbers of vertices of polygons but no good results about upper and lower complexity bounds for the transformations.

References

[1] Brouwer, L.E.J.:
     Historical background, principles, and methods of intuitionism, South
     African J.Sc. 49, 139 - 146 (1952)

[2] Heyting, A.:
     Intuitionism, an introduction, North-Holland, Amsterdam, 1956 (revised
     1972)

[3] Troelstra, A.S.:
     Principles of intuitionism, Springer–Verlag, Berlin, Heidelberg, 1969

[4] Bishop,E.; Bridges, D.S.:
     Constructive Analysis, Springer-Verlag, Berlin, Heidelberg, 1985

[5] Ceitin, G.S.:
     Algorithmic operators in constructive complete separable metric spaces
     (in Russian), Doklady Akad, Nauk 128, 49 - 52 (1959)

[6] Markov, A.A.:
     On constructive mathematics (in Russian), Trudy Mat. Inst. Stektov
     67, 8 - 14 (1962)

[7] Kushner, B.A.:
     Lectures on constructive mathematical logic and foundations of mathe-
     matics, Izdat. "Nauka", Moscow, (1973)

[8] Aberth, O.:
     Computable analysis, McGraw-Hill, New York, (1980)

[9] Mazur, S.:
     Computable analysis, Rozprawy Matematyczne XXXIII (1963)

[10] Grzegorczyk, A.:
     On the definition of computable real continuous functions, Fund.Math.
     44, 61 - 71 (1957)

[11] Mostowski, A.:
     On computable sequences, Fund.Math. 44, 37 - 51 (1955)

[12] Lacombe, D.:
     Quelques procedes de definition en topologie recursive. In: Construc-
     tivity in mathematics (A.Heyting, ed.), North–Holland, Amsterdam,1959

[13] Klaua, D.:
     Konstruktive Analysis, Deutscher Verlag der Wissenschaften, Berlin,
     1961

[14] Hauck, J.:
      Berechenbare reelle Funktionen, Zeitschrift f. math. Logik und Grdl.
      Math. 19, 121 - 140 (1973)
[15] Scott, D.:
      Outline of a mathematical theory of computation Science, Proc. 4th
      Princeton Confernce onInform. Sci., 1970
[16] Brent, R.P.:
      Fast multiple precision evaluation of elementary functions, J. ACM 23,
      242 - 251 (1976)
[17] Ko, K.; Friedman, H.:
      Computational complexity of real functions, Theoret. Comput. Sci. 20,
      323 - 352 (1982)
[18] Müller, N.Th.:
      Subpolynomial complexity classes of real functions and real numbers. In:
      Lecture notes in Computer Science 226, Springer–Verlag, Berlin, Heidel-
      berg, 284 - 293, 1986
[19] Müller, N.Th.:
      Uniform computational complexity of Taylor series. In: Lecture notes in
      Computer Science 267, Springer–Verlag, Berlin, Heidelberg, 435 - 444,
      1987
[20] Beeson, M.J.:
      Foundations of constructive mathematics, Springer–Verlag, Berlin, Hei-
      delberg, 1985
[21] Kreitz,Ch.; Weihrauch,K.:
      Compactness in constructive analysis revisited, Informatik–Berichte Nr.
      49, Fernuniversität Hagen (1984) and Annals of Pure and Applied Logic
      36, 29 - 38 (1987)
[22] Kreitz,Ch.; Weihrauch,K.:
      A unified approach to constructive and recursive analysis. In: Compu-
      tation and proof theory, (M.M. Richter et al., eds.), Springer– Verlag,
      Berlin, Heidelberg, 1984
[23] Kreitz,Ch.; Weihrauch,K.:
      Theory of representations, Theoretical Computer Science 38, 35 - 53
      (1985)
[24] Weihrauch, K.:
      Type 2 recursion theory, Theoretical Computer Science 38, 17 - 33 (1985)
[25] Weihrauch, K.; Kreitz, Ch.:
      Representations of the real numbers and of the open subsets of the set
      of real numbers, Annals of Pure and Applied Logic 35, 247 - 260 (1987)

[26] Weihrauch, K.:
Computability, Springer–Verlag, Berlin, Heidelberg, 1987

[27] Weihrauch, K.:
On natural numberings and representations, Informatik–Berichte Nr.29, Fernuniversität Hagen, 1982

[28] Weihrauch, K.; Kreitz, Ch.:
Type 2 computational complexity of functions on Cantor's space, Theoretical Computer Science 82, 1 - 18 (1991)

[29] Weihrauch, K.:
Towards a general effectivity theory for computable metric spaces (to appear in Theoretical Computer Science)

[30] Weihrauch, K.:
The complexity of online computations of real functions (to appear in Journal of Complexity)

[31] Deil, Th.:
Darstellungen und Berechenbarkeit reeller Zahlen, Informatik–Berichte Nr.51, Fernuniversität Hagen, 1984

[32] Hinman, P.G.:
Recursion–theoretic Hierachies, Springer–Verlag, Berlin, Heidelberg, 1978

[33] Pour–El,M.B.; Richards, J.I.:
Computability in Analysis and Physics, Springer–Verlag, Berlin, Heidelberg, 1989

[34] Bishop, E.:
Foundations of Constructive Analysis, Mc-Graw–Hill, New York, 1967

[35] Egli, H.; Constable, R.L.:
Computability concepts for programming language semantics, Theoretical Computer Science 2, 133 – 145 (1976)

[36] Van Wesep, R.:
Wadge degrees and descriptive set theory, in Cabal Seminar 76 – 77, Lecture Notes in Mathematics 689, Springer–Verlag, Berlin, Heidelberg, 1978

[37] Weihrauch, Klaus:
The lowest Wadge degrees of subsets of the Cantor Space, Informatik–Berichte Nr. 107, Fernuniversität Hagen, 1991

[38] von Stein, Thorsten:
Vergleich nicht konstruktiv lösbarer Probleme in der Analysis, Diplomarbeit, Fernuniversität Hagen, 1989

Author Index

Lecture Notes in Computer Science

For information about Vols. 1–529
please contact your bookseller or Springer-Verlag

Vol. 569: A. Beaumont, G. Gupta (Eds.), Parallel Execution of Logic Programs. Proceedings, 1991. VII, 195 pages. 1991.

Vol. 570: R. Berghammer, G. Schmidt (Eds.), Graph-Theoretic Concepts in Computer Science. Proceedings, 1991. VIII, 253 pages. 1992.

Vol. 571: J. Vytopil (Ed.), Formal Techniques in Real-Time and Fault-Tolerant Systems. Proceedings, 1992. IX, 620 pages. 1991.

Vol. 572: K. U. Schulz (Ed.), Word Equations and Related Topics. Proceedings, 1990. VII, 256 pages. 1992.

Vol. 573: G. Cohen, S. N. Litsyn, A. Lobstein, G. Zémor (Eds.), Algebraic Coding. Proceedings, 1991. X, 158 pages. 1992.

Vol. 574: J. P. Banâtre, D. Le Métayer (Eds.), Research Directions in High-Level Parallel Programming Languages. Proceedings, 1991. VIII, 387 pages. 1992.

Vol. 575: K. G. Larsen, A. Skou (Eds.), Computer Aided Verification. Proceedings, 1991. X, 487 pages. 1992.

Vol. 576: J. Feigenbaum (Ed.), Advances in Cryptology - CRYPTO '91. Proceedings. X, 485 pages. 1992.

Vol. 577: A. Finkel, M. Jantzen (Eds.), STACS 92. Proceedings, 1992. XIV, 621 pages. 1992.

Vol. 578: Th. Beth, M. Frisch, G. J. Simmons (Eds.), Public-Key Cryptography: State of the Art and Future Directions. XI, 97 pages. 1992.

Vol. 579: S. Toueg, P. G. Spirakis, L. Kirousis (Eds.), Distributed Algorithms. Proceedings, 1991. X, 319 pages. 1992.

Vol. 580: A. Pirotte, C. Delobel, G. Gottlob (Eds.), Advances in Database Technology – EDBT '92. Proceedings. XII, 551 pages. 1992.

Vol. 581: J.-C. Raoult (Ed.), CAAP '92. Proceedings. VIII, 361 pages. 1992.

Vol. 582: B. Krieg-Brückner (Ed.), ESOP '92. Proceedings. VIII, 491 pages. 1992.

Vol. 583: I. Simon (Ed.), LATIN '92. Proceedings. IX, 545 pages. 1992.

Vol. 584: R. E. Zippel (Ed.), Computer Algebra and Parallelism. Proceedings, 1990. IX, 114 pages. 1992.

Vol. 585: F. Pichler, R. Moreno Díaz (Eds.), Computer Aided System Theory – EUROCAST '91. Proceedings. X, 761 pages. 1992.

Vol. 586: A. Cheese, Parallel Execution of Parlog. IX, 184 pages. 1992.

Vol. 587: R. Dale, E. Hovy, D. Rösner, O. Stock (Eds.), Aspects of Automated Natural Language Generation. Proceedings, 1992. VIII, 311 pages. 1992. (Subseries LNAI).

Vol. 588: G. Sandini (Ed.), Computer Vision – ECCV '92. Proceedings. XV, 909 pages. 1992.

Vol. 589: U. Banerjee, D. Gelernter, A. Nicolau, D. Padua (Eds.), Languages and Compilers for Parallel Computing. Proceedings, 1991. IX, 419 pages. 1992.

Vol. 590: B. Fronhöfer, G. Wrightson (Eds.), Parallelization in Inference Systems. Proceedings, 1990. VIII, 372 pages. 1992. (Subseries LNAI).

Vol. 591: H. P. Zima (Ed.), Parallel Computation. Proceedings, 1991. IX, 451 pages. 1992.

Vol. 592: A. Voronkov (Ed.), Logic Programming. Proceedings, 1991. IX, 514 pages. 1992. (Subseries LNAI).

Vol. 593: P. Loucopoulos (Ed.), Advanced Information Systems Engineering. Proceedings. XI, 650 pages. 1992.

Vol. 594: B. Monien, Th. Ottmann (Eds.), Data Structures and Efficient Algorithms. VIII, 389 pages. 1992.

Vol. 595: M. Levene, The Nested Universal Relation Database Model. X, 177 pages. 1992.

Vol. 596: L.-H. Eriksson, L. Hallnäs, P. Schroeder-Heister (Eds.), Extensions of Logic Programming. Proceedings, 1991. VII, 369 pages. 1992. (Subseries LNAI).

Vol. 597: H. W. Guesgen, J. Hertzberg, A Perspective of Constraint-Based Reasoning. VIII, 123 pages. 1992. (Subseries LNAI).

Vol. 598: S. Brookes, M. Main, A. Melton, M. Mislove, D. Schmidt (Eds.), Mathematical Foundations of Programming Semantics. Proceedings, 1991. VIII, 506 pages. 1992.

Vol. 599: Th. Wetter, K.-D. Althoff, J. Boose, B. R. Gaines, M. Linster, F. Schmalhofer (Eds.), Current Developments in Knowledge Acquisition - EKAW '92. Proceedings. XIII, 444 pages. 1992. (Subseries LNAI).

Vol. 600: J. W. de Bakker, K. Huizing, W. P. de Roever, G. Rozenberg (Eds.), Real-Time: Theory in Practice. Proceedings, 1991. VIII, 723 pages. 1992.

Vol. 601: D. Dolev, Z. Galil, M. Rodeh (Eds.), Theory of Computing and Systems. Proceedings, 1992. VIII, 220 pages. 1992.

Vol. 602: I. Tomek (Ed.), Computer Assisted Learning. Proceedigs, 1992. X, 615 pages. 1992.

Vol. 603: J. van Katwijk (Ed.), Ada: Moving Towards 2000. Proceedings, 1992. VIII, 324 pages. 1992.

Vol. 604: F. Belli, F.-J. Radermacher (Eds.), Industrial and Engineering Applications of Artificial Intelligence and Expert Systems. Proceedings, 1992. XV, 702 pages. 1992. (Subseries LNAI).

Vol. 605: D. Etiemble, J.-C. Syre (Eds.), PARLE '92. Parallel Architectures and Languages Europe. Proceedings, 1992. XVII, 984 pages. 1992.

Vol. 606: D. E. Knuth, Axioms and Hulls. IX, 109 pages. 1992.

Vol. 607: D. Kapur (Ed.), Automated Deduction – CADE-11. Proceedings, 1992. XV, 793 pages. 1992. (Subseries LNAI).

Vol. 608: C. Frasson, G. Gauthier, G. I. McCalla (Eds.), Intelligent Tutoring Systems. Proceedings, 1992. XIV, 686 pages. 1992.

Vol. 609: G. Rozenberg (Ed.), Advances in Petri Nets 1992. VIII, 472 pages. 1992.

Vol. 610: F. von Martial, Coordinating Plans of Autonomous Agents. XII, 246 pages. 1992. (Subseries LNAI).

Vol. 612: M. Tokoro, O. Nierstrasz, P. Wegner (Eds.), Object-Based Concurrent Computing. Proceedings, 1991. X, 265 pages. 1992.

Vol. 613: J. P. Myers, Jr., M. J. O'Donnell (Eds.), Constructivity in Computer Science. Proceedings, 1991. X, 247 pages. 1992.